Albertus Johannes Malan

Passivity-Based Stabilisation and Coordination in Networked Multi-Energy Systems

Karlsruher Beiträge zur
Regelungs- und Steuerungstechnik
Karlsruher Institut für Technologie

Band 27

Passivity-Based Stabilisation and Coordination in Networked Multi-Energy Systems

by

Albertus Johannes Malan

KIT Scientific Publishing

Karlsruher Institut für Technologie
Institut für Regelungs- und Steuerungssysteme

Passivity-Based Stabilisation and Coordination in Networked Multi-Energy Systems

Zur Erlangung des akademischen Grades eines Doktors der Ingenieurswissenschaften von der KIT-Fakultät für Elektrotechnik und Informationstechnik des Karlsruher Instituts für Technologie (KIT) genehmigte Dissertation

von Albertus Johannes Malan, M.Sc.

Tag der mündlichen Prüfung: 25. März 2025
Hauptreferent: Prof. Dr.-Ing. Sören Hohmann
Korreferent: Prof. Dr.-Ing. Johannes Schiffer

Impressum

KIT Scientific Publishing

Karlsruher Institut für Technologie (KIT)
KIT Scientific Publishing
Straße am Forum 2
D-76131 Karlsruhe

KIT Scientific Publishing is a registered trademark
of Karlsruhe Institute of Technology.
Reprint using the book cover is not allowed.

www.bibliothek.kit.edu/ksp.php | E-Mail: info@ksp.kit.edu | Shop: www.ksp.kit.edu

Print on Demand 2026 – Gedruckt auf FSC-zertifiziertem Papier

ISSN 2511-6312
ISBN 978-3-7315-1453-4
DOI 10.5445/KSP/1000184150

Preface

Although a PhD is an intrinsically individual pursuit, we are not isolated as we stand upon the shoulders of giants, peering into the unexplored. I would therefore like to start by expressing my sincerest gratitude for those who have accompanied, guided, challenged and supported me in this pursuit.

Firstly, my deepest thanks go to Prof. Dr.-Ing. Sören Hohmann for being a key enabler of my work and for years of supervision, support and trust. It has been a joy and a privilege to discuss the finer technicalities of control systems, energy and the world at large with you. Similarly, my sincerest thanks go to Prof. Dr.-Ing. Johannes Schiffer for his interest and his assessment of my work. I greatly appreciate our discussions and your genuine interest in my work.

Furthermore, my sincerest appreciation goes to my colleagues at the IRS for adding spice and camaraderie to my PhD pursuit through our many discussions, travels, and laughs together. In particular, thank you, Felix, Pol, Armin, and the rest of the NMES group for delving into the depths of theoretical mysteries with me and for meticulously providing feedback on my work. My earnest thanks also go out the Joel Ferguson and Michele Cucuzzella for our productive collaboration over the last two years and to my students who assisted my work with great enthusiasm.

I finally want to thank my parents, my family, and my friends for their enduring support and for providing balance throughout this journey. And lastly, utmost gratitude goes out to my Father, my inspiration, confidant and friend.

Karlsruhe, March 2025

Abstract

The drive towards net-zero emissions is transforming energy networks. This manifests as an increasing reliance on volatile renewable energy sources, even as new technologies enable greater flexibility through couplings between energy networks and smart loads. Fully realising the flexibility benefits of such future networked multi-energy systems (NMESs), however, presupposes a dynamically stable operation in light of the higher degree of volatility and coupling.

This thesis addresses the challenge of achieving dynamical stability in DC networks, gas networks, and NMESs. Controllers are designed for the decentralised stabilisation and distributed coordination in DC networks with nonlinear loads and which lack pervasive power availability at the loads. Similarly, decentralised stabilisation controllers are presented for compressor-equipped gas networks which may experience an oversupply of gas locally. The controlled DC and gas networks are combined into NMESs via coupling components. A stability result for a dynamical NMES is constructed using the stability results of its constitutive networks in isolation. A scalability for networks and NMESs of arbitrary sizes is obtained through the use of equilibrium-independent passivity (EIP). Furthermore, EIP ensures a modularity which allows any component or controller to be substituted with different versions that have equivalent EIP properties.

The stability results are shown to be robust against parameter and network topology changes. This is validated through comprehensive simulation studies. The simulations further illustrate how transient effects can propagate between networks in an NMES. Collectively, the proposed controllers and the stability-analysis methods open the door for achieving enhanced coordination in future NMESs with guaranteed stability.

Kurzfassung

Das Ziel, sogenannte Netto-Null-Emissionen zu erreichen, führt zu einer Transformation der Energienetze mit einer zunehmenden Abhängigkeit von volatilen, erneuerbaren Energiequellen. Gleichzeitig bieten neue Technologien die Chance, durch die Kopplung von Energienetzen und intelligenten Lasten, eine höhere Flexibilität zu erzielen. Um die Flexibilitätsvorteile solcher zukünftiger vernetzter Multienergiesysteme mit ihrem hohen Maß an Volatilität und Verkopplung voll ausschöpfen zu können, ist jedoch ein dynamisch stabiler Betrieb erforderlich.

Die vorliegende Dissertation befasst sich mit den Herausforderungen der dynamischen Stabilität für Gleichstromnetze, Gasnetze und vernetzte Multienergiesysteme. Es werden Regler für die dezentrale Stabilisierung und verteilte Koordination in Gleichstromnetzen mit nichtlinearen Lasten und ohne flächendeckende Leistungsverfügbarkeit an den Lasten entworfen. Des Weiteren werden Regleransätze für eine dezentrale Stabilisierung von Gasnetzen mit Kompressoren präsentiert, die potentiell lokale Überangebote an Gas aufweisen. Die derart geregelten Gleichstrom- und Gasnetze werden über Kopplungskomponenten zu vernetzten Multienergiesystemen zusammengeführt. Die Stabilität eines solchen dynamischen vernetzten Multienergiesystems lässt sich aus den Stabilitätsergebnissen seiner konstitutiven Netze ableiten. Die Skalierbarkeit der Ergebnisse für Energienetze und vernetzte Multienergiesysteme ist durch die Verwendung von ruhelangenunabhängiger Passivität sichergestellt. Darüber hinaus gewährleistet die ruhelangenunabhängige Passivität eine Modularität, welche die Substitution jeder Komponente oder jeden Reglers durch andere Versionen mit gleichwertigen Passivitätseigenschaften ermöglicht.

Die Stabilitätsergebnisse sind nachweislich robust gegenüber Netztopologie- und Parameteränderungen. Diese Erkenntnisse werden durch umfangreiche Simulationsstudien belegt. Zudem zeigen die Simulationen, wie sich transiente Effekte zwischen den enthaltenen Einzelnetzen in einem vernetzten Multienergiesystem ausbreiten können. Die präsentierten Regler sowie die Methoden der Stabilitätsanalyse eröffnen somit die Möglichkeit einer optimierten Koordination in zukünftigen vernetzten Multienergiesystemen mit garantierter Stabilität.

Contents

1 Introduction

Modern energy networks are some of the largest and most complex systems created by mankind, built with the singular goal of meeting our energy needs. The continual rise in energy demand has been a prime driver for the rapid expansion of the power and gas transmission and distribution networks.[1] Although the growth in energy consumption has fuelled and been driven by vast economic growth globally [HHY08; Zha+17; HMJ22], [Smi21, p. 10ff.], an over-reliance on fossil fuels risks causing severe damage on a global scale [Arm+22; Lee+23]. This has spurred individuals, companies, and countries to aim for net-zero emissions by increasing their energy efficiency and the share of renewable energy sources (RESs) [Höh+21; Eur23; FS23]. A further pillar of net-zero strategies is the development and widespread adoption of new technologies, including electric vehicles, heat pumps, and renewable gas production [Per+18; Abb+21; Ozc+23].

These measures for achieving net-zero emissions are systematically changing modern energy systems. Driven by greater efficiency, the total energy consumption in developed nations like Germany, France, and the United States has plateaued or decreased slightly in the last decade [EKK23, p. 8], [Int23b]. Indeed, the decrease of 13.0 % in the German public net electricity generation (2002 to 2023) comes despite the net installed electrical generation capacity *increasing* by 111.1 % in the same period [Fra23, p. 14, p. 88], highlighting the volatile and intermittent nature of RESs. Moreover, decarbonising district heating, district cooling, and gas networks creates a greater reliance on the predominantly electric RESs in these networks [VvH18; Qad+19] and in energy systems which are increasingly interconnected.[2]

The coupling of previously independent energy domains into a networked multi-energy system (NMES) thus makes a holistic consideration increasingly important. Such NMESs, if not properly controlled, can see the volatility of RESs propagating to other energy networks, possibly lowering the network resilience (see [LES08; CM16; Tay+22; Dvo+22]). The potential transient effects and stability concerns arising from

[1] E.g., the size of the low voltage distribution network in Germany increased by 26.5 % between 2011 and 2022. The low and medium pressure gas networks grew by 18.3 % in the same period. [BB12, p. 19, p. 173], [BB23, p. 108, p. 223]

[2] Although there are different views on how best to decarbonise the various energy networks (see [vRe20; LLC21; Bil+24]), an all-electric approach would require a significant expansion of the electric grid. Replacing the energy provided by gas with electric heat pumps in Germany would more than double the peak electricity demand in winter [BJ21]. This roughly translates to a 25 % base-load increase in summer. However, including hydrogen in the energy mix can potentially reduce the total system cost by between 14 % and 16 % in a European context, compared to the all-electric case [Sas+21].

dynamically coupled energy networks in NMESs, however, remain largely unexplored in the literature [Gue+19]. This is especially important in light of the overlapping timescales of energy network transients.[3] Nevertheless, NMESs offer great potential in providing short- and long-term flexibility, storage capabilities, and transport capacity[4] for RES-dominated electrical grids, which can help alleviate network strain and prevent RES curtailment [Man14; BS18; Qad+20; Dvo+22]. This thesis thus aims to investigate the dynamical coupling effects in NMESs while presenting controllers for the constitutive energy networks which yield stability at the network and the NMES level. Specifically, a focus is placed on RES-dominated NMESs which comprise DC and gas networks.

DC networks are steadily gaining attention for their efficiency and practicality, despite the historical dominance of AC [Jus+13; Men+17]. Indeed, many practical applications use DC, including electric vehicles, consumer electronics, telecommunication components, data centres, electric train networks, and LED lighting. DC is also a natural interface for various RESs, battery storage systems, and sector-coupling technologies like electrolysis-based power to gas (P2G) and fuel cells [BS18]. Moreover, the efficiency gained from partially or fully switching over to a DC distribution network for such applications can yield significant savings [GAH16; Ger+18].

Gas networks, despite also undergoing significant changes, are expected to remain relevant in a world transitioning to net-zero and beyond. While natural gas usage is projected to decrease to reach net-zero [Qad+20; Int23a], significant investments are being made towards producing renewable gas for gas-dependent sectors and to help counteract RES volatility.[5] The combination of lower demand and local production, however, can cause regions to vary between being net gas consumers and producers. Unlike in power networks, where transformers enable bidirectional power flows, conventional gas networks with a top-down supply must be retrofitted with compressors to allow the regional distribution or storage of excess gas [Eur+20].

From a control systems perspective, the changes in energy networks and their interconnection in particular necessitate new approaches for controlling and managing NMESs [Gue+19]. Due to the large number of components with variable availability (e.g., RESs and flexible loads), controllers are required which can guarantee stable network operation under changing conditions and network configurations. For the individual networks and NMESs in general, control solutions and analysis methods

---

3 While the propagation speed of electricity is very high ($\sim 1.6 \cdot 10^8$ m/s) [Pap21, p. 281], the controlled network dynamics can exhibit transients in the range from microseconds to hours [OW15; Cvi+17]. The propagation dynamics in other domains range from hundreds of milliseconds to hours, as characterised by the speed of sound, e.g., in methane ($\sim$430 m/s), hydrogen ($\sim$1300 m/s), and water ($\sim$1480 m/s) [Ben+02, p. 409f.]. Thus, there are transients in electrical networks which act within similar timescales as the dynamics in other domains.

4 High-power electric lines can transmit up to 2 GW, whereas gas pipelines can transport 20 GW at a tenth to a twentieth of the construction cost [vRe20].

5 Europe alone plans on producing 10 million tonnes of renewable hydrogen domestically and importing a further 10 million tonnes by 2030 [Eur22].

are thus required which are *scalable* and *modular*. Scalable approaches ensure a practicality when considering the ever-growing size and complexity of the networks. A modularity, on the other hand, enables controllers, components, or entire networks to be substituted with alternative variants as long as certain requirements for the interface of the connection are met.

Beyond scalability and modularity, the specific control requirements in DC and gas networks are shaped by their respective technical contexts. In the DC case, power supplying components must *form* the grid while ensuring voltage stability [Dra+16a; Men+17; KAA19]. However, the destabilising effect of constant power loads and a lack of power availability at all network buses complicate this task. Furthermore, where several components cooperate to form the grid, an *average* or *global voltage regulation* along with *load-sharing* is typically desired [Nas+15; ZD15; Tuc+18; Dou+22]. This prevents unequal component strain and degradation while improving network resilience.

In gas networks, on the other hand, the pressure regulation requirements are less stringent compared to the voltage regulation in DC networks. Instead, the regulatory frameworks typically require a *balancing* of the network inflows and outflows whilst simultaneously managing the storage capacity of the network, termed the linepack [Koc+15, p. 51ff.], [GasNZV10, § 3, § 22], [Qad+20]. Control systems in gas networks must therefore contend with the inherently nonlinear nature of gas network components[6] along with increasing volatility in demand and supply as a consequence of interconnecting various domains. Nevertheless, compared to the fast dynamics in DC networks, the slower dynamics and higher network capacities in gas networks lessen the need for fast-transient load-sharing between multiple regulating components.

Therefore, NMESs require control solutions that:

- guarantee the *stabilisation* of dynamical NMESs,
- provide a fast-transient, load-sharing *coordination* where required.

The control solutions must also be scalable and modular to ensure that implementations in large networks remain manageable w.r.t. the computational and communication complexity, and w.r.t. the regulatory overhead to ensure stability.

These requirements preclude centralised control schemes which present single points of failure for the network, restrict the reaction speed against disturbances, and strain the communication infrastructure in large networks. Instead, *distributed* controllers—where information is shared only with a limited number of neighbours—should be implemented where communication is indispensable, and *decentralised* controllers, which only use locally available information, should be utilised otherwise.[7]

6 These include nonlinearities from pipelines [Koc+15], valves [MF19] and compressors [GE99b].

7 The definitions of decentralised and distributed control systems are chosen in line with [Loh+14, p. 83f.], where more detail is provided.

1.1 Objective and Research Contributions

The main objective of this thesis is to **design controllers for the decentralised stabilisation and distributed coordination** in DC networks, gas networks, and NMESs.

Despite such control schemes being extensively studied in a DC context, limitations remain w.r.t. the inclusion of nonlinear loads and sparse power availability. On the other hand, the stability of gas networks containing dynamical compressors has scarcely been investigated, reinforcing the need for controllers which yield scalable and modular stability results. Furthermore, although NMESs are gaining prominence in the literature, there is a conspicuous absence of a substantiated, interconnected stability analysis which considers the dynamics of every energy network. These limitations relating to DC networks, gas networks, and NMESs are discussed in greater detail in the next chapter.

In light of the above, this thesis makes the following contributions towards addressing these needs of future energy systems:

1. Controllers ensuring **decentralised stabilisation in DC and gas networks**: Decentralised control laws are provided for so-called distributed generation units (DGUs)[8] in DC networks, and compressors and valves in gas networks. Through the use of passivity theory, stability results are provided for these networks. These results scale to arbitrary network sizes and are robust against various factors such as load and network parameter changes, topology changes, and components connecting or disconnecting.
2. A control scheme for fast-transient **proportional power-sharing through distributed coordination** in DC networks: A multi-stage controller is presented that ensures each component participating in the control injects the same amount of power, proportional to a chosen component-specific factor. By employing dynamic distributed averaging (DDA), the controller achieves average voltage regulation using all available voltage measurements in the network. The passivity-based stability result for the distributed coordination is independent of the network size, network topology, connecting or disconnecting components, and changes in the network and load parameters.
3. **Stability conditions for NMESs** comprising dynamical DC and gas networks: Components coupling DC and gas networks are shown to be compatible with the stability achieved through the designed decentralised and distributed controllers. This allows for a constructive stability result for an NMES which builds on the stability results of its constitutive networks in isolation.

[8] RESs and energy storage components typically connect to DC networks via a DGU.

1.2 Outline and Notation

To provide clarity and improve readability, an overview of the thesis structure is provided in Figure 1.1 below. Subsequently, the notation used in this thesis is introduced.

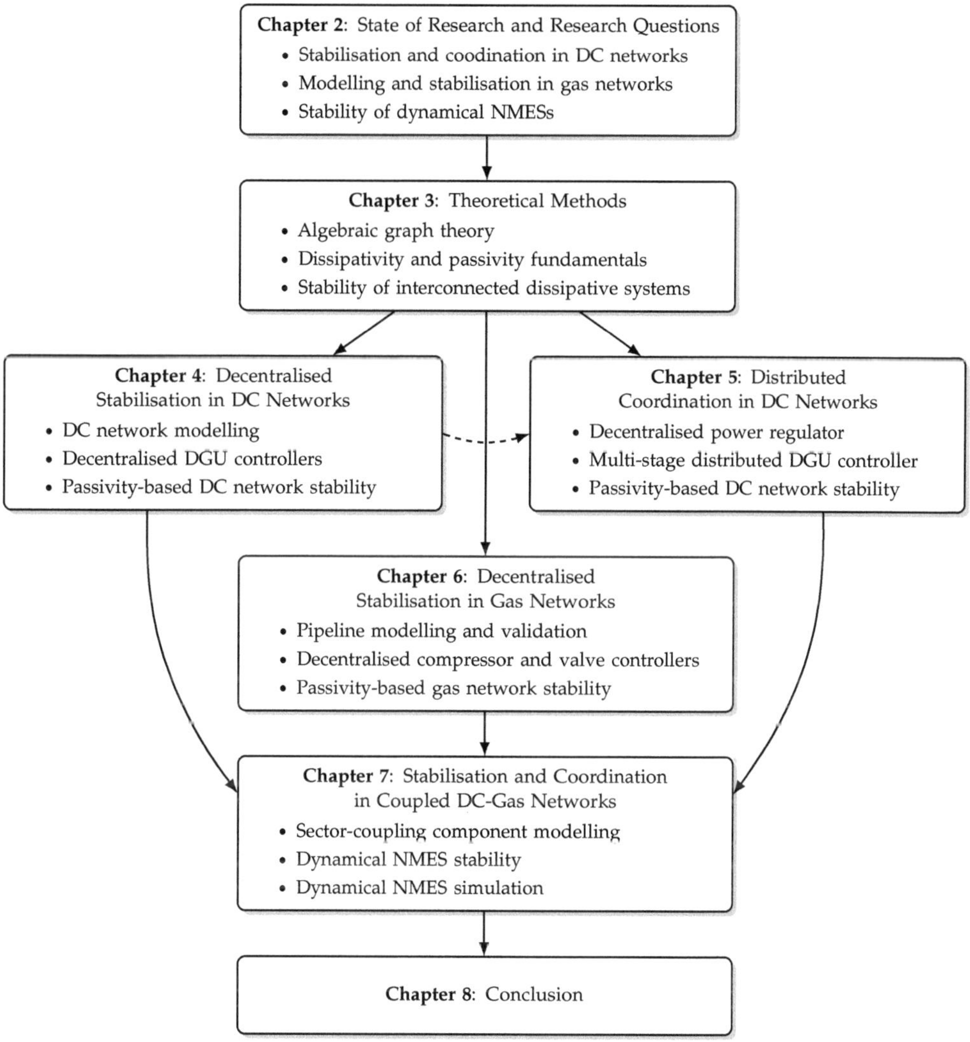

Figure 1.1: Outline of the thesis.

Notation

Denote by $\mathbb{N}$ the set of natural numbers, $\mathbb{R}$ the set of real numbers, $\mathbb{R}_{\geq 0}$ the set of positive real numbers, and $\mathbb{R}_{>0}$ the set of strictly positive real numbers. All other sets are denoted using calligraphic letters, with $|\mathcal{N}|$ denoting the cardinality of a set $\mathcal{N}$. Subtracting sets is achieved via the $\setminus$ operator, i.e., $\mathbb{R}_{>0} = \mathbb{R}_{\geq 0} \setminus \{0\}$.

Define as a vector $\boldsymbol{a} = (a_n), \boldsymbol{a} \in \mathbb{R}^k$ and a matrix $\boldsymbol{A} = (a_{n,m}), \boldsymbol{A} \in \mathbb{R}^{k \times l}$. All vectors are column vectors unless explicitly stated otherwise. Let $\mathbb{1}_k$ be a k-dimensional vector of ones and $\boldsymbol{I}_k$ be the identity matrix of dimension k. The k-dimensional vector and $k \times l$-dimensional matrix of zeros are $\boldsymbol{0}_k$ and $\boldsymbol{0}_{k,l}$, respectively. Where clear from context, subscripts indicating the dimensions of $\mathbb{1}$, $\boldsymbol{I}$, and $\boldsymbol{0}$ are omitted.

Consider a function $f(\boldsymbol{a})$, with $f\colon \mathbb{R}^k \to \mathbb{R}$. Then, $f(\boldsymbol{a})$ is positive definite in $\boldsymbol{a}$ if $f(\boldsymbol{a}) > 0$ for $\boldsymbol{a} \neq \boldsymbol{0}$ and $f(\boldsymbol{a}) = 0$ for $\boldsymbol{a} = \boldsymbol{0}$. Additionally, $f(\boldsymbol{a})$ is positive semi-definite in $\boldsymbol{a}$ if $f(\boldsymbol{a}) \geq 0$ for $\boldsymbol{a} \neq \boldsymbol{0}$ and $f(\boldsymbol{a}) = 0$ for $\boldsymbol{a} = \boldsymbol{0}$. Moreover, $f(\boldsymbol{a})$ is negative (semi-)definite in $\boldsymbol{a}$ if $-f(\boldsymbol{a})$ is positive (semi-)definite in $\boldsymbol{a}$. Given $\boldsymbol{b} = \boldsymbol{h}(\boldsymbol{a})$, $f(\boldsymbol{a})$ is positive definite in $\boldsymbol{b}$ if $f(\boldsymbol{a}) > 0$ in the set $\{\boldsymbol{a} \in \mathbb{R}^k | \boldsymbol{a} \neq \boldsymbol{0}, \boldsymbol{b} \neq \boldsymbol{0}, \boldsymbol{b} = \boldsymbol{h}(\boldsymbol{a})\}$.[9] A function $f(c)$ with $f\colon \mathbb{R} \to \mathbb{R}$ is a class C^p function if its pth derivative is continuous in the domain of c. For notational convenience, the following shorthand is used for the Jacobian of a vector function $\boldsymbol{f}\colon \mathbb{R}^k \to \mathbb{R}^l$,

$$\nabla \boldsymbol{f}(\boldsymbol{a}) = \frac{\partial \boldsymbol{f}(\boldsymbol{a})}{\partial \boldsymbol{a}} = \begin{bmatrix} \frac{\partial f_1(\boldsymbol{a})}{\partial a_1} & \cdots & \frac{\partial f_1(\boldsymbol{a})}{\partial a_k} \\ \vdots & \ddots & \vdots \\ \frac{\partial f_l(\boldsymbol{a})}{\partial a_1} & \cdots & \frac{\partial f_l(\boldsymbol{a})}{\partial a_k} \end{bmatrix} = \begin{bmatrix} \nabla_{a_1} f_1(\boldsymbol{a}) & \cdots & \nabla_{a_k} f_1(\boldsymbol{a}) \\ \vdots & \ddots & \vdots \\ \nabla_{a_1} f_l(\boldsymbol{a}) & \cdots & \nabla_{a_k} f_l(\boldsymbol{a}) \end{bmatrix}. \tag{1.1}$$

The operator $\dim(\cdot)$ returns the dimension of a vector or a matrix. The operator $\mathrm{stack}(\cdot)$ stacks vectors into a single column vector, e.g., $\mathrm{stack}_{k\in\mathcal{N}}(\boldsymbol{a}_k) = [\boldsymbol{a}_1^T, \ldots, \boldsymbol{a}_{|\mathcal{N}|}^T]^T$. The operator $\mathrm{diag}(\cdot)$ constructs a (block-)diagonal matrix from the given vector (or matrices) and the matrix determinant of a matrix $\boldsymbol{A}$ is given by $\det(\boldsymbol{A})$. Let $\boldsymbol{A} \succ 0$ and $\boldsymbol{A} \succcurlyeq 0$ denote a positive definite and a positive semi-definite matrix, respectively, where $\boldsymbol{A} - \boldsymbol{B} \succ 0 \iff \boldsymbol{A} \succ \boldsymbol{B}$ for matrices of the same dimensions. The Kronecker product of two matrices is denoted as $\boldsymbol{A} \otimes \boldsymbol{D}$.

For a given variable $x(t)$, $\hat{x}$ denotes its (potentially unknown) steady state, $\tilde{x}(t) := x(t) - \hat{x}$ is the error w.r.t. its steady state, and $x^\star(t)$ is the desired control setpoint. Let $x(t) \equiv 0$ denote that $x(t)$, along with all its derivatives, are zero for all $t \geq 0$. Note that the time dependence of variables is omitted where clear from context.

9 As an example, given $\boldsymbol{b} = \boldsymbol{h}(\boldsymbol{a})$, $f(\boldsymbol{a})$ is negative definite in $\boldsymbol{b}$ if $f(\boldsymbol{a}) < -\boldsymbol{b}^T\boldsymbol{b} < 0$ for all $\boldsymbol{a} \neq \boldsymbol{0}$ and $\boldsymbol{b} \neq \boldsymbol{0}$.

2 State of Research and Research Questions

This chapter presents an overview of the state of research for DC networks, gas networks, and NMESs. The state of research is used to derive research questions pertaining to the main objective of achieving decentralised stabilisation and distributed coordination, as established in Section 1.1.

The chapter commences with DC networks in **Section 2.1**, where a description of the state of the art is followed by a literature review of the approaches for achieving decentralised stabilisation and distributed coordination in DC networks. Next, gas networks are considered in **Section 2.2**, where the state of the art is followed by literature reviews on the modelling of gas pipelines and the control of gas compressors. Thereafter, a literature review on NMESs comprising electrical and gas networks is presented in **Section 2.3**. In each of these sections, the literature reviews are used to derive research questions which are answered in the subsequent chapters. A summary of these research questions is given in **Section 2.4**.

2.1 DC Network Stabilisation and Coordination: State of the Art

The standard setup in a DC network comprises one or more *buses*, denoting points of equal voltage in the network, interconnected via transmission lines. Each bus may be equipped with a load representing the aggregate user demand at that bus along with a DGU connecting the bus to RESs, storage systems or other energy domains. While a central requirement for DC networks pertains to ensuring the voltage quality at the buses, this has not yet been formalised in a common standard [PPC23; Dra+16b].[1] Following the pertinent literature on power systems [Kun+04; Hat+21], the following definition for the stability of a DC network will thus be used in this thesis.

1 Certain standards such as the power supply for telecommunications [Eur12] define voltage specifications, e.g., a normal operating range of $365\,\mathrm{V} \pm 15\,\mathrm{V}$ and a service range of $260\,\mathrm{V}$ to $400\,\mathrm{V}$. Others like [Car+22] consider a voltage operating range of $8.57\,\%$ above or below the nominal voltage while transients must remain within $-28.57\,\%$ and $+54.29\,\%$ of the nominal voltage.

Definition 2.1 (DC Network Stability)
The steady-state voltage of each bus in a stable DC network is constant and within a tolerable range from the reference voltage.[2] In the event of changes or non-critical faults, the voltage at each bus converges to a steady state within the tolerable range.

Remark 2.1. *While the stability of power systems is often understood in the context of Definition 2.1, this term can various other interpretations including in a control systems sense (the stability of equilibria) or in a feasibility sense (e.g., the static stability considered in [Liu+21]. In this thesis, a stability in of the system equilibria are considered with the aim of achieving a system stability as per Definition 2.1.)*

Two common problems typically complicate or inhibit a stable network operation as per Definition 2.1: the properties of realistic loads and the availability of power with which control action can be achieved. In general, loads in AC and DC networks are modelled by static[3] functions relating the bus voltage to the current of the load [vV98; MBB08]. While any function may be used to describe the loads, the use of constant impedance (Z), constant current (I), constant power (P) (ZIP) loads is especially prevalent due to their wide range of applicability and their interpretability. Any load function which exhibits a negative incremental impedance, i.e., one that does not increase monotonically, has the potential to cause voltage oscillations and potentially a voltage collapse in the network. While such phenomena, which often result from including P loads [Jus04; SGF17; AN+17], could be compensated by local controllers, many loads are not situated directly at a DGU. Furthermore, the volatility of RESs means that the power needed for the DGU control action is not necessarily available at all times. Stability in DC networks must thus be achieved in the context of destabilising load functions and a sparsity of steady-state control action in the network.

Beyond guaranteeing the stability of a DC network, several additional objectives can be defined for operating the network. Starting with lower level objectives and moving up, this includes: achieving a coordinated load-sharing between the DGUs [Kar+21; NTFT22; Dou+22]; managing the energy storage levels in the network [OCK11; JXG13; Vu+17]; operating the network economically [Li+18; ABD22; JS+23a]; and implementing market functions [Li+17; Wer+18; Zha+20]. While these objectives can be separated into hierarchical levels with separate controllers for each level [Pla+15; Men+17; KAA19; Gao+19; AI21; Mod+23], there are also controllers which aim to fulfil

2 A tolerable voltage range is not yet standardised and can be application specific (see [PPC23; Dra+16b; Eur12; Car+22]). Since this thesis primarily considers the dynamic convergence to some network and load-dependent steady state, a tolerable range of 10 % above and below the nominal voltage is considered.

3 Although extensions to general dynamical or time-varying loads are made in the literature, e.g., [MBB08, p. 112], [vV98, p. 97f.], [Sil+22], DC loads are typically represented by static functions in the pertinent literature (see the references and discussions in Sections 2.1.1 and 2.1.2).

multiple objectives (see, e.g., [NFT20; Kar+21; ABD22; JS+23a]) and thus fall outside a strict hierarchy.

A further classification for controllers in DC networks pertains to the implementation schemes of the controllers, i.e., whether the controllers are centralised as in [OCK11; JXG13; Li+17; Li+18; Zha+20], distributed as in [Vu+17; Wer+18; NTFT22; Dou+22; ABD22; JS+23a] or decentralised as in [Nah+20; Kar+21; CKS23]. Since the objective of a controller does not strictly dictate its implementation scheme, it is useful to consider both the objective and the implementation scheme as independent classifications.

In line with the central objective of this thesis, **Sections 2.1.1 and 2.1.2** present an overview of the pertinent literature on the stabilisation and coordination objectives in DC networks, respectively. Specific focus is placed on decentralised schemes for stabilisation and distributed schemes for coordination.[4] Thereafter, the results of the literature review are discussed and research questions are formulated for the control of DC networks.

Remark 2.2. *The term microgrid is frequently used in the literature relating to DC networks. Although not standardised in its meaning, the term introduced in [Las01] is usually understood to mean a subsection of an electrical grid which connects to the main grid at a single point, which contains DGUs and storage elements, and which can operate in a grid-connected or disconnected (islanded) mode. Furthermore, a microgrid can supply its own energy demand at least for some amount of time and behaves as a single source or load when connected to the grid (see [Oli+14; Far+20][Sch15, p. 52]). The methods presented in this thesis are not limited to small networks and do not require a single point of connection to the main grid. To this end, the more generic term* DC network *is used instead of* DC microgrid *in this thesis.*

2.1.1 Literature Review: Decentralised Stabilisation

The task of stabilising the voltage as per Definition 2.1 in a DC network is typically achieved by designing controllers for the DGUs which can regulate the local bus voltage to a desired setpoint. Since this requires regulating the fast transients of the DGUs, which often drop below the millisecond range [OW15; Cvi+17], distributed or centralised controllers which suffer from communication delays are ill-suited for voltage setpoint regulation.

Numerous methods have been proposed to solve the decentralised voltage regulation problem, including: plug-and-play-capable controllers[5] [Tuc+16; SSK18; TRFT18]; linearisation controllers [Per+20; Mac+21]; energy-based controllers [Kos+21; CKS23];

[4] For a broader overview of DC controllers beyond the scope of this thesis, the interested reader is directed towards the review papers [Dra+16a; Men+17; KAA19; Gao+19; AI21; Mod+23] and the sources therein.

[5] The plug-and-play capability refers to a control design in which controllers only require information only from itself or its neighbours, in which components can automatically join or leave the network, and in which controllers can self-configure [Sto09; RFFT13].

integral-augmented state-feedback controllers [FCS21; Sil+22; FCS23]; and passivity-based controllers [Nah+20; Str+20b; Str24].

A plug-and-play-capable controller is proposed in [Tuc+16], which achieves voltage regulation for a quasi-stationary network with I loads. Stability is assured by solving a linear matrix inequality (LMI) and updating any neighbouring DGUs before allowing a new DGU to connect. This approach is improved upon in [Tuc+18], where updating the neighbouring DGUs is no longer required. A different plug-and-play-capable controller is designed in [SSK18] for Z loads and static lines, with a robustness against P loads being demonstrated through linearisation.

In [Per+20], a feedback linearisation is used to provide voltage regulation with an input-to-state stability (ISS) property providing robustness against disturbances. The controllers are designed for a single bus network with linear loads operating in a smart railway system. An input-output linearisation is proposed in [Mac+21], where a single bus with a P load is stabilised by adding a converter parallel to a voltage source and estimating the load parameters.

The Brayton-Moser framework, which models the system power as opposed to the system energy, is utilised in [Kos+21] to design a controller through input and output shaping. A Krasovskii-type storage function[6] is used to show stability in a network with linear loads. This approach is extended in [CKS23] to accommodate nonlinear ZIP loads connected at controlled DGUs.

Among the state-feedback controllers with augmented integration states, [FCS21] presents a controller that ensures exponential stability for monotonically increasing ZIP loads located at buses equipped with DGUs. Their results also give an ISS property which yields disturbance rejection capabilities against time-varying loads. In [Sil+22], a controller is designed which provides highly accurate disturbance rejection capabilities against time-varying loads described by linear exosystems, although non-local network parameters are required for the controllers. The augmented state-feedback controller in [FCS23] ensures voltage regulation for non-monotone ZIP loads located at controlled DGUs. Despite requiring local parameter values for the control, an ISS property grants resilience against external disturbances.

Integral-augmented state-feedback controllers have also been used in conjunction with passivity theory to achieve voltage regulation, as in [Nah+20], where stability is assured for monotonically increasing ZIP loads at controlled DGUs. A similar result is found in [Str+20b], where the controller is designed using interconnection and damping assignment passivity-based control. This result is extended in [Str24] to allow for monotonically increasing loads at buses without a DGU.

[6] See [KKS19; KKS21] for more detail on Krasovskii-type storage functions and Krasovskii passivity.

Discussion

Despite the extensive research on decentralised voltage regulation, many approaches restrict the types or properties of loads allowed in the network. This includes allowing only linear Z or I loads as in [Tuc+16; TRFT18; SSK18; Per+20; Kos+21] or restricting the load function to a monotonically increasing relation [Nah+20; Str+20b; FCS21; Str24], thus preventing the use of purely P or dominantly P loads.[7] More recent results in [Mac+21; CKS23; FCS23] present controllers that can circumvent these restrictions, although this requires an estimate of the constant power load [Mac+21], a numerical derivative of a measured state to be calculated [CKS23], or which require exact DGU parameter values [FCS23].

Furthermore, many approaches can only guarantee stability under restrictive network descriptions. This might entail a network where *each* bus is equipped with a DGU which has sufficient power available for control [Tuc+16; SSK18; Nah+20; Str+20b; FCS21] or which comprises only a single bus [Per+20; Mac+21]. Alternatively, some approaches allow buses without DGUs by analysing the Kron-reduced network [TRFT18; Sil+22]. The Kron reduction eliminates buses without DGUs, yielding an electrically equivalent network from the perspective of the remaining DGU-equipped buses [DB13]. However, since it is computed with the Schur complement, the Kron reduction can fail for negative load parameters[8] and its extension to nonlinear loads remains an open research topic. The cases where buses may consist of only a load invariably require loads which are monotonically increasing functions or maps [Kos+21; Str24].

Remark 2.3. *Droop-based control approaches are often grouped together with other controllers dealing with voltage stability. However, the inner-loop voltage and current controllers are typically responsible for the voltage regulation, with the droop controllers modifying the setpoints of the inner-loop controllers based on the output current or power [Men+17]. As such, droop-based approaches are considered in the next section on coordination.*

2.1.2 Literature Review: Distributed Coordination

After ensuring voltage stability, average voltage regulation and load sharing are used to improve the resilience of the network. Since DC power is transferred through voltage differences and voltage-regulating DGUs might not be available at each bus, it is desirable to ensure that the *average* voltage in the network is equal to a desired reference (see [Nas+15; Tri+19; CKS07; Cuc+19; Fan+20; SSB21]). Load sharing can be

---

[7] Since electronics are represented as P loads [IEEE22, Table 7], such P loads have a large and growing importance in electrical networks.

[8] Recent results have shown that *some* negative loads can yield a valid Kron reduction (see [Che+21]), but such cases are highly topology and parameter dependent.

implemented either as (proportional) current or power sharing, where participating components inject the same (proportional) current or power, thus sharing the burden of supplying the loads. Although power sharing ensures a balanced energy usage, the linearity of current sharing simplifies the process of deriving theoretical guarantees.

Many controllers have been proposed for average voltage regulation and load sharing in DC networks, including: droop controllers as in [Gao+19; Mod+23]; secondary controllers with underlying droop mechanisms [ZD15; DL18; Iov+18]; consensus-based controllers [Nas+15; Tuc+18; DWD18; Cuc+19; Fan+20; NFT20; NTFT22]; controllers using distributed averaging [Tri+19; SSB21]; other distributed controllers [Vu+17; Han+18; Han+19; CKS07]; and model predictive control (MPC) approaches [NTG18; Kar+21; JS+23b]. Furthermore, several other consensus-based and current sharing controllers, which exhibit similar properties to the publications above, are given in [Gao+19].

In droop control, load sharing is achieved with a decentralised control scheme using local measurements as a proxy for communication. However, the proportional nature of standard droop approaches forces a trade-off between load sharing and voltage regulation. This necessitates a higher-level distributed or centralised controller to achieve both objectives simultaneously [Gao+19; Mod+23]. The requirement extends to other droop approaches, including inversed, nonlinear, and dead-band droop (see [Gao+19; Mod+23] and the sources therein).

In [ZD15], a droop-based secondary controller using consensus is proposed for current sharing with I loads in a resistive grid. Similarly, [DL18] uses consensus and droop for proportional power sharing in a grid with static lines and Z loads. In [Iov+18], a decentralised state-feedback proportional-integral (PI) controller for a single bus system is proposed for power sharing with Z loads.

Several consensus-based controllers have been proposed in which each DGU is an agent in a multi-agent system (MAS).[9] The authors of [Nas+15] and [Tuc+18] use consensus for current sharing with Z loads and I loads, respectively. In [DWD18], a consensus-based power sharing is achieved if a network- and load-parameter-dependent LMI can be solved for a grid with ZIP loads, static lines, and ideal voltage sources. In [Cuc+19], a weighted consensus on the injected DGU currents is implemented for a Kron-reduced grid with I loads. The approach is then extended by a sliding mode controller, which steers the network to a desired manifold, yielding current sharing. Consensus-based controllers have also been presented for monotonically increasing ZIP loads located at DGUs, with [Fan+20] and [NFT20; NTFT22] achieving power sharing and current sharing, respectively.

By making use of distributed averaging of available measurements, controllers are presented in [Tri+18] for current sharing and power sharing with ZIP loads. This approach achieves voltage regulation without requiring any voltage measurements

9 See [Lun19, Chapter 3] for an overview of the standard consensus theory for MASs.

and verifies stability through a practical experiment. The stability is then proven in [Tri+19] for a Kron-reduced network containing I loads. The authors of [SSB21] propose a proportional current sharing for Z loads that is resilient to data-injection and man-in-the-middle attacks.

Other distributed controllers include the distributed PI controller in [Vu+17], which achieves power sharing and energy management, and a current-sharing controller using event-triggered communication in [Han+18]. Making use of MAS theory, a leader-based current-sharing controller is presented in [Han+19], while [CKS07] introduces a distributed passivity-based controller for proportional current sharing with no voltage measurements. The latter result is achieved for monotone ZIP loads and uses Krasovskii-type storage functions.

MPC has also been used for voltage regulation and load sharing, although such controllers are more common for higher-level objectives operating on slower dynamics, e.g., optimal power flow in [JS+23b]. In [NTG18], a finite-step control-Lyapunov-based MPC is used to achieve optimal centralised voltage regulation and current sharing under constrained inputs in a linear system. A decentralised MPC for current sharing a single bus network with P loads is proposed in [Kar+21].

Discussion

As is the case for decentralised voltage stabilisation, many controllers aiming for a distributed coordination restrict the usable loads or topologies. Indeed, linear loads [Nas+15; ZD15; Vu+17; DL18; Iov+18; Tuc+18; Tri+18; NTG18; Cuc+19; Tri+19; SSB21] or monotone loads [DWD18; Han+18; CKS07; Fan+20; NFT20; NTFT22] are typically required to ensure stability using the various controllers. Moreover, only a few publications consider loads not connected at DGUs [Nas+15; DL18; DWD18; Tri+18; SSB21], whereas others require DGUs to be present at the loads [Han+18; Tuc+18; NTG18; Fan+20; NFT20; NTFT22], consider Kron-reduced networks [ZD15; Cuc+19; CKS07; Tri+19], or are limited to single bus networks [Vu+17; Iov+18; Kar+21]. Placing limits on where loads can be connected in the network restricts the practicality of the network, especially when the volatility of DGU-connected RESs is considered.

Furthermore, the approaches which allow loads to be connected at buses without a DGU operate using typical consensus algorithms. In standard consensus approaches, only the measurements of agents that also participate in the consensus are considered. Voltage measurements at unpowered DGUs are therefore discarded and not considered for the global voltage regulation. This can have a significant performance and resilience impact in networks with few active DGUs, in networks where the active DGU measurements are not representative of the average network voltage, or when several DGUs suddenly gain or lose power locally.

2.1.3 Gaps and Research Questions

In the context of rapidly growing and changing energy networks, controllers are required which can ensure stable voltage regulation or coordinated load sharing for DC networks in a modular and scalable way. This is especially important in light of the volatile power availability of DGU-interfaced RESs and the need to supply non-monotone P loads in an increasingly digital world.[10]

As per the literature review in Section 2.1.1, it can be seen that a variety of works have been considered for the decentralised stabilisation in DC networks. Yet despite much progress and many novel ideas, many restrictions persist, especially w.r.t. which loads may be included in the network (e.g., linear, monotonically increasing, or more general continuous static functions) and where they may connect in the network (only at a single bus, only at DGUs, or anywhere in the network). Indeed, a *single* non-monotone P load at a bus without a voltage regulating DGU foils the stability proofs of all controllers considered in Section 2.1.1.

To overcome this challenge, the following two research questions (RQs) are considered for the decentralised stabilisation in DC networks:

RQ 1.1: How can voltage regulation be achieved in the presence of non-monotone load functions at buses where power is available while enabling a scalable[11] stability result?

RQ 1.2: How can the stability of a DC network, in which only a subset of buses have power available locally, be guaranteed in a way that is scalable, modular, and robust against parameter and network topology changes?

Moving from stabilisation to the distributed coordination in DC networks, the literature review in Section 2.1.2 demonstrates similar restrictions w.r.t. the load functions and load locations. Indeed, the power-sharing approaches either consider only linear loads [Vu+17; DL18; Iov+18], ignore line dynamics [Fan+20], require a poorly scaling and network-dependent LMI to be solved [DWD18], or provide no proof of stability [Tri+18]. Moreover, in networks where the power availability is sparse or where loads might be far away from the supply, the average network voltage might differ substantially from the average bus voltage at powered DGUs. The usage of voltage measurements should thus be determined by the availability of measurement hardware and not by control system restrictions resulting from the availability of power at a bus.

Two RQs are thus formulated for the distributed coordination in DC networks:

RQ 2.1: How can average voltage regulation and proportional power sharing be implemented in a scalable manner while considering all voltage measurements in a network where steady-state power is not available at every bus?

---

10 Electronic devices are typically represented by P loads [IEEE22, p. 47].

11 The complexity of a scalable stability result is largely or fully independent of the size of the system.

RQ 2.2: How can the stability of a DC network with distributed coordination be guaranteed in a way that is robust against parameter and network topology changes while promoting modularity?

2.2 Gas Network Stabilisation: State of the Art

Gas networks have a similar setup to DC networks. In this setup, points of equal pressure in the gas network (nodes) are interconnected via gas pipelines, compressors and valves (edges). Loads representing the aggregate user demand at a given network location are situated at the corresponding node. Control of the gas network is achieved through compressors and valves, which can effect pressure increases or decreases, respectively.[12] Compressors and valves are often combined into compressor units, groups, or substations, to enable greater granularity in the pressure control, routing, and throughput [Koc+15, Section 2.4], [Men05, p. 139f., p. 173], [Dom+21, p. 23]. The direction of supply in gas networks conventionally follows a strictly top-down setup, with gas being transported over long distances in high-pressure pipelines before being distributed to lower-pressure networks. Compressors are used to maintain the high pressure needed for the transport of gas and to compensate for pressure losses from friction and elevation differences. At the distribution level, control valves and pressure reducing valves are used to maintain an acceptable pressure level and to route gas as required [Koc+15, p. 10f.], [Qad+20, p. 9f].

Remark 2.4. *Similar to electrical networks, gas networks are categorised according to their pressure levels. In Germany, the categories comprise low-pressure (up to* 100 mbar*) and medium-pressure (between* 100 mbar *and* 1 bar*) networks for gas distribution and high-pressure (between* 1 bar *and* 100 bar*) networks for gas transmission [BB23, p. 221]. Similar categories with different thresholds are also used in other countries [Qad+20, p. 11]. Note that the stated pressure ranges are gauge pressures, i.e., relative to the atmospheric pressure.*

A central requirement for gas network operation is the balancing of the inflows and outflows of gas in the network under consideration for the linepack [GasNZV10; Koc+15; Qad+20]. In the electrical analogy, this equates to balancing the current inflows and outflows under consideration of the accumulated charge. While the capacity to store accumulated charge is small in electrical networks, meaning the loads can exhaust the accumulated charge almost instantaneously if there is a power supply disruption, the linepack capacity in large gas networks can supply the gas demand for sevaral minutes to several hours during supply disruptions (see the linepack and gas demand in [Tra+18; WR19; Dvo+22]). Continuing the analogy, stabilising the gas pressure can be used as a proxy for ensuring balancing in a gas network in the same way that a

[12] While realistic valves can exhibit non-zero response times (see [Eme23]), valves are generally considered to be purely static components [Koc+15, pp. 29–31], [Men05, Chapter 9], [Qad+20, pp. 15–18].

constant DC voltage corresponds to a balance of current inflows and outflows. Thus, the following definition which mirrors the DC case (see Definition 2.1) is used for the stability of gas networks in this thesis.

Definition 2.2 (Gas Network Stability)
The steady-state pressure of each node in a stable gas network is constant and within a tolerable range from the reference pressure.[13] *In the event of changes or non-critical faults, the pressure at each node converges to a steady state within the tolerable range.*

Beyond ensuring the stable operation, additional objectives can be defined for gas networks. These include achieving load sharing between the compressors in the same compressor station [Mil+20; ZST20; Zag+23], and optimally managing the valves and compressors to lower operational costs [Dom+11; RMBS15; Dvo+22]. In [OC20], the management of a gas network is divided into the operation (the lower-level control ranging from seconds to hours), the balancing (i.e., matching the inflows and outflows on a scale of hours to seasons), and the design (where the temporal aspect is fully static) of the network. Interestingly, for the operation and balancing tasks, the control of compressors exhibits a similar hierarchy as seen for DGUs in DC networks (see Section 2.1). This involves low-level control of a single compressor (as in [GEV02; BG08; BGS16]) as a primary layer, followed by a secondary layer tasked with load sharing in a compressor station [Mil+20; ZST20; Zag+23], and a tertiary controller which seeks to manage the compressor stations in the network efficiently [Dom+11; RMBS15; Dvo+22].

Challenging the status quo in gas networks is the combination of the expected decrease in gas demand and the increase in renewable gas production as countries seek to decarbonise [Qad+19; Qad+20; Eur22], [Int23a, p. 30]. In such future networks where local supply can outstrip the local demand, additional compressors will thus be required to prevent a local pressure build-up and to enable a decentralised bottom-up supply [Eur+20]. The control of compressors and control values also gains additional relevance in light of the volatility from RESs permeating between domains in NMESs [LES08; CM16; Tay+22; Dvo+22].

In line with the central objective of ensuring the stable operation in dynamical NMESs, the following two subsections present an overview of the pertinent literature on the dynamical modelling of gas pipelines for the purposes of control and stability analysis (**Section 2.2.1**), and on the classification and control of the compressors used in gas networks (**Section 2.2.2**). An emphasis is placed on models and control methods that allow or yield modular and scalable results. Thereafter, the literature review is

[13] Since this thesis primarily considers the dynamic convergence to some network and load-dependent steady state, a tolerable range of 10 % above and below the nominal pressure is considered.

discussed and research questions are formulated for the decentralised stabilisation of gas networks.

2.2.1 Literature Review: Pipelines Modelling

Unlike in power systems, where electrical lines are often considered static and linear, gas pipelines exhibit highly nonlinear behaviour with transients ranging from seconds to hours (see [Ke00]). These effects stem from the pressure losses due to friction and elevation differences, as well as the small propagation-rate-to-capacity ratio in gas networks. A further complicating aspect is the partial differential nature of the pipeline dynamics, which arises from the conservation of mass, momentum and energy for the gas. A pipeline model is thus required which is accessible to control and analysis methods while accurately representing the complex physical phenomena.

Many different models have been proposed for dealing with the nonlinear set of partial differential equations (PDEs) representing the flow of gas in a pipeline. The existing approaches can broadly be categorised into discretised models [HG+09; DF11; PBLD16; BS19; WLC20; OC20], static models [HMS10; WM11; SSW15; Ekh+19], linearised models [ABNG12; Zho+17; Wen+18], analytical models [MVZ20; Dom+21], and models exploiting an electrical analogy [Ke00; THM17; Yan+20].

A common approach for dealing with the PDE nature of the gas pipeline equations is to discretise the model in time and space. As shown in [HG+09; WLC20], this yields a model suitable for MPC with a good fidelity. [PBLD16] similarly presents a discretised simulation model which shows a similar fidelity as the commercial SIMONE simulation software package. In [BS19], the pipeline PDEs are discretised with a finite element approach, yielding an efficient method for calculating the linepack. Finally, [OC20] provides both transient and steady-state pipeline models while also considering cases of non-homogenous gas mixtures. To reduce the large size associated with discretised models, quasi-static [HMS10] or static models [WM11; SSW15; Ekh+19] have been suggested which neglect some or all temporal effects, respectively.

To make gas network models more accessible to standard control theory, discretisation is often combined with linearisation. In [ABNG12; Wen+18], separate fourth-order state-space models are proposed for a pipeline, while similar steps are followed in [Zho+17] to obtain a transfer function for pipeline segments. Other authors have sought to align the pipeline dynamics with control theory methods by investigating and exploiting the properties of the pipeline PDEs. In [MVZ20], the authors investigate the monotonicity of a quasi-static gas network model to ensure the tractability of robust optimisation problems for gas networks. On the other hand, the technical report [Dom+21] shows that the pipeline PDEs make up an infinite-dimensional port-Hamiltonian system.[14]

[14] See [vJ14; Dui+09] for comprehensive overviews on the port-Hamiltonian modelling framework.

Apart from using the port-Hamiltonian framework to provide model structure, some authors have proposed electrical analogies for gas pipelines. In [Ke00], a pipeline is modelled as a parallel RC branch connected in series with an inductor and an energy source. A similar approach is presented in [THM17] for the quasi steady state. Finally, [Yan+20] propose cascaded RC π-structures with linearised resistances for the friction.

Discussion

In the literature, a variety of gas pipeline models have been proposed for tasks ranging from control and optimisation to simulation and analysis. Although the approaches in [HG+09; DF11; PBLD16; BS19; WLC20; OC20; MVZ20] provide a high model fidelity by discretising the gas pipeline PDEs, these models are less accessible to analytical control or analysis methods. In contrast, the static or quasi-static models in [HMS10; WM11; SSW15; Ekh+19; Yan+20] allow the gas models to be simplified at the cost of the temporal fidelity of the models. While the linearised models proposed by [ABNG12; Zho+17; Wen+18] allow for a straightforward application of conventional control and stability methods, the fidelity of these models differs materially from the models obtained purely through discretisation when compared in a standard benchmark system. In contrast, the port-Hamiltonian approach in [Dom+21] provides structured insights into the properties of gas pipelines, which may be incorporated into control design and system analysis methods. Nevertheless, the infinite-dimensional nature of these models requires a discretisation before they can be implemented in a numerical simulation. Finally, the approaches in [Ke00; THM17; Yan+20] drawing on the electrical analogy for gas present an interesting idea in exploiting the similarities between many of the thermodynamic domains (see [Dui+09, p. 24]), yet the fidelity of the proposed models as demonstrated in benchmark simulations only approximately match the fidelity achieved by the linearised models.

When considering the results of the various gas pipeline models covered in this section, it is interesting to note that approximating the pipeline PDEs through a linearisation has a much greater effect on the model fidelity compared to the coarseness of the discretisation. Indeed, while the approaches in [HG+09; DF11; BS19] use 20 or more discrete elements or sections in their implementations, it has also been shown in [PBLD16; WLC20] that good performance can be achieved using only one or two segments for the pipeline, respectively. Notably, this is not significantly different from the linearisation-based approaches in [ABNG12; Zho+17; Wen+18], where between two and five sections are used for the pipelines. Furthermore, while the electrically inspired models in [Ke00; THM17; Yan+20] do not show a similar fidelity level compared to [PBLD16], the structured port-Hamiltonian model presented in [Dom+21] hints at an electrical equivalent circuit for a gas pipeline which conforms to the common thermodynamic framework and which might result in an increased model fidelity.

2.2.2 Literature Review: Gas Network Compressor

Pressure increases in gas networks are required for overcoming pressure losses from elevation changes or friction and for injecting gas into a higher-pressure network or gas storage. Such increases are realised using compressors, which are categorised into positive displacement (e.g., piston and rotary compressors) and non-positive displacement (e.g., centrifugal and axial compressors) types [Qad+20, p. 11]. Both operation types are common in gas networks [Koc+15, p. 31]. Piston compressors generally achieve higher compression ratios at a lower, discontinuous throughput, whereas centrifugal compressors yield greater efficiency and a continuous flow rate [SSW15, p. 154], [Men05, p. 167], [Qad+20, p. 11f.], [Koc+15, p. 35]. In this thesis, centrifugal compressors are considered since these are more commonly used for transporting gas via pipelines [GE99b, p. 11], [Men05, p. 167].

A central requirement for centrifugal compressors is preventing instability arising from compressor surge [GE99b, p. 14]. Surge occurs at low flow rates where the incremental flow-rate-to-pressure-ratio becomes positive. This causes positive feedback where a lower flow rate yields a lower pressure ratio between the compressor inlet and outlet. This in turn causes a further decrease of the flow rate [GE99b, p. 14f], [Gra+00].

Approaches to combat surge in centrifugal compressors can be grouped into active surge control [GE99b; GE99a; GEV02; BG05; BG06; BG08; SAA09; BGS16; Gha+18; ZM20; KH21] and anti-surge control [Cor+15; Ben+15; Tor+17; Mil+20; Mei+22]. In active surge control, controllers ensure stable operation within the surge-capable region. In contrast, anti-surge controllers actively prevent the compressor from entering the surge region. Both controller types use a variety of control inputs, including changing the angle of the inlet guide vanes (IGV) to affect the flow rate, adding a close-coupled valve (CCV) after the compressor to lower the pressure, using recycle or backflow valves to increase the required flow rate through the compressor, or using a throttle control valve (TCV) at the outlet of the entire system.

Starting with the active surge controllers, [GE99b; GE99a] develop backstepping and passivity-based controllers for a CCV-equipped compressor. These ideas are expanded in [GEV02], where surge prevention is achieved without a CCV for a constant inlet pressure. Similar controllers are proposed in [BG05; BG06; BG08], where the compressor speed is explicitly considered. In contrast, [SAA09] proposes single-input-single-output feedback controllers for a CCV- and TCV-equipped constant speed compressor.

More recent active surge controllers include the controller using state feedback linearisation in [BGS16] for a CCV-equipped compressor, where robustness is demonstrated via the circle criterion. Similarly, [Gha+18] proposes an adaptive controller combined with backstepping which is robust against uncertainties in the compressor characteristic. In [ZM20], a backstepping design is used for a CCV- and TCV-equipped compressor and a robust Lyapunov stability is demonstrated. Finally, in [KH21], a nonlinear feedback controller is derived for a wet-gas centrifugal compressor using a Lyapunov analysis in combination with optimisation.

Several MPC-based anti-surge controllers have also been proposed for centrifugal compressors. In [Cor+15], a linearised MPC is used to control a recycle valve and the compressor torque, which is shown to yield good results compared to a gain-scheduled PI controller while being robust against input pressure disturbances. A linearised MPC approach is also used in [Ben+15] to control the IGV, torque, and a recycle valve while being robust against inlet and outlet pressure changes. In [Tor+17], linear, sequential quadratic programming, and nonlinear pressure-regulating MPCs are compared, with the nonlinear variants ensuring a larger distance to the surge region. [Mil+20] uses a nonlinear MPC for optimal load sharing in serial and parallel compressor configurations. Finally, [Mei+22] proposes a nonlinear MPC for pressure regulation under non-ideal and non-isothermal gas flow.

Discussion

The instability associated with surge behaviour in centrifugal compressors occurs when the incremental change in pressure ratio w.r.t. the flow rate becomes positive. Approaches to deal with this behaviour include adding a CCV which adds damping to the flow-rate state, using the compressor drive torque which indirectly affects the flow rate, or avoiding the instability by using a recycle valve to increase the flow rate. Interestingly, this closely mirrors the problem of stabilising the bus voltages[15] in Section 2.1.1, as shown in Table 2.1.[16] While the underlying physical phenomena of these systems differ greatly, many similarities can also be seen in the control structures used to combat the instabilities of the respective systems.

Similar to the DC case before, it is desirable to ensure that the regulated compressors will work correctly in arbitrary network configurations. The current approaches in the literature, however, only consider the stability of operating points [GE99b; GE99a; BG05; BG06; BG08; SAA09], or a robust stability subject to model uncertainties and/or inlet/outlet disturbances [GEV02; Cor+15; Ben+15; BGS16; Tor+17; Gha+18; ZM20; Mil+20; KH21; Mei+22].[17] While some authors determine bounds for the disturbances like in [GEV02], the robustness of the controllers w.r.t. inlet and outlet disturbances is only demonstrated via simulations.

15 Voltage and flow rates may both be considered as generalised effort variables in the thermodynamic framework [Dui+09, p. 24].

16 The use of a recycle valve corresponds electrically to a parallel current-controlled source connecting to a bus. This, however, is not a common solution in the electrical domain, since the regions where the loads exhibit negative incremental impedance often corresponds with the desirable region of operation.

17 The passivity-based controllers such as those in [GE99b; BG06] use passivity theory to derive the respective stabilising controller. However, these approaches do not yield a controlled system that is passive w.r.t. the inlet and outlet ports.

Table 2.1: Gas Compressor Versus DC Bus Control

Aspect	Gas Compressor Control	DC Bus Control
Unstable variable	Compressor flow rate.	Bus voltage.
Cause of instability	Positive incremental flow-rate-to-pressure-ratio of the compressor characteristic.	Negative incremental impedance of the static load function.
Direct damping injection	Adding a CCV to dampen the flow rate while decreasing the efficiency.	Adding a shunt resistance to dampen the bus voltage while decreasing the efficiency.
Indirect control action	Control the compressor drive torque to influence the first-order speed dynamics of the compressor.	Control the converter voltage to influence the first-order DGU filter current dynamics.

Finally, while the various MPC approaches in [Cor+15; Ben+15; Tor+17; Mil+20; Mei+22] yield robust stability results and incorporate additional factors, including the operational efficiency and actuator saturation, they all employ anti-surge control which avoids the regions of instability. Active surge control, which brings with it the ability to operate the compressor in the surge region, is currently limited to the direct feedback controllers. This is due to the computational time required for the MPC approaches currently resulting in the actuation being too slow to counteract surge once it occurs.

2.2.3 Gaps and Research Questions

In future gas networks, the conventional top-down supply structure of gas networks will be challenged by the combination of lower gas demand, local gas production, and volatility propagating through from other energy networks. This can give rise to situations of local gas oversupply, which require controllers for valves and compressors to ensure the stable operation of the gas networks in a scalable and modular manner.

Guaranteeing the stability of an entire gas network necessitates a model for gas pipelines that accurately reflects their complex nonlinear dynamics while being accessible for use in the prevalent analytical stability analysis methods. From the literature review in Section 2.2.1, it can be seen that a variety of pipeline models have been proposed in the literature. Despite the quantity and the variety of approaches used to deal with the nonlinear set of PDEs describing the flow of gas in a pipeline, none of the models provides a combination of a verified high model fidelity together with a model structure conducive to analytical control and analysis methods.

The addition of compressors in gas networks with surplus gas production also has the potential to exacerbate pressure fluctuations, leading to potential instability. While several state-of-the-art approaches demonstrate a regulated compressor stability that is robust w.r.t. model uncertainties and inlet or outlet disturbances, as discussed in Section 2.2.2, no approach analyses and demonstrates the stable operation of a gas network which contains surge-capable compressors. Furthermore, no approaches consider the extraction of excess gas in a network which requires the regulation of the compressor inlet pressure.

In this thesis, the following two RQs are therefore formulated for the decentralised stabilisation in gas networks:

RQ 3.1: How can gas pipelines be modelled in a way that is conducive to analytical control design and stability analysis methods while retaining a high fidelity?

RQ 3.2: How can centrifugal compressors be controlled to achieve pressure regulation of the inlet in a decentralised manner while ensuring that compressor-equipped gas networks exhibit a scalable and modular stability that is robust against parameter and network topology changes?

2.3 Dynamical Networked Multi-Energy Systems

In a future where there is greater interconnection and codependence between various energy networks, the question of how such NMESs influence each other becomes increasingly relevant. Of special interest is the extent to which the dynamics of the respective networks influence each other. In this work, NMESs comprising an interconnection of DC and gas networks are considered more closely. This section builds upon the state of the art and the literature overview of the respective networks in Sections 2.1 and 2.2, while dealing explicitly with cases where an interconnection of these network types is considered. Note that since a majority of publications consider AC systems, both types of electric power systems are considered in the review below. Thus, following an overview of the literature and a discussion thereof in **Section 2.3.1**, gaps relating to NMESs comprising DC and gas networks are identified and the corresponding research questions are formulated in **Section 2.3.2**.

2.3.1 Literature Review: Interconnected Electric and Gas Networks

A large body of research is dedicated to the interconnection of electric and gas networks, as indicated in the review papers [He+18; Gue+19]. Broadly speaking, this body of research can be grouped into publications dealing with the topics of analysis, simulation and feasibility [Erd+14; Pam+17; Pam18; Liu+21; Zha+23], or which perform a coordination or optimisation to obtain operational or economic benefits [Liu+09; Ala+15; CPSM15; CM16; He+17; Fan+18; Bel+20; Xu+20; OMa21].

Starting with the category of analysis, simulation and feasibility, [Erd+14] proposes a simulation model for gas-electric systems coupled via gas-fired power plants (GFPPs) and the compressor power requirements. The model is used to analyse the static coupling effect on network loading and efficiency. Similarly, [Pam+17; Pam18] propose a simulation framework with dynamical gas networks for analysing the security of the energy supply. In [Liu+21], the static stability of electric-gas networks coupled via GFPPs is investigated using the steady-state power flow and gas flow equations. Furthermore, the authors of [Zha+23] present a simulation model using static AC power flow and linearised gas flow equations for integrated electric and gas systems coupled by GFPPs and P2G components.

A number of publications use interconnected electric-gas networks as a basis for obtaining an operational or economic benefit. In [Liu+09], the short-term scheduling and unit commitment of GFPPs are considered such that the gas supply is minimised for the case that the electric network has no slack node. In [Ala+15], the operational costs of gas turbines are minimised in the presence of intermittent wind power while accounting for component start-up times. A similar study is conducted in [CPSM15], where the optimal power flow and linearised gas flow equations are used to minimise the operation cost in a GFPPs-coupled gas-electric system. In [CM16], a two-stage optimisation is proposed in which the power flow equations are used to check whether curtailment is needed before optimally coordinating the use of coupling P2G components. The authors of [He+17] similarly consider a cost-minimising co-optimisation for the scheduling of a coupled electric-gas system while accounting for load and wind power uncertainties.

Continuing the list of optimisation-based approaches, [Fan+18] minimise the cost of operation while reducing network congestion and the curtailment of RESs. The authors also demonstrate the necessity of using a dynamical gas network model while using a static power system consideration. Similarly, [TWZ19; Bel+20] minimise the operational costs of gas-electric systems using linearised gas flow models. In [Xu+20], an NMES comprising gas, electric and district heating networks is considered with the authors proposing a method for an optimal consensus-based coordination of the energy hubs which interconnect the various networks. Finally, the author of [OMa21] uses optimal gas and power flow equations to solve the dispatch problem in GFPPs-coupled gas-electric networks. This is done by defining a flexibility set for the GFPPs such that a feasible electric solution is guaranteed.

Discussion

The literature dealing with interconnected gas and electric systems, while varied in their goals and approaches, overwhelmingly deals with timescales in the range of hours or longer. To solve optimisation and coordination problems in a large NMES comprising complex, nonlinear dynamics, many simplifying assumptions are made for the networks. Indeed, while some authors consider and demonstrate the necessity of the dynamical effects of gas networks [CPSM15; Pam+17; Pam18; Fan+18;

Bel+20; OMa21; Zha+23], several other publications focus purely on static gas networks [Liu+09; Erd+14; Ala+15; CM16; He+17; TWZ19; Xu+20; Liu+21].

The consideration of the electric power systems, however, is generally constrained to the static case. While many authors use the AC power flow equations to relate the nodal voltage to the injected active and reactive powers [Erd+14; Pam+17; Pam18; TWZ19; Xu+20; Liu+21; Zha+23], many others simplify these equations even further by employing the DC power flow equations [Liu+09; Ala+15; CPSM15; CM16; He+17; Fan+18; Bel+20]. Not only do these equations fully omit the dynamics of the electric power systems, they implicitly restrict the components in the network to constant power or constant impedance loads or sources. Restricting the electrical networks to AC or DC power flow equations is thus also problematic due to the current lack of stability results for networks dominated by the non-monotone incremental impedances associated with constant power loads (see the discussions in Section 2.1).

Beyond the static versus dynamic considerations of the respective networks in an NMES, the research gaps pertaining to the dynamical stability of the independent networks, as identified in Sections 2.1 and 2.2, remain relevant here. This is especially important in light of the fact that the investigated literature on interconnected networks assumes a stable operation of the networks in isolation. However, neither analytical stability studies nor extensive dynamical simulation studies have been performed on the interconnection of possibly unstable power and gas networks to verify the assumed stable operation. Indeed, the coupling of networks introduces further disturbances which could adversely affect stability. Thus, as also remarked in the review paper [Gue+19], there is a need for the "analysis of network integration into multi-energy models with the aim of understanding the effects of dynamics on sudden conversion of energy from one form to another".

2.3.2 Gaps and Research Questions

Interconnecting various energy networks allows potential economic and operational benefits to be realised by exploiting the flexibility and the unique characteristics of the energy networks. Nevertheless, such higher level objectives rely on networks which exhibit a dynamically stable operation along with the assumption that the effects of the coupling components do not degrade the stability assumption. Building on the RQs formulated in Sections 2.1 and 2.2, which aim to achieve a robust stability for individual DC and gas networks, the pertinent question is if and how these stability results transfer to the interconnected NMES case.

To this end, two RQs are formulated for an NMES comprising DC and gas networks:

RQ 4.1: How can DC and gas networks be coupled such that the stability of the networks is conserved while also ensuring a scalable and modular stability for the NMES?

RQ 4.2: To what extent are the transients and steady states of DC and gas networks affected by the coupling in an NMES?

2.4 Summarised Research Questions

Concerning the main objective of this thesis of providing controllers for the decentralised stabilisation and the distributed coordination in DC networks, gas networks and NMESs, an investigation of the literature is presented. Important gaps in the existing literature are identified, both for individual DC and gas networks as well as for NMESs. Based on these gaps, research questions are formulated which are summarised as follows:

- DC networks require decentralised controllers which ensure a scalable, modular, and robust stability for a network with non-monotone loads and without pervasive steady-state power availability. To this end, RQ 1.1 and RQ 1.2 are addressed in Chapter 4.
- DC networks require a distributed coordination scheme in which proportional power sharing is achieved along with an average voltage regulation using all measured bus voltages while yielding a stability result that is scalable, modular, and robust. The corresponding RQ 2.1 and RQ 2.2 are addressed in Chapter 5.
- Gas networks require high-fidelity pipeline models which are conducive to analytical design and stability analysis methods. Furthermore, decentralised controllers are required for compressors which ensure that the pressure regulation and stability in gas networks are achieved in a modular, scalable and robust manner. To address these gaps, answers to RQ 3.1 and RQ 3.2 are presented in Chapter 6.
- NMESs comprising DC and gas networks require descriptions of coupling components which enable a scalable and modular stability analysis for the NMESs. Additionally, an investigation into the effects that coupling components have on the transients and steady states of DC and gas networks in dynamical NMESs is required. Answers to the corresponding RQ 4.1 and RQ 4.2 are provided in Chapter 7.

3 Theoretical Methods

DC networks, gas networks, and NMESs are large and complex networks comprising a variety of different components. To facilitate answering the RQs formulated in Chapter 2, this chapter provides theoretical methods pertaining to graph theory and dissipativity theory. Whereas graph theory is helpful for representing and analysing networks, dissipativity theory allows the stability of complex and nonlinear systems to be investigated in a modular fashion.

The chapter starts of with **Section 3.1**, in which a brief introduction to algebraic graph theory is presented. Beyond introducing the requisite notation for graphs, important matrices derived from graphs are recalled along with their properties.

Thereafter, in **Section 3.2**, the theory of dissipativity and the special case of passivity are introduced. Equilibrium-independent variants for dissipativity and passivity are also introduced and links to the Lyapunov stability of systems are recalled.

Finally, in **Section 3.3**, new dissipativity results are presented for systematically deriving asymptotic stability results for cases not covered by the existing literature. Two useful lemmas are also provided for investigating the equilibrium independent passivity properties of static and nonlinear systems.

3.1 Algebraic Graph Theory

Graphs play a key role in the description of large networks. In this thesis, graphs are used to describe both the physical and the communication interconnections between buses or nodes in a network. This section introduces the graph notation and lists some properties required in the sequel.

Let $\mathcal{G} = (\mathcal{N}, \mathcal{E})$ define a finite graph without self loops in which the vertices $\mathcal{N} = \{1, \ldots, |\mathcal{N}|\}$, $|\mathcal{N}| \geq 1$ are interconnected by the edges $\mathcal{E} = \{1, \ldots, |\mathcal{E}|\}$, $|\mathcal{E}| \geq 0$. Let each edge be assigned an arbitrarily chosen direction. For each vertex $n \in \mathcal{N}$, denote the sets of edges $\mathcal{E}_n^+$ and $\mathcal{E}_n^-$ which have vertex n as a sink node or a source node, respectively. Similarly, the source node and sink node of an edge $o \in \mathcal{E}$ are defined by the sets $\mathcal{N}_o^+$ and $\mathcal{N}_o^-$, respectively.

In graphs describing physical connections, it is useful to assign a positive flow direction in the direction of the edge, even though bidirectional flow can occur. Using the directed edges, the *incidence matrix* of a graph $\mathcal{G}$ can be described as

$$\boldsymbol{E} = (e_{no}) \in \mathbb{R}^{|\mathcal{N}|\times|\mathcal{E}|}, \quad e_{no} = \begin{cases} +1 & \text{if vertex } n \text{ is the sink of edge } o, \\ -1 & \text{if vertex } n \text{ is the source of edge } o, \\ 0 & \text{otherwise.} \end{cases} \tag{3.1}$$

The incidence matrix $\boldsymbol{E}$ may also be divided into row vectors $\boldsymbol{e}_{\text{row},n} \in \mathbb{R}^{1\times|\mathcal{E}|}$ describing the incoming and outgoing edges associated with a node n, or column vectors $\boldsymbol{e}_{\text{col},o} \in \mathbb{R}^{|\mathcal{N}|\times 1}$ specifying the sink and source nodes associated with an edge o, i.e.,

$$\boldsymbol{E} = \begin{bmatrix} \boldsymbol{e}_{\text{row},1} \\ \hdashline \vdots \\ \hdashline \boldsymbol{e}_{\text{row},|\mathcal{N}|} \end{bmatrix} = \begin{bmatrix} \boldsymbol{e}_{\text{col},1} & \cdots & \boldsymbol{e}_{\text{col},|\mathcal{E}|} \end{bmatrix}. \tag{3.2}$$

For graphs where the edges are associated with the positive weights $R_o > 0$, $\boldsymbol{R} = \text{diag}(R_o)$, the *Laplacian matrix* can be obtained from $\boldsymbol{\mathcal{L}} = \boldsymbol{ERE}^T \in \mathbb{R}^{|\mathcal{N}|\times|\mathcal{N}|}$ [Che+21]. For a connected graph, the eigendecomposition of $\boldsymbol{\mathcal{L}}$ is [Che+21]:

$$\boldsymbol{\mathcal{L}} = \boldsymbol{\Xi}\boldsymbol{\Lambda}\boldsymbol{\Xi}^{-1}, \qquad \boldsymbol{\Xi} = [\mathbb{1}_{|\mathcal{N}|}, \boldsymbol{\Xi}_{\text{Im}}], \tag{3.3a}$$

$$\boldsymbol{\Lambda} = \text{diag}(\lambda_n), \qquad 0 = \lambda_1 < \lambda_2 \leq \cdots \leq \lambda_{|\mathcal{N}|}, \tag{3.3b}$$

i.e., $\boldsymbol{\mathcal{L}}$ has a kernel comprising the vector of ones with the eigenvalue zero, and an image $\boldsymbol{\Xi}_{\text{Im}}$ associated with $|\mathcal{N}| - 1$ strictly positive eigenvalues. The smallest strictly positive eigenvalue $\lambda_2(\boldsymbol{\mathcal{L}})$ is known as the *algebraic connectivity* of the graph, with larger values indicating a greater vertex connectivity. For certain graph types—like star, cycle, or path graphs—the algebraic connectivity has a known value [BH12, p. 8], [Fie73]. In more general cases, bounds for $\lambda_2(\boldsymbol{\mathcal{L}})$ can be computed (see [BH12, p. 52], [dAb07]) or $\lambda_2(\boldsymbol{\mathcal{L}})$ can be determined using the Courant-Fischer theorem [Fie73; dAb07]

$$\lambda_2(\boldsymbol{\mathcal{L}}) = \min_{\boldsymbol{\xi}\neq\boldsymbol{0},\,\boldsymbol{\xi}\perp\mathbb{1}} \frac{\boldsymbol{\xi}^T\boldsymbol{\mathcal{L}}\boldsymbol{\xi}}{\boldsymbol{\xi}^T\boldsymbol{\xi}}. \tag{3.4}$$

3.2 Dissipativity Theory Fundamentals

An important aspect when analysing the stability of large networked systems pertains to the scalability of the methods. By extending and generalising concepts like the internal stability associated with Lyapunov theory and input-output measures such as L_2-gain and ISS, *dissipativity theory* allows the stability of large systems to be investigated modularly. This modularity is achieved by leveraging the property that interconnecting dissipative systems again yields a dissipative system. Since the equilibria of large networks, however, are not necessarily equal to the equilibria of the unconnected network components, it is useful to consider the equilibrium-independent dissipativity (EID) for large networks. This allows a subsequent stability result for a network to hold for any possible equilibrium.

Dissipativity theory, originally introduced in [Wil72], is only presented in an abbreviated form in this section, with the interested reader being directed to [SJK97; BL07; AMP16; vdS17; Bro+20] for more expansive overviews on the topic. The introduction of this section continues with the dissipativity and passivity definitions. Dissipativity is then extended to EID in **Section 3.2.1**. Thereafter, **Section 3.2.2** provides an overview of how the stability of dissipative systems can be investigated.

Consider the general nonlinear system

$$\dot{\boldsymbol{x}} = \boldsymbol{f}(\boldsymbol{x}, \boldsymbol{u}), \tag{3.5a}$$

$$\boldsymbol{y} = \boldsymbol{h}(\boldsymbol{x}, \boldsymbol{u}), \tag{3.5b}$$

where $\boldsymbol{x} \in \mathcal{X} \subseteq \mathbb{R}^{\mathrm{n}}$, $\boldsymbol{u} \in \mathcal{U} \subseteq \mathbb{R}^{\mathrm{m}}$, $\boldsymbol{y} \in \mathcal{Y} \subseteq \mathbb{R}^{\mathrm{m}}$, and $\boldsymbol{f}\colon \mathcal{X} \times \mathcal{U} \to \mathbb{R}^{\mathrm{n}}$ and $\boldsymbol{h}\colon \mathcal{X} \times \mathcal{U} \to \mathcal{Y}$ are class C^1 functions. Furthermore, consider a continuous function $w\colon \mathcal{U} \times \mathcal{Y} \to \mathbb{R}$ which is called the supply rate.

Definition 3.1 (Dissipativity, see [SJK97, p. 27], [AMP16, p. 4], [vdS17, p. 34])
A system (3.5) *is dissipative w.r.t. a supply rate* $w(\boldsymbol{u}, \boldsymbol{y})$ *if there exists a positive semi-definite class* C^1 *storage function* $S\colon \mathcal{X} \to \mathbb{R}_{\geq 0}$ *with* $S(\boldsymbol{0}) = 0$ *such that* $\dot{S} \leq w(\boldsymbol{u}, \boldsymbol{y})$ *holds for all* $(\boldsymbol{x}, \boldsymbol{u}) \in \mathcal{X} \times \mathcal{U}$. *Strict dissipativity follows if the inequality holds strictly.*

Remark 3.1. *Dissipativity is typically introduced in its integral form (see [SJK97, Definition 2.1], [AMP16, Definition 1.1], and [vdS17, Definition 3.1.2]) which also considers the initial states and is valid for system dynamics which are not* C^1. *However, compared to the differential form in Definition 3.1, integral dissipativity is harder to verify since it involves computing the state trajectories [vdS17, p. 43].*

By specifying a form for w, various concepts in control theory can be related to dissipativity theory. Passivity theory, in particular, allows for a modular stability result to be established for a system by considering only the passivity properties of the constitutive subsystems and their interconnection structure.

Definition 3.2 (Dissipative Supply Rates)
A system (3.5) that is dissipative w.r.t. a supply rate $w(\boldsymbol{u}, \boldsymbol{y})$

1. *is passive if $w = \boldsymbol{u}^T\boldsymbol{y}$ [SJK97, Definition 2.2],*
2. *is input-feedforward passive (IFP) if $w = \boldsymbol{u}^T\boldsymbol{y} - \nu\boldsymbol{u}^T\boldsymbol{u}$, with $\nu \in \mathbb{R}$ [SJK97, Definition 2.12],*
3. *is input-strictly passive (ISP) if $w = \boldsymbol{u}^T\boldsymbol{y} - \nu\boldsymbol{u}^T\boldsymbol{u}$, with $\nu \in \mathbb{R}_{>0}$ [vdS17, Definition 2.2.8],*
4. *is output-feedback passive (OFP) if $w = \boldsymbol{u}^T\boldsymbol{y} - \rho\boldsymbol{y}^T\boldsymbol{y}$, with $\rho \in \mathbb{R}$ [SJK97, Definition 2.12],*
5. *is output-strictly passive (OSP) if $w = \boldsymbol{u}^T\boldsymbol{y} - \rho\boldsymbol{y}^T\boldsymbol{y}$, with $\rho \in \mathbb{R}_{>0}$ [vdS17, Definition 2.2.8],*
6. *is input-feedforward output-feedback passive (IFOFP) if $w = (1+\nu\rho)\boldsymbol{u}^T\boldsymbol{y} - \nu\boldsymbol{u}^T\boldsymbol{u} - \rho\boldsymbol{y}^T\boldsymbol{y}$, with $\nu, \rho \in \mathbb{R}$ [BL07, p. 27],*
7. *is ISS if $w = \beta(\|\boldsymbol{u}\|) - \alpha(\|\boldsymbol{y}\|)$, where $\alpha, \beta \in \mathcal{K}_\infty$ [Son08, p. 178],*[1]
8. *is integral ISS if $w = \beta(\|\boldsymbol{u}\|) - \alpha(\|\boldsymbol{y}\|)$, where $\beta \in \mathcal{K}_\infty$ and $\alpha\colon \mathbb{R}_{\geq 0} \to \mathbb{R}_{\geq 0}$ is positive definite, i.e., $\alpha(r) > 0$ for $r > 0$ and $\alpha(0) = 0$ [Son08, p. 184],*
9. *has an L_2-gain $\leq \gamma_{L_2}$ if $w = \gamma_{L_2}^2\boldsymbol{u}^T\boldsymbol{u} - \boldsymbol{y}^T\boldsymbol{y}$, with $\gamma_{L_2} > 0$ [vdS17, Definition 8.1.1].*

Remark 3.2. *In the passivity framework, the input passivity index ν and the output passivity index ρ may be used to classify the excess (> 0) or lack (< 0) of passivity of a system. This is particularly relevant for the interconnection of passive systems, where an excess passivity in one system may compensate for a lack of passivity in another system (see [SJK97, Theorem 2.34], [Kha02, Theorem 6.2]).*

Remark 3.3. *Systems with ISS or finite L_2-gain properties are well known for their input disturbance rejection properties. Moreover, since OSP systems have an L_2-gain of $\frac{1}{\rho}$ [AMP16, p. 3], such systems also exhibit disturbance rejection properties. Indeed, as discussed in [MJSH22], a system with a finite L_2-gain also admits an IFOFP supply rate with $\rho > 0$ and $\nu < 0$. An OSP system, which is IFOFP with $\rho > 0$ and $\nu = 0$, is therefore a stronger condition than having a finite L_2-gain or being ISS.*

Remark 3.4. *A supply rate $w(\boldsymbol{u}, \boldsymbol{y})$ is defined by the input and output signals of a system. Given the dissipation inequality $\dot{S} \leq w_1(\boldsymbol{u}, \boldsymbol{y})$, it therefore follows that $\dot{S} \leq w_2(\boldsymbol{u}, \boldsymbol{y})$ if*

[1] A function $\alpha \in \mathcal{K}_\infty$ with $\alpha\colon \mathbb{R}_{\geq 0} \to \mathbb{R}_{\geq 0}$ is continuous, strictly increasing, unbounded and has $\alpha(0) = 0$ [Kha02, p. 144].

$w_1(\boldsymbol{u}, \boldsymbol{y}) \leq w_2(\boldsymbol{u}, \boldsymbol{y})$ for all $\boldsymbol{u} \in \mathcal{U}$ and $\boldsymbol{y} \in \mathcal{Y}$. Since this supply rate comparison is independent of $\dot{S}$ and must also hold for the steady state with $\dot{S} = 0$, it is sufficient to evaluate $\dot{S} = 0 \leq w_1(\boldsymbol{u}, \boldsymbol{y}) \leq w_2(\boldsymbol{u}, \boldsymbol{y})$. For the quadratic supply rates in Definition 3.2 for which sector conditions can be established [Kha02, Definition 6.2], $w_1 \leq w_2$ if the sector $0 \leq w_1$ lies within the sector $0 \leq w_2$.

3.2.1 Equilibrium-Independent Dissipativity

In many cases, the exact operation point of a system is not known a priori. To derive stability from dissipativity theory in such cases, the required dissipativity must hold not only for an equilibrium, but for a range of valid equilibria. This can be done by investigating the EID of a system [AMP16; SP19] or, in specialised cases, its equilibrium-independent passivity (EIP) [HAP11] or shifted passivity [vdS17, p. 96]. A central requirement in these cases pertains to the properties of the state equilibrium $\hat{\boldsymbol{x}}$.

Assumption 3.1 (Equilibrium Input-Output Uniqueness, see [AMP16, p. 24], [SP19])

For the system (3.5), there exists a non-empty set $\hat{\mathcal{X}} \subset \mathbb{R}^n$ where there is a unique input $\hat{\boldsymbol{u}} \in \hat{\mathcal{U}} \subset \mathbb{R}^m$ and a unique output $\hat{\boldsymbol{y}} \in \hat{\mathcal{Y}} \subset \mathbb{R}^m$ for every equilibrium $\hat{\boldsymbol{x}} \in \hat{\mathcal{X}}$ such that (3.5) verifies $\boldsymbol{f}(\hat{\boldsymbol{x}}, \hat{\boldsymbol{u}}) = \boldsymbol{0}$ and $\hat{\boldsymbol{y}} = \boldsymbol{h}(\hat{\boldsymbol{x}}, \hat{\boldsymbol{u}})$.

For a system where the conditions in Assumption 3.1 hold, a dissipativity independent of the equilibrium can be defined.

Definition 3.3 (EID [AMP16, Definition 3.1])

A system (3.5) is (strictly) EID if there exists a positive semi-definite class C^1 storage function $S(\boldsymbol{x}, \hat{\boldsymbol{x}})$, $S\colon \mathcal{X} \times \hat{\mathcal{X}} \to \mathbb{R}_{\geq 0}$, with $S(\hat{\boldsymbol{x}}, \hat{\boldsymbol{x}}) = 0$, that is (strictly) dissipative w.r.t. a supply rate $w(\boldsymbol{u} - \hat{\boldsymbol{u}}, \boldsymbol{y} - \hat{\boldsymbol{y}})$ for all $(\boldsymbol{x}, \hat{\boldsymbol{x}}, \boldsymbol{u}) \in \mathcal{X} \times \hat{\mathcal{X}} \times \mathcal{U}$.

Remark 3.5 (Simplified EID storage function notation). *For notational convenience, let $S(\tilde{\boldsymbol{x}}) := S(\boldsymbol{x}, \hat{\boldsymbol{x}})$, where $S(\tilde{\boldsymbol{x}})$ comprises a family of storage functions defined for each $\hat{\boldsymbol{x}} \in \hat{\mathcal{X}}$ (see also [SP19, Definition 3.2]).*

It should be noted that the EID supply rates are independent of the equilibrium and are thus uniform for all $\hat{\boldsymbol{x}} \in \hat{\mathcal{X}}$ [SP19]. Nevertheless, if Assumption 3.1 holds, any of the supply rates in Definition 3.2 may be converted to their equilibrium-independent versions by changing $w(\boldsymbol{u}, \boldsymbol{y})$ to $w(\tilde{\boldsymbol{u}}, \tilde{\boldsymbol{y}}) = w(\boldsymbol{u} - \hat{\boldsymbol{u}}, \boldsymbol{y} - \hat{\boldsymbol{y}})$. The following non-exhaustive list of EID supply rates is defined for use in this thesis.

Definition 3.4 (EID supply rates, see [AMP16, p. 25], [HAP11])
A system (3.5) *that is EID w.r.t. a supply rate* $w(\tilde{\boldsymbol{u}}, \tilde{\boldsymbol{y}})$ *is*

1. *EIP if* $w = \tilde{\boldsymbol{u}}^T \tilde{\boldsymbol{y}}$*,*
2. *input-feedforward equilibrium-independent passive (IF-EIP) if* $w = \tilde{\boldsymbol{u}}^T \tilde{\boldsymbol{y}} - \nu \tilde{\boldsymbol{u}}^T \tilde{\boldsymbol{u}}$*, with* $\nu \in \mathbb{R}$*,*
3. *input-strictly equilibrium-independent passive (IS-EIP) if* $w = \tilde{\boldsymbol{u}}^T \tilde{\boldsymbol{y}} - \nu \tilde{\boldsymbol{u}}^T \tilde{\boldsymbol{u}}$*, with* $\nu \in \mathbb{R}_{>0}$*,*
4. *output-feedback equilibrium-independent passive (OF-EIP) if* $w = \tilde{\boldsymbol{u}}^T \tilde{\boldsymbol{y}} - \rho \tilde{\boldsymbol{y}}^T \tilde{\boldsymbol{y}}$*, with* $\rho \in \mathbb{R}$*,*
5. *output-strictly equilibrium-independent passive (OS-EIP) if* $w = \tilde{\boldsymbol{u}}^T \tilde{\boldsymbol{y}} - \rho \tilde{\boldsymbol{y}}^T \tilde{\boldsymbol{y}}$*, with* $\rho \in \mathbb{R}_{>0}$*,*
6. *input-feedforward output-feedback equilibrium-independent passive (IFOF-EIP) if* $w = (1 + \nu\rho)\tilde{\boldsymbol{u}}^T \tilde{\boldsymbol{y}} - \nu \tilde{\boldsymbol{u}}^T \boldsymbol{u} - \rho \tilde{\boldsymbol{y}}^T \tilde{\boldsymbol{y}}$*, with* $\nu, \rho \in \mathbb{R}$.

3.2.2 Stability of Equilibrium-Independent Dissipative Systems

Investigating the stability of the equilibria of systems via dissipativity theory requires a description of the system input $\boldsymbol{u}$. In the simplest case, setting $\boldsymbol{u} = \boldsymbol{0}$ yields an autonomous system along with the simplified dissipativity inequality $\dot{S} \leq w(\boldsymbol{0}, \boldsymbol{y})$. For the case that S is positive definite in $\boldsymbol{x}$ and $\hat{\boldsymbol{x}} = \boldsymbol{0}$ is an equilibrium of the system, this directly yields a Lyapunov stability result [SJK97, Theorem 2.20], with $\dot{S} \leq 0$, for a system dissipative w.r.t. any of the supply rates in Definition 3.2 (except for OFP and IFOFP systems, where $\rho \geq 0$ must also hold).[2,3]

The stability of autonomous EID systems can similarly be investigated by setting $\tilde{u} = \boldsymbol{0}$, i.e., where $\hat{\boldsymbol{u}}$ may be a non-zero constant. As in Lyapunov theory, any states in which $\dot{S}$ is negative definite in an otherwise Lyapunov-stable system will converge to the equilibrium associated with the specific $\hat{\boldsymbol{u}}$. Thus, an OS-EIP system (or an IFOF-EIP system with $\rho > 0$) will ensure that the states reflected in $\tilde{\boldsymbol{y}}$ converge to their equilibria if $\tilde{\boldsymbol{u}} \equiv \boldsymbol{0}$. However, since not all states are necessarily directly present in the output $\tilde{y}$, additional properties are required to ensure the asymptotic stability (AS) or global asymptotic stability (GAS) for the equilibria of autonomous EID systems. These definitions for dissipative systems include zero-input detectability [SJK97,

[2] If S is only positive semi-definite in $\boldsymbol{x}$, additional criteria for Lyapunov stability as described in [SJK97, Theorem 2.24] is required.

[3] Other stability analysis methods which can be written as Lyapunov functions—including the small gain, positivity, circle and Popov theorems [HB93]—are therefore also naturally compatible with the dissipativity framework.

Definition 2.32], and zero-state observability [vdS17, Definition 3.2.11] and zero-state detectability [vdS17, Definition 3.2.15], which are adapted for EID systems.

Definition 3.5 (Equilibrium-Independent-Input Detectable (EIID))
A system (3.5) is EIID if, for every $\hat{\boldsymbol{x}} \in \hat{\mathcal{X}}$ and its associated $\hat{\boldsymbol{u}}$ and $\hat{\boldsymbol{y}}$, $\tilde{\boldsymbol{y}} \equiv \mathbf{0}$ implies $\lim_{t\to\infty} \tilde{\boldsymbol{u}} = \mathbf{0}$.

Definition 3.6 (Equilibrium-Independent-State Observable (EISO), c.f. [SP19, Definition 4.1])
A system (3.5) is EISO if, for every $\hat{\boldsymbol{x}} \in \hat{\mathcal{X}}$ and its associated $\hat{\boldsymbol{u}}$ and $\hat{\boldsymbol{y}}$, $\tilde{\boldsymbol{u}} \equiv \mathbf{0}$ and $\tilde{\boldsymbol{y}} \equiv \mathbf{0}$ imply $\tilde{\boldsymbol{x}} = \mathbf{0}$.

Definition 3.7 (Equilibrium-Independent-State Detectable (EISD))
A system (3.5) is EISD if, for every $\hat{\boldsymbol{x}} \in \hat{\mathcal{X}}$ and its associated $\hat{\boldsymbol{u}}$ and $\hat{\boldsymbol{y}}$, $\tilde{\boldsymbol{u}} \equiv \mathbf{0}$ and $\tilde{\boldsymbol{y}} \equiv \mathbf{0}$ imply $\lim_{t\to\infty} \tilde{\boldsymbol{x}} = \mathbf{0}$.

Note that an EISO system is automatically also EISD. Moreover, in an EISO system, the rates of convergence of $\tilde{\boldsymbol{x}}$ and $\tilde{\boldsymbol{y}}$ are linked, whereas $\tilde{\boldsymbol{x}}$ in an EISD system might show a slower convergence than $\tilde{\boldsymbol{y}}$. Using the above definitions, the stability of EID systems can be investigated.

Proposition 3.1 (Equilibrium Stability of Autonomous EID Systems, see [AMP16, Theorem 3.1], [SP19, Lemma 4.3])
Consider a system (3.5) for which Assumption 3.1 holds and which has a storage function $S(\tilde{\boldsymbol{x}})$ that is positive definite in $\tilde{\boldsymbol{x}}$, for every $\hat{\boldsymbol{x}} \in \hat{\mathcal{X}}$. Let the system be EID w.r.t. a supply rate $w(\tilde{\boldsymbol{u}}, \tilde{\boldsymbol{y}})$. If $\tilde{\boldsymbol{u}} \equiv \mathbf{0}$, then any equilibrium of the system is:

1. *stable if, for every $\hat{\boldsymbol{x}} \in \hat{\mathcal{X}}$, $\dot{S}$ is negative semi-definite in $\tilde{\boldsymbol{x}}$;*
2. *AS if, for every $\hat{\boldsymbol{x}} \in \hat{\mathcal{X}}$, $\dot{S}$ is negative definite in $\tilde{\boldsymbol{x}}$;*
3. *AS if, for every $\hat{\boldsymbol{x}} \in \hat{\mathcal{X}}$, $\dot{S}$ is negative definite in $\tilde{\boldsymbol{y}}$ and the system is EISD;*
4. *AS if it is stable and, for every $\hat{\boldsymbol{x}} \in \hat{\mathcal{X}}$, $\dot{S} \equiv 0$ implies $\lim_{t\to\infty} \tilde{\boldsymbol{x}} = \mathbf{0}$;*
5. *GAS if it is AS and, for every $\hat{\boldsymbol{x}} \in \hat{\mathcal{X}}$, $S(\tilde{\boldsymbol{x}})$ is radially unbounded in $\tilde{\boldsymbol{x}}$.*

Remark 3.6. *By Statement 3 in Proposition 3.1, any equilibrium of an OS-EIP and EISD system with a storage function $S(\tilde{\boldsymbol{x}})$ that is positive definite in $\tilde{\boldsymbol{x}}$ is AS for $\tilde{\boldsymbol{u}} = \mathbf{0}$.*

Remark 3.7. *Whereas input-output stability measures can be considered system properties, the Lyapunov stability of the internal dynamics of a system is a property of the individual equilibria. The line between these properties is blurred in dissipativity theory, where the internal stability represented by the storage function* $S(\boldsymbol{x})$ *is related to the supply rate* $w(\boldsymbol{u}, \boldsymbol{y})$ *representing an input-output measure. Moreover, through the extension to EID, Lyapunov stability results obtained via Proposition 3.1 hold for any equilibrium of the system, allowing such results again to be considered a system property which is independent of the equilibrium. This is useful for achieving a stability according to Definition 2.1 as discussed in Remark 2.1.*

3.3 Dissipativity Theory Revisited

Dissipativity theory is a versatile tool for the stability analysis of a system comprising many interconnected and nonlinear subsystems. However, the established theory culminating in Proposition 3.1 presents no systematic method of verifying AS if $\dot{S}$ is only negative semi-definite in $\tilde{\boldsymbol{y}}$ (e.g., via Statement 4) or for verifying the EIP or OS-EIP for nonlinear systems. Therefore, the stability analysis of interconnected EID systems is extended in **Section 3.3.1** with a focus on providing a systematic method for investigating Statement 4 in Proposition 3.1. Thereafter, in **Section 3.3.2**, methods are provided by which EIP properties can be established for classes of static and dynamical nonlinear systems.

3.3.1 Stability of Interconnected EID Systems

Investigating the stability of the equilibria for systems of interconnected EID subsystems is achieved by combining the subsystem storage functions and applying Proposition 3.1, e.g., to obtain stability via Statement 1 (see [vdS17, Proposition 3.3.3], [SP19]), or AS via Statement 2 (see [AMP16, Theorem 3.1]) or Statement 3 (see [SP19]). Furthermore, it is shown in [SJK97, Theorem 2.33] that combining GAS and zero-input detectability conditions for feedback-interconnected passive subsystems can yield AS. In this section, this idea is extended to EID systems and generalised for arbitrary interconnections. An example is provided which demonstrates the systematic approach for deriving the AS result. Thereafter, this approach is formalised and simplifications for interconnected EIP systems are presented.

To use the asymptotic conditions in Definitions 3.5 and 3.7 in an interconnected system, the following assumption is made. Note that this assumption typically holds for practical systems.

Assumption 3.2 (Asymptotic Convergence)
For any EIID system and for any GAS or EISD system, let the variable convergence of $\lim_{t\to\infty}\tilde{\boldsymbol{u}} = \boldsymbol{0}$ *and* $\lim_{t\to\infty}\tilde{\boldsymbol{x}} = \boldsymbol{0}$ *be such that* $\tilde{\boldsymbol{u}} \approx \boldsymbol{0}$ *and* $\tilde{\boldsymbol{x}} \approx \boldsymbol{0}$ *may be considered after some finite time, respectively.*

Example 3.1:
Consider the system in Figure 3.1 comprising five interconnected EID and EISD subsystems of the form (3.5) with a storage function

$$S_\Sigma(\tilde{\boldsymbol{x}}_\Sigma) = \sum_{s=1}^{5} S_s(\tilde{\boldsymbol{x}}_s), \quad \tilde{\boldsymbol{x}}_\Sigma = \mathrm{stack}(\tilde{\boldsymbol{x}}_1, \tilde{\boldsymbol{x}}_2, \tilde{\boldsymbol{x}}_3, \tilde{\boldsymbol{x}}_4, \tilde{\boldsymbol{x}}_5), \tag{3.6}$$

such that S_Σ *is positive definite in* $\tilde{\boldsymbol{x}}_\Sigma$ *and* $\dot{S}_\Sigma$ *is negative semi-definite in* $\tilde{\boldsymbol{x}}_\Sigma$*, i.e., the system is stable via Proposition 3.1. Furthermore, let* $\dot{S}_\Sigma$ *be negative definite in* $\tilde{\boldsymbol{y}}_1$ *and* $\tilde{\boldsymbol{y}}_2$*, let subsystem* $s = 2$ *be EIID, and let subsystems* $s = \{4, 5\}$ *be GAS if* $\tilde{\boldsymbol{u}}_4 = \boldsymbol{0}$ *and* $\tilde{\boldsymbol{u}}_5 = \boldsymbol{0}$*, respectively.*

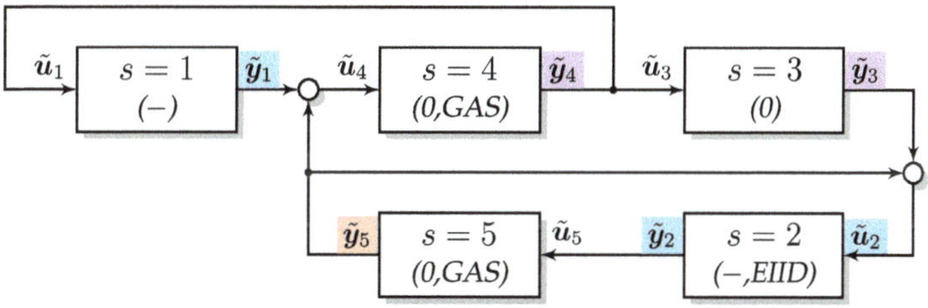

Figure 3.1: Example system comprising five interconnected EID and EISD subsystems.

Assuming the system has equilibria, the AS of its equilibria can be established in the following steps. 1) Since $\dot{S}_\Sigma$ *is negative definite in* $\tilde{\boldsymbol{y}}_1$ *and* $\tilde{\boldsymbol{y}}_2$*,* $\lim_{t\to\infty}\tilde{\boldsymbol{y}}_1 = \boldsymbol{0}$ *and* $\lim_{t\to\infty}\tilde{\boldsymbol{y}}_2 = \boldsymbol{0}$*. Moreover,* $\lim_{t\to\infty}\tilde{\boldsymbol{u}}_2 = \boldsymbol{0}$ *follows from the EIID property. 2) Since subsystem* $s = 5$ *is GAS,* $\lim_{t\to\infty}\tilde{\boldsymbol{y}}_5 = \boldsymbol{0}$ *follows after* $\tilde{\boldsymbol{u}}_5 = \tilde{\boldsymbol{y}}_2 = \boldsymbol{0}$*. 3) Similarly,* $\lim_{t\to\infty}\tilde{\boldsymbol{y}}_4 = \boldsymbol{0}$ *results after* $\tilde{\boldsymbol{u}}_4 = \tilde{\boldsymbol{y}}_1 + \tilde{\boldsymbol{y}}_5 = \boldsymbol{0}$*. Additionally, since* $\tilde{\boldsymbol{u}}_2 = \tilde{\boldsymbol{y}}_3 + \tilde{\boldsymbol{y}}_5$*,* $\tilde{\boldsymbol{y}}_3 = \boldsymbol{0}$ *holds after* $\tilde{\boldsymbol{u}}_2 = \boldsymbol{0}$ *and* $\tilde{\boldsymbol{y}}_5 = \boldsymbol{0}$*. The equilibrium of the system is thus AS, since the output of each subsystem converges to zero and since each system is EISD.*

From Example 3.1, it can be seen that the definiteness of $\dot{S}_\Sigma$ in the subsystem outputs, the interconnection structure, and the EIID and GAS properties of the subsystems contributed towards an AS which corresponds to Statement 4 of Proposition 3.1. Consider now the generalised case in which a system comprising a set of $\mathcal{J} = \{1, \ldots, |\mathcal{J}|\}$ subsystems, each with the form (3.5), is linearly interconnected according to

$$\tilde{\boldsymbol{u}}_\Sigma = \boldsymbol{H}\tilde{\boldsymbol{y}}_\Sigma, \quad \tilde{\boldsymbol{u}}_\Sigma = \mathrm{stack}_{s\in\mathcal{J}}(\tilde{\boldsymbol{u}}_s), \quad \tilde{\boldsymbol{y}}_\Sigma = \mathrm{stack}_{s\in\mathcal{J}}(\tilde{\boldsymbol{y}}_s), \tag{3.7}$$

and where a storage function of the system

$$S_\Sigma(\tilde{\boldsymbol{x}}_\Sigma) = \sum_{s \in \mathcal{J}} p_s S_s(\tilde{\boldsymbol{x}}_s), \quad \tilde{\boldsymbol{x}}_\Sigma = \mathrm{stack}_{s \in \mathcal{J}}(\tilde{\boldsymbol{x}}_s), \quad p_s > 0, s \in \mathcal{J}, \tag{3.8}$$

is the weighted sum of the subsystem storage functions. Moreover, the following assumption ensures that a valid interconnected equilibrium exists.

Assumption 3.3 (Non-Empty Interconnected Equilibrium Set)
The system comprising the set $\mathcal{J}$ of subsystems of the form (3.5) *interconnected via* (3.7) *has a non-empty equilibrium set $\hat{\mathcal{X}}_\Sigma$, with $\hat{\boldsymbol{x}}_\Sigma \in \hat{\mathcal{X}}_\Sigma$.*

The following theorem formalises the steps leading to AS in Example 3.1 by using an algorithm to successively apply the EIID and GAS subsystem properties through the interconnection matrix $\boldsymbol{H}$ to determine which outputs converge to zero. To assist in this process, let

$$\mathfrak{y}_i = \begin{cases} 0 & \text{if } \lim_{t\to\infty} \tilde{y}_i = 0, \\ 1 & \text{otherwise,} \end{cases} \tag{3.9}$$

indicate the AS-status of the outputs, which changes as the algorithm progresses.

Theorem 3.2 (AS Equilibria of Interconnected EID Subsystems)
Consider a system of $|\mathcal{J}|$ subsystems of the form (3.5) *where each subsystem $s \in \mathcal{J}$ has the dynamics* (3.5) *along with a storage function $S_s(\tilde{\boldsymbol{x}}_s)$ that is positive definite in $\tilde{\boldsymbol{x}}_s$, for every $\hat{\boldsymbol{x}}_s \in \hat{\mathcal{X}}_s$. Let each subsystem be EID and EISD, and let Assumptions 3.1, 3.2 and 3.3 hold for the subsystems and the system, respectively. Let the storage function S_Σ in* (3.8) *be positive definite in $\tilde{\boldsymbol{x}}_\Sigma$ and let $\dot{S}_\Sigma$ be negative semi-definite in $\tilde{\boldsymbol{x}}_\Sigma$. Furthermore, let each subsystem $s \in \mathcal{J}$ belong to exactly one of the following sets*

- *$s \in \mathcal{J}_-$: ($\dot{S}_\Sigma$ is negative definite in $\tilde{\boldsymbol{y}}_s$) or ($\dot{S}_\Sigma$ is negative semi-definite in $\tilde{\boldsymbol{y}}_s$ and $\dot{S}_\Sigma \equiv 0$ implies $\lim_{t\to\infty} \tilde{\boldsymbol{y}}_s = \boldsymbol{0}$);*
- *$s \in \mathcal{J}_{-,\mathrm{EIID}}$: $\big[$($\dot{S}_\Sigma$ is negative definite in $\tilde{\boldsymbol{y}}_s$) or ($\dot{S}_\Sigma$ is negative semi-definite in $\tilde{\boldsymbol{y}}_s$ and $\dot{S}_\Sigma \equiv 0$ implies $\lim_{t\to\infty} \tilde{\boldsymbol{y}}_s = \boldsymbol{0}$)$\big]$ and subsystem s is EIID;*
- *$s \in \mathcal{J}_0$: $\dot{S}_\Sigma$ is negative semi-definite in $\tilde{\boldsymbol{y}}_s$;*
- *$s \in \mathcal{J}_{0,\mathrm{GAS}}$: $\dot{S}_\Sigma$ is negative semi-definite in $\tilde{\boldsymbol{y}}_s$ and subsystem s is GAS if $\tilde{\boldsymbol{u}}_s = \boldsymbol{0}$;* [4]

[4] For $s \in \mathcal{J}_0 \cup \mathcal{J}_{0,\mathrm{GAS}}$, $\dot{S}_\Sigma$ might be negative definite in some element of $\tilde{\boldsymbol{y}}_s$ even though it is only negative semi-definite in $\tilde{\boldsymbol{y}}_s$ as a whole.

with $\mathcal{J} = \mathcal{J}_- \cup \mathcal{J}_{-,\mathrm{EIID}} \cup \mathcal{J}_0 \cup \mathcal{J}_{0,\mathrm{GAS}}$. *Partition the interconnection* (3.7) *according to*

$$\underbrace{\begin{bmatrix} \tilde{\boldsymbol{u}}_- \\ \tilde{\boldsymbol{u}}_{-,\mathrm{EIID}} \\ \tilde{\boldsymbol{u}}_0 \\ \tilde{\boldsymbol{u}}_{0,\mathrm{GAS}} \end{bmatrix}}_{\tilde{\boldsymbol{u}}_\Sigma} = \underbrace{\begin{bmatrix} \boldsymbol{H}_{11} & \boldsymbol{H}_{12} & \boldsymbol{H}_{13} & \boldsymbol{H}_{14} \\ \boldsymbol{H}_{21} & \boldsymbol{H}_{22} & \boldsymbol{H}_{23} & \boldsymbol{H}_{24} \\ \boldsymbol{H}_{31} & \boldsymbol{H}_{32} & \boldsymbol{H}_{33} & \boldsymbol{H}_{34} \\ \boldsymbol{H}_{41} & \boldsymbol{H}_{42} & \boldsymbol{H}_{43} & \boldsymbol{H}_{44} \end{bmatrix}}_{\boldsymbol{H}} \underbrace{\begin{bmatrix} \tilde{\boldsymbol{y}}_- \\ \tilde{\boldsymbol{y}}_{-,\mathrm{EIID}} \\ \tilde{\boldsymbol{y}}_0 \\ \tilde{\boldsymbol{y}}_{0,\mathrm{GAS}} \end{bmatrix}}_{\tilde{\boldsymbol{y}}_\Sigma}, \tag{3.10a}$$

$$\begin{cases} \tilde{\boldsymbol{u}}_- = \mathrm{stack}_{s\in\mathcal{J}_-}(\tilde{\boldsymbol{u}}_s), & \tilde{\boldsymbol{y}}_- = \mathrm{stack}_{s\in\mathcal{J}_-}(\tilde{\boldsymbol{y}}_s), \\ \tilde{\boldsymbol{u}}_{-,\mathrm{EIID}} = \mathrm{stack}_{s\in\mathcal{J}_{-,\mathrm{EIID}}}(\tilde{\boldsymbol{u}}_s), & \tilde{\boldsymbol{y}}_{-,\mathrm{EIID}} = \mathrm{stack}_{s\in\mathcal{J}_{-,\mathrm{EIID}}}(\tilde{\boldsymbol{y}}_s), \\ \tilde{\boldsymbol{u}}_0 = \mathrm{stack}_{s\in\mathcal{J}_0}(\tilde{\boldsymbol{u}}_s), & \tilde{\boldsymbol{y}}_0 = \mathrm{stack}_{s\in\mathcal{J}_0}(\tilde{\boldsymbol{y}}_s), \\ \tilde{\boldsymbol{u}}_{0,\mathrm{GAS}} = \mathrm{stack}_{s\in\mathcal{J}_{0,\mathrm{GAS}}}(\tilde{\boldsymbol{u}}_s), & \tilde{\boldsymbol{y}}_{0,\mathrm{GAS}} = \mathrm{stack}_{s\in\mathcal{J}_{0,\mathrm{GAS}}}(\tilde{\boldsymbol{y}}_s). \end{cases} \tag{3.10b}$$

Then, any equilibrium of the interconnected system is AS if Algorithm 1 terminates with $\boldsymbol{\mathfrak{y}}_0 = \boldsymbol{0}$ *and* $\boldsymbol{\mathfrak{y}}_{0,\mathrm{GAS}} = \boldsymbol{0}$.

Algorithm 1 Test for output convergence, with $\mathfrak{y}_i$ defined in (3.9).

$\boldsymbol{\mathfrak{y}}_0 \leftarrow \mathbb{1}_{\dim(\tilde{\boldsymbol{y}}_0)}$
$\boldsymbol{\mathfrak{y}}_{0,\mathrm{GAS}} \leftarrow \mathbb{1}_{\dim(\tilde{\boldsymbol{y}}_{0,\mathrm{GAS}})}$
for each $\tilde{y}_{0,i}$ in $\tilde{\boldsymbol{y}}_0$ **do**
    if $\dot{S}_\Sigma$ is negative definite in $\tilde{y}_{0,i}$ **then**
        $\mathfrak{y}_{0,i} \leftarrow 0$
for each $\tilde{y}_{0,\mathrm{GAS},i}$ in $\tilde{\boldsymbol{y}}_{0,\mathrm{GAS}}$ **do**
    if $\dot{S}_\Sigma$ is negative definite in $\tilde{y}_{0,\mathrm{GAS},i}$ **then**
        $\mathfrak{y}_{0,\mathrm{GAS},i} \leftarrow 0$
repeat
    unchanged ← true
    ▷ *Test if a single EIID input maps uniquely to a single non-converged output.* ◁
    $\tilde{\boldsymbol{u}}_{-,\mathrm{EIID}} \leftarrow \boldsymbol{H}_{23}\boldsymbol{\mathfrak{y}}_0 + \boldsymbol{H}_{24}\boldsymbol{\mathfrak{y}}_{0,\mathrm{GAS}}$
    for each $\tilde{u}_{-,\mathrm{EIID},i}$ in $\tilde{\boldsymbol{u}}_{-,\mathrm{EIID}}$ **do**
        if $\tilde{u}_{-,\mathrm{EIID},i}$ equals a single non-zero summand $h_{ij}\mathfrak{y}_j$ **then**
            $\mathfrak{y}_j \leftarrow 0$
            unchanged ← false
    ▷ *Test if all inputs to a GAS subsystem have converged to zero.* ◁
    $\tilde{\boldsymbol{u}}_{0,\mathrm{GAS}} \leftarrow \boldsymbol{H}_{43}\boldsymbol{\mathfrak{y}}_0 + \boldsymbol{H}_{44}\boldsymbol{\mathfrak{y}}_{0,\mathrm{GAS}}$
    for each $s \in \mathcal{J}_{0,\mathrm{GAS}}$ **do**
        if $\tilde{\boldsymbol{u}}_s = \boldsymbol{0}$ **then**
            $\boldsymbol{\mathfrak{y}}_s \leftarrow \boldsymbol{0}$
            unchanged ← false
until unchanged

Proof:
Due to the negative definiteness of $\dot{S}_\Sigma$ or the LaSalle-like condition for the outputs of $s \in \mathcal{J}_- \cup \mathcal{J}_{-,\mathrm{EIID}}$, it follows that $\lim_{t\to\infty} \tilde{\boldsymbol{y}}_- = \boldsymbol{0}$ and $\lim_{t\to\infty} \tilde{\boldsymbol{y}}_{-,\mathrm{EIID}} = \boldsymbol{0}$. Note that the convergence of these outputs does not presuppose that $\tilde{\boldsymbol{u}}_-$ and $\tilde{\boldsymbol{u}}_{-,\mathrm{EIID}}$ converge to zero. To complete the AS proof, the convergence of $\tilde{\boldsymbol{y}}_0$ and $\tilde{\boldsymbol{y}}_{0,\mathrm{GAS}}$ must be established, which is done through Algorithm 1. Even though $\dot{S}_\Sigma$ is only negative semi-definite in $\tilde{\boldsymbol{y}}_s$, for $s \in \mathcal{J}_- \cup \mathcal{J}_{-,\mathrm{EIID}}$, any scalar output in $\tilde{\boldsymbol{y}}_s$ will converge to zero if $\dot{S}_\Sigma$ is negative definite in that output, yielding Lines 3–8 of Algorithm 1. Next, the subsystem interconnection and the EIID and GAS subsystem properties are utilised. Consider now (3.10a) using the output convergence notation in (3.9) with $\mathfrak{y}_- = \boldsymbol{0}$ and $\mathfrak{y}_{-,\mathrm{EIID}} = \boldsymbol{0}$,

$$\tilde{\boldsymbol{u}}_{-,\mathrm{EIID}} = \boldsymbol{H}_{23}\mathfrak{y}_0 + \boldsymbol{H}_{24}\mathfrak{y}_{0,\mathrm{GAS}}, \tag{3.11a}$$

$$\tilde{\boldsymbol{u}}_{0,\mathrm{GAS}} = \boldsymbol{H}_{43}\mathfrak{y}_0 + \boldsymbol{H}_{44}\mathfrak{y}_{0,GAS}. \tag{3.11b}$$

Since $\mathfrak{y}_{-,\mathrm{EIID}} = \boldsymbol{0}$, $\tilde{\boldsymbol{u}}_{-,\mathrm{EIID}}$ also converges to zero through the EIID property. Thus, if any scalar input in (3.11a) maps exclusively to a single unconverged output $\mathfrak{y}_j$ through the interconnection matrix $\boldsymbol{H}$, the EIID property also ensures that $\mathfrak{y}_j = 0$. This process is executed in Lines 12–16 of Algorithm 1. Next, the subsystems with GAS properties are checked in Lines 18–22. If the input vector of any GAS subsystem calculated from (3.11b) is zero, the output vector of that system converges to zero. Since both the EIID and GAS properties can modify $\mathfrak{y}_0$ and $\mathfrak{y}_{0,\mathrm{GAS}}$ in (3.11b) and (3.11b), these properties are repeatedly applied until no further changes occur. If Algorithm 1 terminates with $\mathfrak{y}_0 = \boldsymbol{0}$ and $\mathfrak{y}_{0,\mathrm{GAS}} = \boldsymbol{0}$, then any equilibrium of the system is AS since $\lim_{t\to\infty} \tilde{\boldsymbol{y}}_\Sigma = \boldsymbol{0}$ and since every subsystem is EISD. ■

By exploiting the interconnection of the subsystems along with the subsystem GAS and EIID properties, Theorem 3.2 presents a systematic means by which the AS of any equilibrium of interconnected EID and EISD subsystems can be investigated, in the case that the $\dot{S}_\Sigma$ is only negative semi-definite in $\tilde{\boldsymbol{y}}_\Sigma$ (see Statement 4 of Proposition 3.1). To demonstrate the use of Theorem 3.2, the system in Figure 3.1 is again investigated in the following example.

Example 3.2:

Consider the setup in Example 3.1, which is now investigated using Theorem 3.2. Since $\dot{S}_\Sigma$ is negative definite in $\tilde{\boldsymbol{y}}_s$, with $s = \{1, 2\}$, AS is assured if Algorithm 1 terminates with $\mathfrak{y}_s = \boldsymbol{0}$ for $s = \{3, 4, 5\}$. Since no additional information on the definiteness of $\dot{S}_\Sigma$ in the elements of $\tilde{\boldsymbol{y}}_s$, $s = \{3, 4, 5\}$ is available, Lines 3–8 make no changes to $\mathfrak{y}_0 = \mathfrak{y}_3$ and $\mathfrak{y}_{0,\mathrm{GAS}} = \mathrm{stack}(\mathfrak{y}_4, \mathfrak{y}_5)$. Over two iterations, the repeat loop makes changes to $\mathfrak{y}_3$, $\mathfrak{y}_4$, and $\mathfrak{y}_5$ as listed in Table 3.1.

Table 3.1: Repeat-Loop Iterations of Algorithm 1 for Example 3.1

Iteration 1			*Iteration 2*		
$\tilde{\boldsymbol{u}}_2 = \mathfrak{y}_3 + \mathfrak{y}_5$		$\mathfrak{y}_3 = \mathbb{1}$	$\tilde{\boldsymbol{u}}_2 = \mathfrak{y}_3$	*Line 15 (EIID)* $\Longrightarrow$	$\mathfrak{y}_3 = \mathbf{0}$
$\tilde{\boldsymbol{u}}_4 = \mathfrak{y}_5$		$\mathfrak{y}_4 = \mathbb{1}$	$\tilde{\boldsymbol{u}}_4 = \mathbf{0}$	*Line 21 (GAS)* $\Longrightarrow$	$\mathfrak{y}_4 = \mathbf{0}$
$\tilde{\boldsymbol{u}}_5 = \mathbf{0}$	*Line 21 (GAS)* $\Longrightarrow$	$\mathfrak{y}_5 = \mathbf{0}$	$\tilde{\boldsymbol{u}}_5 = \mathbf{0}$		$\mathfrak{y}_5 = \mathbf{0}$

In the third repeat-loop iteration, no changes occur and Algorithm 1 terminates with $\mathfrak{y}_0 = \mathbf{0}$ and $\mathfrak{y}_{0,\mathrm{GAS}} = \mathbf{0}$, thus ensuring AS according to Theorem 3.2.

Remark 3.8. *Theorem 3.2 can be extended by splitting the sets $\mathcal{J}_0$ and $\mathcal{J}_{0,\mathrm{GAS}}$ into EIID and non-EIID subsets. This opens further options for inferring convergence using the EIID property, similar to Lines 12–16 of Algorithm 1.*

Remark 3.9. *The GAS property required by a subsystem $s \in \mathcal{J}_{0,\mathrm{GAS}}$ can be relaxed to an AS property if it is guaranteed that the state trajectories $\boldsymbol{x}_s(t)$ stay within the region of attraction of the equilibrium. Practically, this includes the case where the equilibria of a subsystem s are AS in a region $\mathcal{X}_s \subset \mathbb{R}^n$ and where $\boldsymbol{x}_s(t) \in \mathcal{X}$ for all $t \geq 0$.*

Although Theorem 3.2 offers a systematic approach for investigating the AS via Statement 4 of Proposition 3.1, this approach requires information about the interconnected system storage function S_Σ. By restricting the subsystems to be EIP[5] and by requiring a skew-symmetric interconnection of the subsystems, the need for investigating the properties of S_Σ may be relaxed. A stability analysis can thus be performed by only considering the local subsystem storage functions S_s.

Theorem 3.3 (AS Equilibria of Interconnected EIP Subsystems)
Consider a system of $|\mathcal{J}|$ subsystems in which each subsystem $s \in \mathcal{J}$ has the dynamics (3.5) along with a storage function $S_s(\tilde{\boldsymbol{x}}_s)$ that is positive definite in $\tilde{\boldsymbol{x}}_s$, for every $\hat{\boldsymbol{x}}_s \in \hat{\mathcal{X}}_s$. Let each subsystem be EIP and EISD, and let Assumptions 3.1, 3.2 and 3.3 hold for the subsystems and the system, respectively. Furthermore, let each subsystem $s \in \mathcal{J}$ belong to exactly one of the following sets

- *$s \in \mathcal{J}_-$: ($\dot{S}_s - \tilde{\boldsymbol{u}}_s^T \tilde{\boldsymbol{y}}_s$ is negative definite in $\tilde{\boldsymbol{y}}_s$) or ($\dot{S}_s - \tilde{\boldsymbol{u}}_s^T \tilde{\boldsymbol{y}}_s$ is negative semi-definite in $\tilde{\boldsymbol{y}}_s$ and $\dot{S}_s \equiv 0$ implies $\lim_{t\to\infty} \tilde{\boldsymbol{y}}_s = \mathbf{0}$);*

[5] This includes subsystems that are OS-EIP or IFOF-EIP with $\nu \geq 0$ and $\rho \geq 0$.

- $s \in \mathcal{J}_{-,\mathrm{EIID}}$: $\big[(\dot{S}_s - \tilde{\boldsymbol{u}}_s^T \tilde{\boldsymbol{y}}_s$ *is negative definite in* $\tilde{\boldsymbol{y}}_s)$ *or* $(\dot{S}_s - \tilde{\boldsymbol{u}}_s^T \tilde{\boldsymbol{y}}_s$ *is negative semi-definite in* $\tilde{\boldsymbol{y}}_s$ *and* $\dot{S}_s \equiv 0$ *implies* $\lim_{t\to\infty} \tilde{\boldsymbol{y}}_s = \boldsymbol{0})\big]$ *and subsystem* s *is EIID;*
- $s \in \mathcal{J}_0$: $\dot{S}_s - \tilde{\boldsymbol{u}}_s^T \tilde{\boldsymbol{y}}_s$ *is negative semi-definite in* $\tilde{\boldsymbol{y}}_s$*;*
- $s \in \mathcal{J}_{0,\mathrm{GAS}}$: $\dot{S}_s - \tilde{\boldsymbol{u}}_s^T \tilde{\boldsymbol{y}}_s$ *is negative semi-definite in* $\tilde{\boldsymbol{y}}_s$ *and subsystem* s *is GAS if* $\tilde{\boldsymbol{u}}_s = \boldsymbol{0}$*;*

with $\mathcal{J} = \mathcal{J}_- \cup \mathcal{J}_{-,\mathrm{EIID}} \cup \mathcal{J}_0 \cup \mathcal{J}_{0,\mathrm{GAS}}$. *Let the subsystems be linearly interconnected according to (3.10a) and let*

$$\boldsymbol{P}\boldsymbol{H} + \boldsymbol{H}^T\boldsymbol{P} \succcurlyeq 0, \quad \boldsymbol{P} = \mathrm{diag}(p_s), \tag{3.12}$$

hold with p_s *in (3.8).*[6] *Then, any equilibrium of the system with the storage function* S_Σ *in (3.8) is stable. Furthermore, AS for any equilibrium of the system follows if Algorithm 1 terminates with* $\mathfrak{y}_0 = \boldsymbol{0}$ *and* $\mathfrak{y}_{0,\mathrm{GAS}} = \boldsymbol{0}$.

Proof:
The weighted, skew-symmetric interconnection via $\boldsymbol{H}$, subject to (3.12), renders any equilibrium of the system stable with $\dot{S}_\Sigma$ negative semi-definite in $\boldsymbol{y}_\Sigma$ (see [vdS17, Proposition 4.4.15] and Proposition 3.1). Since each subsystem is at least EIP, $\dot{S}_\Sigma$ is negative (semi-)definite in $\tilde{\boldsymbol{y}}_s$ if $\dot{S}_s$ is negative (semi-)definite in $\tilde{\boldsymbol{y}}_s$. The rest of the proof follows analogously to the proof of Theorem 3.2. ■

Theorem 3.3 differs from Theorem 3.2 in that only the properties of the subsystem storage functions S_s along with a skew-symmetric interconnection are required to yield stability. Moreover, the subsystems are divided into the sets $\mathcal{J}_-$, $\mathcal{J}_{-,\mathrm{EIID}}$, $\mathcal{J}_0$, and $\mathcal{J}_{0,\mathrm{GAS}}$ based on the subsystem storage functions S_s, not the global storage function S_Σ. Replacing EID with the more specific EIP thus allows a higher degree of flexibility in terms of how subsystems may interconnect while automatically yielding a stability result. However, the EIP requirement is also restrictive, e.g., since compensation between subsystems with a lack of passivity and subsystems with an excess of passivity is possible with Theorem 3.2, but not with Theorem 3.3.

Since both Theorem 3.2 and Theorem 3.3 require an evaluation of Algorithm 1, it is useful to consider cases for which the result of the algorithm may be predetermined. One such case is presented in the following corollary, in which additional restrictions are placed on the interconnection matrix $\boldsymbol{H}$.

6 The LMI in (3.12) represents a weighted skew-symmetric interconnection of the subsystems (see [AMP16, p. 17], [vdS17, p. 83]).

Corollary 3.4 (Automatic Computation of Algorithm 1)
Consider Algorithm 1 with the inputs $\tilde{\boldsymbol{u}}_\Sigma$, outputs $\tilde{\boldsymbol{y}}_\Sigma$ and the interconnection matrix $\boldsymbol{H}$ as in (3.10a). *If* $\dim(\tilde{\boldsymbol{u}}_{-,\mathrm{EIID}}) \geq \dim(\tilde{\boldsymbol{y}}_0)$ *and if*

$$\boldsymbol{T}\boldsymbol{H}_{23} = \begin{bmatrix} \star \\ \boldsymbol{H}_\Delta \end{bmatrix}, \qquad \boldsymbol{T}\boldsymbol{H}_{24} = \begin{bmatrix} \star \\ \boldsymbol{0}_{\dim(\tilde{\boldsymbol{y}}_0),\dim(\tilde{\boldsymbol{y}}_{0,\mathrm{GAS}})} \end{bmatrix}, \tag{3.13a}$$

$$\boldsymbol{H}_{44} = \begin{bmatrix} \boldsymbol{0}_{\mu_1,\mu_1} & \cdots & \boldsymbol{0} \\ \star & \ddots & \vdots \\ \vdots \; \ddots & \ddots & \vdots \\ \star \cdots \star & & \boldsymbol{0}_{\mu_{|\mathcal{J}_{0,\mathrm{GAS}}|},\mu_{|\mathcal{J}_{0,\mathrm{GAS}}|}} \end{bmatrix}, \quad \mu_s = \dim(\tilde{\boldsymbol{y}}_s),\, s \in \mathcal{J}_{0,\mathrm{GAS}}, \tag{3.13b}$$

where $\boldsymbol{T}$ is some permutation matrix and $\boldsymbol{H}_\Delta$ is a triangular matrix with $\mathrm{rank}(\boldsymbol{H}_\Delta) = \dim(\tilde{\boldsymbol{y}}_0)$*, then Algorithm 1 terminates with $\mathfrak{y}_0 = \boldsymbol{0}$ and $\mathfrak{y}_{0,\mathrm{GAS}} = \boldsymbol{0}$.*

Proof:
Applying the permutation matrix $\boldsymbol{T}$, Line 12 of Algorithm 1 yields

$$\boldsymbol{T}\tilde{\boldsymbol{u}}_{-,\mathrm{EIID}} = \begin{bmatrix} \star\mathfrak{y}_0 + \star\mathfrak{y}_{0,\mathrm{GAS}} \\ \boldsymbol{H}_\Delta \mathfrak{y}_0 \end{bmatrix}. \tag{3.14}$$

Since $\boldsymbol{H}_\Delta$ is a full rank triangular matrix, $\mathfrak{y}_0 = \boldsymbol{0}$ after at most $\dim(\tilde{\boldsymbol{y}}_0)$ executions of Lines 12–16. After $\mathfrak{y}_0 = \boldsymbol{0}$ is established, Line 18 simplifies to

$$\tilde{\boldsymbol{u}}_{0,\mathrm{GAS}} = \boldsymbol{H}_{44}\mathfrak{y}_{0,\mathrm{GAS}}, \tag{3.15}$$

which allows $\mathfrak{y}_{0,\mathrm{GAS}} = \boldsymbol{0}$ to be achieved after at most $|\mathcal{J}_{0,\mathrm{GAS}}|$ additional executions of Lines 18–22. ∎

Remark 3.10. *In the context of Theorems 3.2 and 3.3, consider the fact that an output $\tilde{y}_{s,i}$, $s \in \mathcal{J}_0$, will only converge to zero if it uniquely connects to an input from an EIID subsystem. Thus,* $\dim(\tilde{\boldsymbol{u}}_{-,\mathrm{EIID}}) \geq \dim(\tilde{\boldsymbol{y}}_0)$ *is necessary for Algorithm 1 to terminate with $\mathfrak{y}_0 = \boldsymbol{0}$. This condition may be relaxed, however, if the EIID properties of subsystems in $\mathcal{J}_0$ and $\mathcal{J}_{0,\mathrm{GAS}}$ are also considered (see Remark 3.8).*

Remark 3.11. *Consider a connected graph $\mathcal{G} = (\mathcal{N}, \mathcal{E})$ with an incidence matrix $\boldsymbol{E}$* (3.1), *with EIP subsystems on the vertices and with OS-EIP and EISO subsystems on the edges. Furthermore, let the vertex subsystems be GAS if the $\tilde{\boldsymbol{u}}_s = \boldsymbol{0}$, $s \in \mathcal{N}$. Then, any equilibrium of the networked system with the interconnection*

$$\begin{bmatrix} \tilde{\boldsymbol{u}}_\mathcal{N} \\ \tilde{\boldsymbol{u}}_\mathcal{E} \end{bmatrix} = \begin{bmatrix} \boldsymbol{0} & -\boldsymbol{E} \\ \boldsymbol{E}^T & \boldsymbol{0} \end{bmatrix} \begin{bmatrix} \tilde{\boldsymbol{y}}_\mathcal{N} \\ \tilde{\boldsymbol{y}}_\mathcal{E} \end{bmatrix} \tag{3.16}$$

is AS through Theorem 3.3 and Corollary 3.4 with $\boldsymbol{P} = \boldsymbol{I}$, $\mathcal{J}_{-,\mathrm{EIID}} = \emptyset$ and $\mathcal{J}_0 = \emptyset$. This mirrors and generalises established results for stability [AMP16, p. 17], passivity [vdS17, Proposition 4.4.1], and AS [Str24, Theorem 3.1].

Remark 3.12. *Any number of subsystems in Theorems 3.2 and 3.3 may be replaced with static subsystems which have equivalent EID or EIP properties. In the context of Theorems 3.2 and 3.3, static subsystems may be considered as being EIID if* $\tilde{\boldsymbol{y}}_s = \mathbf{0} \implies \tilde{\boldsymbol{u}}_s = \mathbf{0}$ *or as being GAS if* $\tilde{\boldsymbol{u}}_s = \mathbf{0} \implies \tilde{\boldsymbol{y}}_s = \mathbf{0}$.

3.3.2 EIP Verification Methods

In order to utilise the EIP-based stability analysis methods presented in Sections 3.2.2 and 3.3.1, it is necessary to verify that a system or a subsystem is EIP. While this process is well established for linear systems[7], similar EIP verification methods are less common for static or nonlinear system. To this end, two lemmas are formulated for use in the subsequent chapters.

Static IFOF-EIP Functions

Recall from [Kha02, Definition 6.2] that a static, sector-bounded nonlinear function exhibits a dissipativity property w.r.t. the supply rate defined by the sector. Moreover, such a function can be categorised as being IFOFP. By considering the dissipativity not only w.r.t. the origin but w.r.t. any other point on the curve, an IFOF-EIP property can be formulated for the nonlinear function.

Consider the single-input-single-output static function along with its definition in the error variables $\tilde{u}$ and $\tilde{y}$:

$$y = h(u), \quad u, \hat{u} \in \mathcal{U} \subseteq \mathbb{R}, \quad y, \hat{y} \in \mathcal{Y} \subseteq \mathbb{R}, \quad h : \mathcal{U} \to \mathcal{Y}, \tag{3.17a}$$

$$\tilde{y} = \tilde{h}(\tilde{u}) := h(u) - h(\hat{u}) = y - \hat{y}, \quad \tilde{u} := u - \hat{u}. \tag{3.17b}$$

Note that since (3.17) is static, Assumption 3.1 is superfluous for an EID property.

7 The passivity of a linear stable system can be verified from its transfer function (see [BL07, p.28]). Moreover, a passive linear system is EIP if Assumption 3.1 holds.

Lemma 3.5 (Static IFOF-EIP Functions)
A static class C^0 and semi-differentiable[8] *function* (3.17) *is IFOF-EIP, with $\nu = \underline{\gamma}$ and $\rho = 1/\overline{\gamma}$, w.r.t. the input-output pair $(\tilde{u}, \tilde{y})$ if there exists a finite $\underline{\gamma} \in \mathbb{R}$ and a finite $\overline{\gamma} \in \mathbb{R}_{>0}$ such that*

$$\underline{\gamma} \leq \frac{dh(u)}{du} \leq \overline{\gamma}, \quad \forall u \in \mathcal{U}. \tag{3.18}$$

Proof:
For the function $h(u)$ in (3.17), consider the slope between any two non-identical points $(u, y), (\hat{u}, \hat{y}) \in \mathcal{U} \times \mathcal{Y}$ defined by the function. This slope is bounded by

$$\underline{\gamma} \leq \frac{y - \hat{y}}{u - \hat{u}} \leq \overline{\gamma}, \quad \forall (u, y), (\hat{u}, \hat{y}) \in \mathcal{U} \times \mathcal{Y}. \tag{3.19}$$

Changing (3.19) into the error variables $\tilde{u}$ and $\tilde{y}$ and multiplying through by $\tilde{u}^2 \geq 0$ gives (see [Kha02, p. 231])

$$\underline{\gamma}\tilde{u}^2 \leq \tilde{u}\tilde{y} \leq \overline{\gamma}\tilde{u}^2 \iff (\tilde{y} - \underline{\gamma}\tilde{u})(\tilde{y} - \overline{\gamma}\tilde{u}) \leq 0 \quad \iff (\tilde{y} - \underline{\gamma}\tilde{u})(1/\overline{\gamma}\,\tilde{y} - \tilde{u}) \leq 0, \tag{3.20}$$

for $\overline{\gamma} > 0$, which describes an IFOFP function (see [Kha02, p. 231]) in the error variables $\tilde{u}$ and $\tilde{y}$. Finally, by application of the mean value theorem, the bounds $\underline{\gamma}$ and $\overline{\gamma}$ in (3.20) may be determined from (3.19) using the global function gradient bounds (3.18). ■

Using Lemma 3.5, the EIP with IFOFP bounds can be derived analytically for a single-input single-output function. Note that if the upper bound $\overline{\gamma} < 0$ in (3.18) for a function $y = h(u)$, Lemma 3.5 can be applied to $y = h(-u)$ or $y = -h(u)$ to obtain an IFOF-EIP property instead.

Remark 3.13 (Symmetrical sector bounds). *Additionally restricting the bounds such that $\underline{\gamma} = -\overline{\gamma}$ in* (3.18) *results in the global Lipschitz continuity of the function $h(u)$. This also implies that the shifted $\tilde{h}(\tilde{u})$ has a finite L_2 gain of $\overline{\gamma}$ (see [AMP16, p. 24]).*

Nonlinear EIP systems

The EIP and OS-EIP for a class of nonlinear input-affine systems can be investigated by the following lemma, which mirrors and extends results presented in the port-Hamiltonian context in [Jay+07; Mon+19] and in a dissipativity context [SP19, Corollary 3.6], [AMP16, p. 26].

[8] A semi-differentiable function is left- or right-differentiable in its entire domain.

Lemma 3.6 (Nonlinear OS-EIP and EIP Systems)
Consider the nonlinear input affine system

$$\boldsymbol{Q}\dot{\boldsymbol{x}} = \boldsymbol{f}(\boldsymbol{x}) + \boldsymbol{B}\boldsymbol{u}, \quad \boldsymbol{y} = \boldsymbol{B}^T\boldsymbol{x}, \tag{3.21}$$

with $\boldsymbol{x} \in \mathcal{X} \subseteq \mathbb{R}^n$, $\boldsymbol{u} \in \mathcal{U} \subseteq \mathbb{R}^m$, $\boldsymbol{y} \in \mathcal{Y} \subseteq \mathbb{R}^m$, *where* $\boldsymbol{Q} \succ 0$ *is diagonal, and* $\boldsymbol{f}\colon \mathcal{X} \to \mathbb{R}^n$ *is a class* C^1 *function. This system is OS-EIP for* $\boldsymbol{x} \in \mathcal{X}$ *with a storage function* $S(\tilde{\boldsymbol{x}}) = \frac{1}{2}\tilde{\boldsymbol{x}}^T\boldsymbol{Q}\tilde{\boldsymbol{x}}$ *if there is a* $\rho > 0$ *such that*

$$-\nabla\boldsymbol{f}(\boldsymbol{x}) - \rho\boldsymbol{B}\boldsymbol{B}^T \succcurlyeq 0, \quad \forall \boldsymbol{x} \in \mathcal{X}. \tag{3.22}$$

If (3.22) *holds for* $\rho = 0$, *then the system* (3.21) *is EIP.*

Proof:
Shifting the nonlinear system (3.21) to the equilibrium described by $\boldsymbol{0} = \boldsymbol{f}(\hat{\boldsymbol{x}}) + \boldsymbol{B}\hat{\boldsymbol{u}}$ and $\hat{\boldsymbol{y}} = \boldsymbol{B}^T\hat{\boldsymbol{x}}$ yields

$$\boldsymbol{Q}\dot{\tilde{\boldsymbol{x}}} = \boldsymbol{f}(\boldsymbol{x}) - \boldsymbol{f}(\hat{\boldsymbol{x}}) + \boldsymbol{B}\tilde{\boldsymbol{u}}, \quad \tilde{\boldsymbol{y}} = \boldsymbol{B}^T\tilde{\boldsymbol{x}}. \tag{3.23}$$

Calculating the time derivative of $S(\tilde{\boldsymbol{x}})$ and substituting in the shifted dynamics (3.23) gives

$$\dot{S}(\tilde{\boldsymbol{x}}) = \tilde{\boldsymbol{x}}^T\boldsymbol{Q}\dot{\tilde{\boldsymbol{x}}} = \tilde{\boldsymbol{x}}^T(\boldsymbol{f}(\boldsymbol{x}) - \boldsymbol{f}(\hat{\boldsymbol{x}})) + \tilde{\boldsymbol{x}}^T\boldsymbol{B}\tilde{\boldsymbol{u}} = \tilde{\boldsymbol{x}}^T(\boldsymbol{f}(\boldsymbol{x}) - \boldsymbol{f}(\hat{\boldsymbol{x}})) + \tilde{\boldsymbol{u}}^T\tilde{\boldsymbol{y}}. \tag{3.24}$$

Evaluating $\dot{S}(\tilde{\boldsymbol{x}}) \leq \tilde{\boldsymbol{u}}^T\tilde{\boldsymbol{y}} - \rho\tilde{\boldsymbol{y}}^T\tilde{\boldsymbol{y}}$ for OS-EIP then results in

$$\begin{aligned} \tilde{\boldsymbol{x}}^T(\boldsymbol{f}(\boldsymbol{x}) - \boldsymbol{f}(\hat{\boldsymbol{x}})) + \tilde{\boldsymbol{u}}^T\tilde{\boldsymbol{y}} &\leq \tilde{\boldsymbol{u}}^T\tilde{\boldsymbol{y}} - \rho\tilde{\boldsymbol{y}}^T\tilde{\boldsymbol{y}} \\ \tilde{\boldsymbol{x}}^T(\boldsymbol{f}(\boldsymbol{x}) - \boldsymbol{f}(\hat{\boldsymbol{x}})) + \rho\tilde{\boldsymbol{x}}^T\boldsymbol{B}\boldsymbol{B}^T\tilde{\boldsymbol{x}} &\leq 0 \\ \tilde{\boldsymbol{x}}^T(\boldsymbol{g}(\boldsymbol{x}) - \boldsymbol{g}(\hat{\boldsymbol{x}})) &\leq 0, \quad \boldsymbol{g}(\boldsymbol{x}) := \boldsymbol{f}(\boldsymbol{x}) + \rho\boldsymbol{B}\boldsymbol{B}^T\boldsymbol{x}. \end{aligned} \tag{3.25}$$

The condition in (3.25) requires $-\boldsymbol{g}(\boldsymbol{x})$ to be monotone (see [RW98, Definition 12.1]). This monotonicity can be determined using $-\nabla\boldsymbol{g}(\boldsymbol{x}) \succcurlyeq 0$ [RW98, Proposition 12.3], which results in (3.22). An EIP result is obtained by setting $\rho = 0$ in (3.22) and (3.25). ■

3.3.3 Discussion

EID and EIP represent scalable and modular frameworks for analysing the stability of large networked systems. Scalability is ensured by the constructive formulation of the storage function for an entire system by using the storage functions of its subsystems. Increasing the number of subsystems therefore does not proportionally increase the difficulty of finding a storage function for the entire system. However, this scalability comes at the cost of more conservative conditions because of the enforced separability of the system storage function. At the same time, EID and EIP allow a modularity by allowing any subsystem to be replaced with a new subsystem, as long as the old and the new subsystems are dissipative w.r.t. the same supply rate. In such cases, the stability result established using the old subsystem remains valid regardless of the size, complexity, or linearity of the new subsystem. For large networked systems, EID and EIP thus allow the internal stability to be investigated by considering the subsystem port properties defined by their supply rates along with the interconnection structure.

Investigating the AS of the system equilibria, however, remains challenging if the system storage function is not negative definite in each subsystem output. By combining the subsystem interconnection with EIID and GAS subsystem properties, the novel results in Theorems 3.2 and 3.3 provide a formalised process by which AS can be investigated. This process, as described in Algorithm 1, can be interpreted as the convergence of variables propagating from subsystems with outputs in which $\dot{S}_\Sigma$ is negative definite towards subsystems with outputs in which $\dot{S}_\Sigma$ is only negative semi-definite.

The provided theorems extend the existing literature by investigating the AS[9] of the equilibria of interconnected EIP and EID subsystems[10] while considering more general topologies[11] and accounting for more system properties.[12] To apply Theorem 3.2 or Theorem 3.3, Algorithm 1 must be executed. Although Algorithm 1 can finish without yielding the state needed for AS, sufficient and necessary conditions for achieving the requisite terminal state are provided in Corollary 3.4 and Remark 3.10, respectively.

In the subsequent chapters, the modularity and scalability provided by the EIP framework along with the newly proposed theorems will be utilised to demonstrate AS results for the decentralised stabilisation and distributed coordination in DC networks, gas networks and NMESs. In this way, the methods in this chapter contribute towards answering RQ 1.2, RQ 2.2, RQ 3.2 and RQ 4.1.

9 E.g., this extends [AMP16, Theorem 3.1], where only stability is considered.
10 E.g., this extends [vdS17, Proposition 4.4.1], where only passivity is considered.
11 E.g., this extends [SJK97, Theorem 2.33], where only passive feedback connections are considered
12 E.g., this extends [Str24, Theorem 3.1], where only the setup in Remark 3.11 is considered.

4 Decentralised Stabilisation in DC Networks

In this chapter, the stabilisation of a DC network is considered. This is done for networks in which loads may be described by arbitrary continuous static functions and where the load locations do not necessarily correspond to the locations where steady-state power is available. The decentralised stabilisation is achieved by dividing the DC network into clusters of buses and designing decentralised controllers for DGUs situated at the buses. By showing that each cluster exhibits an OS-EIP property, the AS of any equilibrium of the DC network is achieved in a scalable and modular fashion using the methods in Chapter 3. In this way, this chapter thus provides an answer to RQ 1.1 and RQ 1.2.

The chapter starts with **Section 4.1**, in which the DC network is partitioned into a cluster of one or more buses and the lines which interconnect them. A differentiation is also made between buses with and without a steady-state supply of power. Models are supplied for the constitutive components of the clusters, including the loads, uncontrolled sources, lines, and DGUs which are then combined into the system description for an entire cluster.

Thereafter, in **Section 4.2**, control objectives are formulated for the buses both with and without steady-state power, and a controller for each bus type is designed. The properties of the buses in closed loop with the proposed controllers are analysed. This leads to OS-EIP conditions for the buses with local steady-state power while it is shown that buses cannot be made OS-EIP using a controller without local steady-state power.

This leads to **Section 4.3**, in which an OS-EIP property for the clusters containing both types of buses is investigated using an LMI. The cluster OS-EIP properties are then combined with the OS-EIP properties of the lines to obtain a GAS result for the entire DC network. The section concludes with an investigation into the robustness of the cluster OS-EIP against various parameter and topology changes.

The control and analysis methods are then applied in a simulation in **Section 4.4** to validate the presented theory and to demonstrate the robustness of the stability result against network topology and parameter changes.

The chapter concludes with a discussion of the methods and results in **Section 4.5** along with a summary of the work and the contributions in **Section 4.6**.

4.1 DC Network Modelling

The DC network considered in this chapter is composed of buses, representing points of equal voltage, interconnected by electrical transmission lines, as depicted in Figure 4.1. Each bus consists of a DGU along with an optional load, representing the aggregated demand of the bus, and an optional uncontrollable P source.[1] The DGU, comprising a buck converter and an LC filter, acts as an interface for DC storage systems, RESs and other components which exhibit a controllable flexibility (e.g., electric vehicles and smart loads). Furthermore, buses within an arbitrarily defined neighbourhood collectively form a cluster, with transmission lines interconnecting the buses in different clusters to form the DC network (see Figure 4.2).

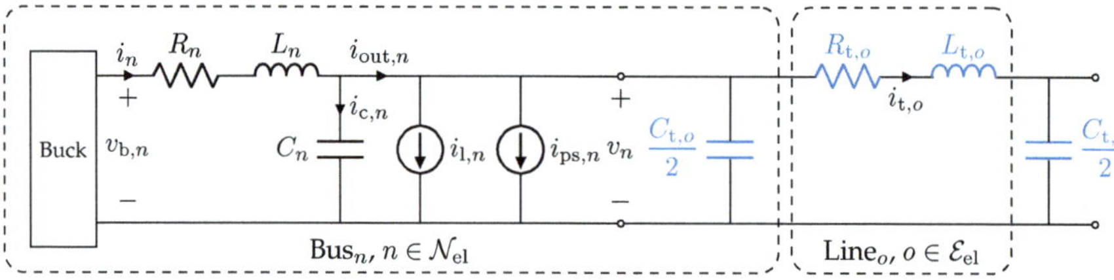

Figure 4.1: A bus n comprising a DC-DC buck converter, an LC filter (R_n, L_n, C_n), a nonlinear load $i_{\text{l},n}$ and a P source $i_{\text{ps},n}$, which connects to a π-model transmission line (blue).

While the typical DGU model assumes an unlimited power supply is available locally for powering the DC network (see [Tuc+16; Tri+19; Str+20b]), many components (e.g., smart loads) might only offer a small amount of control flexibility resulting from a limited amount of power (or stored energy). A distinction between bus types is thus made in this work, with *voltage setting buses* being equipped with an unbounded energy supply for control purposes. Voltage setting buses can thus perform voltage regulation by providing the requisite power during transients and in steady state. On the other hand, *voltage following buses* only have a finite energy supply available locally, restricting their ability to influence the steady state. Voltage following buses can therefore only provide transient stabilisation support, with their steady-state bus voltages being dependent on the rest of the network. In the context of the clustered network, it is thus assumed that each cluster contains at least one voltage setting bus which provides voltage regulation in the cluster.

More formally, the DC network of buses (vertices) and transmission lines (edges) is represented using a graph $\mathcal{G}_{\text{el}} = (\mathcal{N}_{\text{el}}, \mathcal{E}_{\text{el}})$, as depicted in Figure 4.2. The graph is then partitioned into $|\mathcal{M}| \geq 1$ clusters such that $\mathcal{N}_{\text{el}} = \bigcup_{m\in\mathcal{M}} \mathcal{N}_m$ and $\mathcal{E}_{\text{el}} = \mathcal{T} \cup \bigcup_{m\in\mathcal{M}} \mathcal{E}_m$, where $\mathcal{N}_m$ and $\mathcal{E}_m$ denote the buses and lines associated with a cluster $m \in \mathcal{M}$ and $\mathcal{T}$ represents the lines not associated with a cluster, e.g., those

[1] A distinction is made between the load and the P source for components that exhibit different behaviours at low voltages

connecting buses in different clusters. Note that each bus and line appears in exactly one subset: $\mathcal{N}_{m_1} \cap \mathcal{N}_{m_2} = \emptyset$, $\mathcal{E}_{m_1} \cap \mathcal{E}_{m_2} = \emptyset$ and $\mathcal{E}_{m_1} \cap \mathcal{T} = \emptyset$ for any $m_1, m_2 \in \mathcal{M}$ with $m_1 \neq m_2$. Furthermore, each edge is arbitrarily assigned a direction which denotes the positive current flow in the corresponding transmission line. The directed edges $\mathcal{E}_m$ in each cluster also ensure that its buses $\mathcal{N}_m$ are weakly connected. To differentiate between the voltage setting and voltage following buses, the buses in a cluster are partitioned into two sets $\mathcal{N}_m = \mathcal{S}_m \cup \mathcal{F}_m$, with $\mathcal{S}_m \cap \mathcal{F}_m = \emptyset$. The set $\mathcal{S}_m = \{a, \ldots, |\mathcal{S}_m| + a - 1\}$ with $|\mathcal{S}_m| \geq 1$ represents the voltage setting buses, $\mathcal{F}_m = \{|\mathcal{S}_m| + a, \ldots, |\mathcal{S}_m| + |\mathcal{F}_m| + a - 1\}$ with $|\mathcal{F}_m| \geq 0$ represents the voltage following buses, and a is the starting bus index.

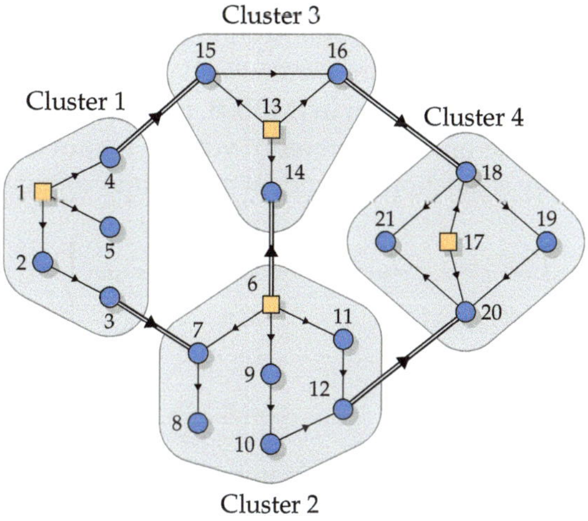

Figure 4.2: A 21-bus DC network partitioned into four clusters. Each cluster m comprises buses with voltage setting controllers ☐ in $\mathcal{S}_m$, buses with voltage following controllers ● in $\mathcal{F}_m$, and lines with arbitrary directions indicating the direction of positive flow →— in $\mathcal{E}_m$. The lines ⇒— in $\mathcal{T}$ are used to interconnect buses in different clusters.

In the subsequent subsections, models which describe the dynamics of the components in the network are provided. These include static load and power supply components (**Section 4.1.1**), transmission lines (**Section 4.1.2**) and DGU-equipped buses (**Section 4.1.3**). These components are then combined into a dynamical system for an entire cluster (**Section 4.1.4**), which serves as the basis for the control and analysis of the network in the subsequent sections.

4.1.1 Static Loads and Sources

In DC networks, uncontrollable loads and sources are typically modelled using static functions (see [vV98; MBB08] and Section 2.1). These static functions describe the voltage-dependent current consumed or injected at a bus.

Remark 4.1. *Since power is transported over lossy inductive lines in DC networks, static components at buses without local control can represent a significant control challenge. Since static components react instantaneously to changing states, they can cause voltage oscillations or a voltage collapse which can be hard to prevent using controllers acting over inductive lines. From a control perspective, it is therefore observed that static loads are generally more problematic than dynamical loads which have stable equilibria.*

Loads

Consider a load at a bus $n \in \mathcal{N}_{\text{el}}$ which is modelled as a static, nonlinear voltage-dependent current source described by a class C^0 (continuous) and globally Lipschitz function. As is typical in the literature (see Section 2.1), the loads are modelled using the ZIP model. Moreover, below a critical voltage $v_{\text{crit},n} \in \mathbb{R}_{>0}$, the loads are modelled as pure resistances (see the elaboration in [MBB08, pp. 110–112]). This critical voltage is usually set to 70 % of the nominal reference voltage v_{Ref} of the network. The ZIP load is then described by a function dependent on the bus voltage $v_n \in \mathbb{R}$:

$$i_{\text{l},n}(v_n) = \left\{ \begin{array}{ll} Z_n^{-1} v_n + I_n + \dfrac{P_n}{v_n}, & v_n \geq v_{\text{crit},n} \\ Z_{\text{crit},n}^{-1} v_n, & v_n < v_{\text{crit},n} \end{array} \right\}, \quad n \in \mathcal{N}_{\text{el}}, \tag{4.1a}$$

$$Z_{\text{crit},n}^{-1} := \frac{i_{\text{l},n}(v_{\text{crit},n})}{v_{\text{crit},n}} = Z_n^{-1} + \frac{I_n}{v_{\text{crit},n}} + \frac{P_n}{v_{\text{crit},n}^2}, \quad v_{\text{crit},n} > 0, \tag{4.1b}$$

where $Z_n, I_n, P_n \in \mathbb{R}$ describe the constant impedance, current and power components of the load, respectively. Note that negative parameters may be used for loads that inject energy.

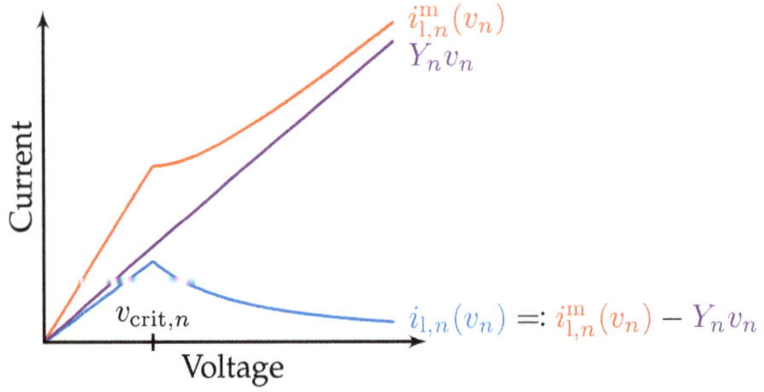

Figure 4.3: The separation of a nonlinear ZIP load (blue) in (4.1) into a monotone increasing nonlinear part (orange) and a linear part (purple), according to (4.2).

As discussed in [Jus04; SGF17; AN+17], the negative incremental impedance associated with P loads can be detrimental to the stability of the network. Naturally, this problem extends to the more general case of ZIP loads with positive and negative load parameters. To simplify the control and analysis tasks in the sequel, the load function (4.1a) is split into a linear part with a gradient $Y_n \in \mathbb{R}$ and a monotone increasing nonlinear part $i^{\mathrm{m}}_{1,n}(v_n)$ such that

$$i_{1,n}(v_n) =: i^{\mathrm{m}}_{1,n}(v_n) - Y_n v_n, \quad n \in \mathcal{N}_{\mathrm{el}}, \tag{4.2}$$

as illustrated in Figure 4.3. Note that Y_n is chosen as small as possible such that $i^{\mathrm{m}}_{1,n}(v_n)$ is guaranteed to be monotone. This means that Y_n is load-dependent, making an exact determination difficult for unknown loads. In such cases, a larger Y_n can be chosen to make the separation in (4.2) robust against load uncertainty. This robustness, however, leads to larger control gains in the sequel.

Remark 4.2. *To ensure that $i^{\mathrm{m}}_{1,n}(v_n)$ is monotone, Y_n must overcome the smallest incremental impedance[2] of the load $i_{1,n}(v_n)$. Thus, from the minimum gradient of the static ZIP load function* (4.1)*, it follows that*

$$Y_n \geq -\min\left(\frac{di_{1,n}(v_n)}{dv_n}\right) = -\min\left(Z_n^{-1}, Z_n^{-1} - \frac{P_n}{v_{\mathrm{crit}}^2}, Z_{\mathrm{crit}}^{-1}\right). \tag{4.3}$$

For the original load $i_{1,n}(v_n)$ to be monotone, i.e., $Y_n = 0$, and if $P_n \geq 0$ and $I_n \geq 0$, (4.3) *yields the load passivity condition $Z^{-1}v_{\mathrm{crit}}^2 \geq P_n$ frequently used in the literature (see [DWD18; CKS07; Str+20b; NFT20; Fan+20]). Furthermore, notice that the smallest Y_n thus corresponds to $-\nu$ for a static, IFOFP, single-input-single-output function as per Lemma 3.5.*

Finally, the monotone part of the load function in (4.2) may be shifted w.r.t. a given constant voltage $\hat{v}_n$. This yields

$$\tilde{i}^{\mathrm{m}}_{1,n}(\tilde{v}_n) := i^{\mathrm{m}}_{1,n}(v_n) - i^{\mathrm{m}}_{1,n}(\hat{v}_n) = i^{\mathrm{m}}_{1,n}(\tilde{v}_n + \hat{v}_n) - i^{\mathrm{m}}_{1,n}(\hat{v}_n), \quad n \in \mathcal{N}_{\mathrm{el}}, \tag{4.4a}$$

$$\tilde{i}^{\mathrm{m}}_{1,n}(0) = 0, \tag{4.4b}$$

as per the static function definition (3.17).

Remark 4.3. *Beyond the ZIP loads considered here, the results in this chapter are compatible with any other continuous and globally Lipschitz static load models which can be described using* (3.17)*, such as exponential load models (see [MBB08, p. 112], [Str+20a]).*

P Sources

Consider a P source which supplies constant power with a maximum current rating $I_{\mathrm{ps,max},n}$. At low voltages, the behaviour of such a P source differs from the constant Z characteristic of the ZIP load representation in (4.1), since the source still seeks to inject the maximum power subject to the current limitation. Thus, the P source shows a

[2] Note that Y_n corresponds to an admittance value ($\Delta i/\Delta v$). Furthermore, the incremental load admittance and impedance have the same sign and monotonicity properties.

constant I behaviour at low voltages. In keeping with the load-oriented representation in Figure 4.1, this gives

$$i_{\mathrm{ps},n}(v_n) = \left\{ \begin{array}{ll} \frac{P_{\mathrm{ps},n}}{v_n}, & v_n \geq \frac{P_{\mathrm{ps},n}}{I_{\mathrm{ps,max},n}}, \\ I_{\mathrm{ps,max},n}, & v_n < \frac{P_{\mathrm{ps},n}}{I_{\mathrm{ps,max},n}}, \end{array} \right\}, \quad n \in \mathcal{N}_{\mathrm{el}}, \tag{4.5}$$

where $P_{\mathrm{ps},n} \in \mathbb{R}_{\leq 0}$ and $I_{\mathrm{ps,max},n} \in \mathbb{R}_{<0}$ are constant.

As is the case for the load function, the P source function (4.5) may be shifted w.r.t. a given constant voltage $\hat{v}_n$, yielding

$$\tilde{i}_{\mathrm{ps},n}(\tilde{v}_n) := i_{\mathrm{ps},n}(v_n) - i_{\mathrm{ps},n}(\hat{v}_n) = i_{\mathrm{ps},n}(\tilde{v}_n + \hat{v}_n) - i_{\mathrm{ps},n}(\hat{v}_n), \quad n \in \mathcal{N}_{\mathrm{el}}, \tag{4.6a}$$

$$\tilde{i}_{\mathrm{ps},n}(0) = 0, \tag{4.6b}$$

as per the static function definition (3.17).

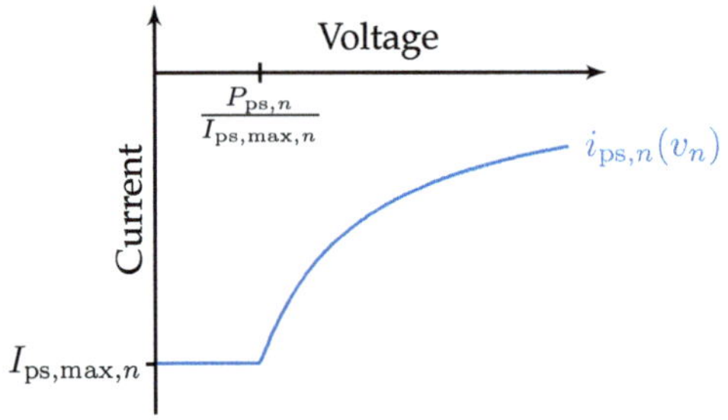

Figure 4.4: A P source function $i_{\mathrm{ps},n}$ (4.5) represented as a load.

Remark 4.4. *The P source function in Figure 4.4 with $P_{\mathrm{ps},n} < 0$ and $I_{\mathrm{ps,max},n} < 0$ is always monotone increasing in v_n, since the derivative of $i_{\mathrm{ps},n}(v_n)$ w.r.t. the voltage is always positive (see Remark 4.2). A separation as in* (4.2) *is thus unnecessary for P sources. This corroborates the observation in [Mac+21] that P sources pose no stability problem for the network.*

4.1.2 Transmission Lines

The same π-model transmission line used in AC networks [Kun94, p. 202, p. 207] [MBB08, p. 65ff.] is also typically employed in DC networks (see the literature sources in Section 2.1). The dynamics of the RL line components connecting a source bus $n_{o,+}$ and a sink bus $n_{o,-}$ are given by

$$L_{\mathrm{t},o}\dot{i}_{\mathrm{t},o} = -R_{\mathrm{t},o}i_{\mathrm{t},o} + v_{n_{o,+}} - v_{n_{o,-}}, \quad o \in \mathcal{E}_{\mathrm{el}}, \tag{4.7}$$

where $i_{\mathrm{t},o} \in \mathbb{R}$ is the line current, $L_{\mathrm{t},o}, R_{\mathrm{t},o} \in \mathbb{R}_{>0}$ are the line inductance and resistance, and $v_{n_{o,+}}, v_{n_{o,-}} \in \mathbb{R}$ are the corresponding bus voltages. Note that the transmission line capacitance is included in the dynamics of the source and sink buses, as shown in the sequel. Furthermore, while the parallel conductances are typically negligible, they may be included as constant impedances in the loads (4.1) for analysis or simulation purposes.

4.1.3 DGU-Equipped Buses

Consider now a bus $n \in \mathcal{N}_{\mathrm{el}}$ which is equipped with a DGU, an optional load (4.1), and an optional P source (4.5) (see Figure 4.1). As is common in the literature (see [Tuc+16; Tri+19; Str+20b]), the DGU is modelled as a time-averaged buck converter which connects an ideal voltage source to the bus through a lossy LC filter. The ideal voltage source represents either an unlimited or a limited source of power for the buses in $\mathcal{S}_m$ and $\mathcal{F}_m$, respectively. The bus dynamics are represented by

$$\begin{bmatrix} L_n \dot{i}_n \\ C_{\mathrm{ef},n} \dot{v}_n \end{bmatrix} = \begin{bmatrix} -R_n & -1 \\ 1 & Y_n \end{bmatrix} \begin{bmatrix} i_n \\ v_n \end{bmatrix} + \begin{bmatrix} v_{\mathrm{b},n} \\ i_{\mathrm{int},n} + i_{\mathrm{ex},n} \end{bmatrix}, \quad n \in \mathcal{N}_{\mathrm{el}}, \tag{4.8a}$$

$$i_{\mathrm{int},n} := \sum_{o \in \mathcal{E}_n^+} i_{\mathrm{t},o} - \sum_{o \in \mathcal{E}_n^-} i_{\mathrm{t},o}, \tag{4.8b}$$

$$i_{\mathrm{ex},n} := \sum_{o \in \mathcal{T}_n^+} i_{\mathrm{t},o} - \sum_{o \in \mathcal{T}_n^-} i_{\mathrm{t},o} - i_{\mathrm{l},n}^{\mathrm{m}}(v_n) - i_{\mathrm{ps},n}(v_n), \tag{4.8c}$$

where $i_n \in \mathbb{R}$ is the filter current, $v_n \in \mathbb{R}$ is the bus voltage, $L_n, R_n \in \mathbb{R}_{>0}$ are the filter inductance and resistance, respectively, and the buck converter voltage $v_{\mathrm{b},n}$ is the control input. The capacitances of the lines which are lumped together at bus n are described by $C_{\mathrm{t},o} \in \mathbb{R}_{>0}$, for $o \in \mathcal{E}_n^+ \cup \mathcal{E}_n^- \cup \mathcal{T}_n^+ \cup \mathcal{T}_n^-$. Adding these to the filter capacitance $C_n \in \mathbb{R}_{>0}$ yields the effective bus capacitance

$$C_{\mathrm{ef},n} = C_n + \sum_{o \in \mathcal{E}_n^+ \cup \mathcal{E}_n^- \cup \mathcal{T}_n^+ \cup \mathcal{T}_n^-} \tfrac{1}{2} C_{\mathrm{t},o}. \tag{4.9}$$

The internal current $i_{\mathrm{int},n} \in \mathbb{R}$ represents the total line currents into and out of bus n from transmission lines within the cluster. Similarly, the external current $i_{\mathrm{ex},n} \in \mathbb{R}$ represents an exogenous input of the bus dynamics (4.8a), which is made up of the transmission lines in $\mathcal{T}$ connecting to or from bus n along with the static monotone part of the load $i_{\mathrm{l},n}^{\mathrm{m}}$ in (4.2) and the static P source (4.5). Note that the linear part of the separated load represented by Y_n is included in the bus dynamics (4.8a).

4.1.4 Clusters

By combining the bus dynamics (4.8) and the line dynamics (4.7), a model for a cluster can be constructed. Recall that the buses in a cluster $m \in \mathcal{M}$ are partitioned into voltage setting buses in $\mathcal{S}_m$ and voltage following buses in $\mathcal{F}_m$. While both types of buses are described using the same dynamics (4.8), they are subject to different conditions related to the availability of power. As such, their associated states are kept separate in the cluster dynamics in vector form

$$\begin{bmatrix} \boldsymbol{L}_{\mathrm{s},m}\dot{\boldsymbol{i}}_{\mathrm{s},m} \\ \boldsymbol{C}_{\mathrm{s},m}\dot{\boldsymbol{v}}_{\mathrm{s},m} \\ \boldsymbol{L}_{\mathrm{f},m}\dot{\boldsymbol{i}}_{\mathrm{f},m} \\ \boldsymbol{C}_{\mathrm{f},m}\dot{\boldsymbol{v}}_{\mathrm{f},m} \\ \boldsymbol{L}_{\mathrm{t},m}\dot{\boldsymbol{i}}_{\mathrm{t},m} \end{bmatrix} = \begin{bmatrix} -\boldsymbol{R}_{\mathrm{s},m} & -\boldsymbol{I}_{|\mathcal{S}_m|} & \boldsymbol{0} & \boldsymbol{0} & \boldsymbol{0} \\ \boldsymbol{I}_{|\mathcal{S}_m|} & \boldsymbol{Y}_{\mathrm{s},m} & \boldsymbol{0} & \boldsymbol{0} & \boldsymbol{E}_{\mathrm{s},m} \\ \boldsymbol{0} & \boldsymbol{0} & -\boldsymbol{R}_{\mathrm{f},m} & -\boldsymbol{I}_{|\mathcal{F}_m|} & \boldsymbol{0} \\ \boldsymbol{0} & \boldsymbol{0} & \boldsymbol{I}_{|\mathcal{F}_m|} & \boldsymbol{Y}_{\mathrm{f},m} & \boldsymbol{E}_{\mathrm{f},m} \\ \boldsymbol{0} & -\boldsymbol{E}_{\mathrm{s},m}^T & \boldsymbol{0} & -\boldsymbol{E}_{\mathrm{f},m}^T & -\boldsymbol{R}_{\mathrm{t},m} \end{bmatrix} \begin{bmatrix} \boldsymbol{i}_{\mathrm{s},m} \\ \boldsymbol{v}_{\mathrm{s},m} \\ \boldsymbol{i}_{\mathrm{f},m} \\ \boldsymbol{v}_{\mathrm{f},m} \\ \boldsymbol{i}_{\mathrm{t},m} \end{bmatrix} + \begin{bmatrix} \boldsymbol{v}_{\mathrm{b,s},m} \\ \boldsymbol{i}_{\mathrm{ex,s},m} \\ \boldsymbol{v}_{\mathrm{b,f},m} \\ \boldsymbol{i}_{\mathrm{ex,f},m} \\ \boldsymbol{0} \end{bmatrix}, \quad m \in \mathcal{M}, \tag{4.10}$$

where $\boldsymbol{i}_{\mathrm{s},m}, \boldsymbol{v}_{\mathrm{s},m}, \boldsymbol{v}_{\mathrm{b,s},m}, \boldsymbol{i}_{\mathrm{ex,s},m}$, and $\boldsymbol{i}_{\mathrm{f},m}, \boldsymbol{v}_{\mathrm{f},m}, \boldsymbol{v}_{\mathrm{b,f},m}, \boldsymbol{i}_{\mathrm{ex,f},m}$ denote the stacked vectors of variables for the buses in $\mathcal{S}_m$ and $\mathcal{F}_m$, respectively. Furthermore, $\boldsymbol{i}_{\mathrm{t},m}$ represents the stacked currents of the edges in $\mathcal{E}_m$. The following matrices are constructed from the bus and line parameters: $\boldsymbol{L}_{\mathrm{s},m} = \mathrm{diag}[(L_j)]$, $\boldsymbol{C}_{\mathrm{s},m} = \mathrm{diag}[(C_{\mathrm{ef},j})]$, $\boldsymbol{L}_{\mathrm{f},m} = \mathrm{diag}[(L_k)]$, $\boldsymbol{C}_{\mathrm{f},m} = \mathrm{diag}[(C_{\mathrm{ef},k})]$, $\boldsymbol{L}_{\mathrm{t},m} = \mathrm{diag}[(L_{\mathrm{t},l})]$, $\boldsymbol{R}_{\mathrm{t},m} = \mathrm{diag}[(R_{\mathrm{t},l})]$ for $j \in \mathcal{S}_m$, $k \in \mathcal{F}_m$, and $l \in \mathcal{E}_m$. For the graph of the cluster $\mathcal{G}_m = (\mathcal{N}_m, \mathcal{E}_m)$, the incidence matrix $\boldsymbol{E}_m$ (see (3.1)) is split according to

$$\boldsymbol{E}_m = \begin{bmatrix} \boldsymbol{E}_{\mathrm{s},m} \\ \boldsymbol{E}_{\mathrm{f},m} \end{bmatrix}, \tag{4.11}$$

where $\boldsymbol{E}_{\mathrm{s},m}$ and $\boldsymbol{E}_{\mathrm{f},m}$ correspond to the first $|\mathcal{S}_m|$ rows and the last $|\mathcal{F}_m|$ rows of $\boldsymbol{E}_m$, respectively. Thus, the internal currents in (4.8b) are included in the dynamics in (4.10) while the external currents in (4.8c) remain exogenous inputs for the cluster.

4.2 Control Problem and Controller Design

Using the component and cluster models in Section 4.1, model-based controllers are designed which achieve voltage regulation in a DC network as specified in RQ 1.1. To this end, the requirements of RQ 1.1 are formalised as a control objective in **Section 4.2.1** and an assumption about the equilibrium of a closed-loop cluster is made. The control objective is stated using the cluster model (4.10) and stipulates the use of decentralised controllers for the different types of buses. Then, in **Section 4.2.2** and **Section 4.2.3**, controllers for the voltage setting and voltage following buses are designed, respectively. Additionally, it is verified in each case that the controllers achieve their local control objectives.

4.2.1 Control Problem Formulation

Recalling Figure 4.2, Statement 3 of Proposition 3.1 and Remark 3.6, the voltage regulation of a DC network as required for RQ 1.1 can be achieved if each cluster is rendered EISD and OS-EIP w.r.t. a suitable power port through the use of decentralised control at the buses. Furthermore, since the voltage setting and voltage following buses differ in their local steady-state power availability, separate sub-objectives are formulated for each bus type.

Objective 4.1 (Cluster OS-EIP)
Design decentralised controllers for each control input $v_{\mathrm{b},n}$, $n \in \mathcal{N}_{\mathrm{el}}$ *such that each cluster* $m \in \mathcal{M}$ *is OS-EIP w.r.t. the input-output pairs* $(\tilde{\boldsymbol{i}}_{\mathrm{ex,s},m}, \tilde{\boldsymbol{v}}_{\mathrm{s},m})$ *and* $(\tilde{\boldsymbol{i}}_{\mathrm{ex,f},m}, \tilde{\boldsymbol{v}}_{\mathrm{f},m})$*. Moreover, the voltage setting and voltage following buses should satisfy the following sub-objectives for each cluster* $m \in \mathcal{M}$*:*

1. *The voltage setting controller at each bus* $j \in \mathcal{S}_m$ *must ensure that, for a constant input* $i_{\mathrm{int},j} + i_{\mathrm{ex},j}$, $\lim_{t\to\infty} v_j = \hat{v}_j = v_j^\star$ *and that the bus is OS-EIP w.r.t. the input-output pair* $(\tilde{i}_{\mathrm{int},j} + \tilde{i}_{\mathrm{ex},j}, \tilde{v}_j)$.
2. *The voltage following controller at each bus* $k \in \mathcal{F}_m$ *must ensure that the equilibrium of the bus is located at* $\hat{i}_k = 0 \iff \hat{v}_k - \hat{v}_{\mathrm{b},k} = 0$.

Note that achieving Objective 4.1 and RQ 1.1 requires the cluster dynamics (4.10) to be controlled to a non-zero equilibrium. Specifically, the voltages at voltage setting buses must converge to the supplied setpoints $\boldsymbol{v}_{\mathrm{s}}^\star$, which then indirectly determine the steady states of the voltages at the voltage following buses $\hat{\boldsymbol{v}}_{\mathrm{f}}$.

Assumption 4.1 (Cluster Equilibrium)
For a cluster m *with the constant input* $(\boldsymbol{v}_{\mathrm{b,s},m}, \boldsymbol{i}_{\mathrm{ex,s},m}, \boldsymbol{v}_{\mathrm{b,f},m}, \boldsymbol{i}_{\mathrm{ex,f},m}) = (\hat{\boldsymbol{v}}_{\mathrm{b,s},m}, \hat{\boldsymbol{i}}_{\mathrm{ex,s},m}, \hat{\boldsymbol{v}}_{\mathrm{b,f},m}, \hat{\boldsymbol{i}}_{\mathrm{ex,f},m})$*, the cluster dynamics* (4.10) *have an equilibrium at some* $(\boldsymbol{i}_{\mathrm{s},m}, \boldsymbol{v}_{\mathrm{s},m}, \boldsymbol{i}_{\mathrm{f},m}, \boldsymbol{v}_{\mathrm{f},m}, \boldsymbol{i}_{\mathrm{t},m}) = (\hat{\boldsymbol{i}}_{\mathrm{s},m}, \boldsymbol{v}_{\mathrm{s},m}^\star, \boldsymbol{0}, \hat{\boldsymbol{v}}_{\mathrm{f},m}, \hat{\boldsymbol{i}}_{\mathrm{t},m})$.

Note that Assumption 4.1 implicitly assumes that the cluster, and thus the DC network as a whole, is well-designed. Specifically, assuming a valid equilibrium exists includes the assumption that the voltage setting buses, the external currents, and any P sources in the cluster can feasibly supply the desired power to the loads over the included transmission lines.

Remark 4.5. *The linearity of the cluster dynamics in* (4.10) *initially seems to suggest that the uniqueness of the equilibrium can easily be determined using the system matrix in* (4.10)*. Nevertheless, the uniqueness of the equilibrium is also dependent on the monotone, nonlinear parts of the static load and the P source functions in* $\boldsymbol{i}_{\mathrm{ex,s},m}$ *and* $\boldsymbol{i}_{\mathrm{ex,f},m}$ *(see* (4.8c)*). While these static functions may be externalised for a stability analysis, their effects cannot be disregarded when calculating the equilibrium of a cluster or the DC network.*

As per Objective 4.1, a bus is equipped with a voltage setting or a voltage following controller depending on the local availability of steady-state power at the bus. A schematic of this implementation is depicted in Figure 4.5, which also provides an overview for the subsequent subsections. Note that equipping a bus with both a voltage setting and a voltage following controller allows the bus to change from a voltage setting bus in $\mathcal{S}_m$ to a voltage following bus in $\mathcal{F}_m$, and vice versa, depending on the local power availability.

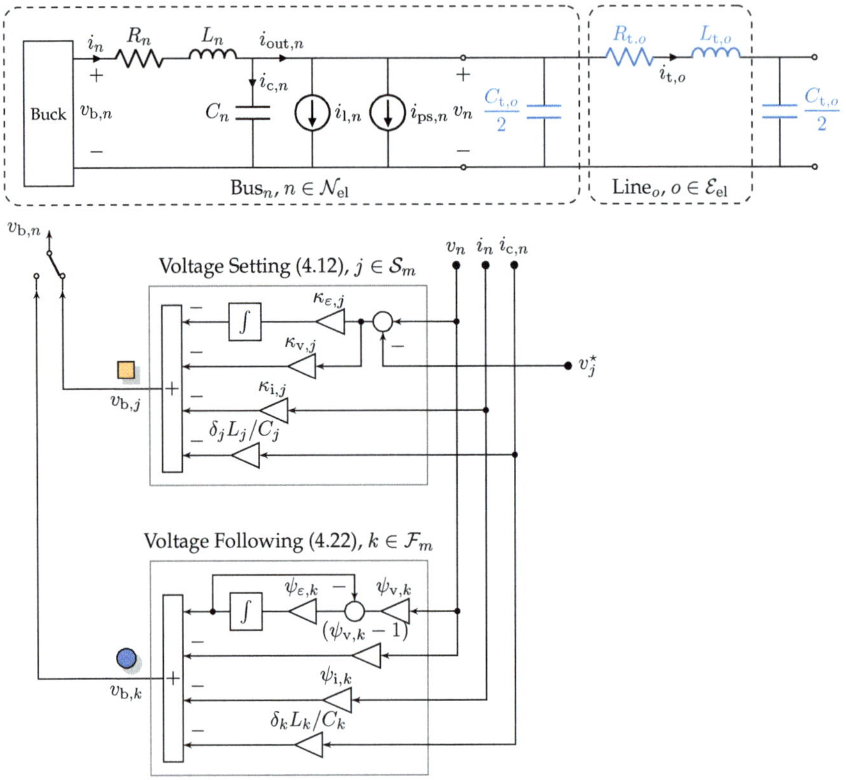

Figure 4.5: Schematic of a bus equipped with a voltage setting controller ($j \in \mathcal{S}_m$) or a voltage following controller ($k \in \mathcal{F}_m$).

4.2.2 Voltage Setting Controller

Consider a voltage setting bus $j \in \mathcal{S}_m$, $m \in \mathcal{N}$, where sufficient steady-state power is available to achieve voltage regulation to a user-defined setpoint $v^\star \in \mathbb{R}_{\geq 0}$. Let this bus be equipped with the decentralised controller

$$\dot{\varepsilon}_j = \kappa_{\varepsilon,j}(v_j - v_j^\star), \tag{4.12a}$$

$$v_{\text{b},j} = -\kappa_{\text{v},j}(v_j - v_j^\star) - \kappa_{\text{i},j} i_j - \varepsilon_j - \frac{\delta_j L_j}{C_j} i_{\text{c},j}, \quad j \in \mathcal{S}_m, \tag{4.12b}$$

where $\kappa_{\varepsilon,j} \in \mathbb{R}_{>0}$ and $\kappa_{\text{v},j} \in \mathbb{R}_{>0}$ are the integral and proportional control gains, $\kappa_{\text{i},j} \in \mathbb{R}_{>0}$ injects damping onto the dynamics of the filter current i_j and $\delta_j \in \mathbb{R}_{>0}$ injects damping onto the dynamics of the capacitor current $i_{\text{c},j}$. Note that (4.12b) only uses the filter capacitance C_j, not the effective bus capacitance $C_{\mathrm{ef},j}$. For simplicity, the subscript j is omitted for the remainder of this subsection.

Next, the bus dynamics in (4.8) equipped with the voltage setting controller (4.12) are analysed. Since the state-dependent current i_c is used in the controller (4.12), it is convenient to write the closed-loop bus dynamics using a new variable

$$i_\mathrm{z} := i + \delta v. \tag{4.13}$$

Calculating the time derivative of (4.13) multiplied by L, and substituting $L\dot{i}$ from (4.8a) along with the filter capacitor equation $C\dot{v} = i_\mathrm{c}$ gives

$$L\dot{i}_\mathrm{z} = L\dot{i} + L\delta\dot{v} = -Ri - v + v_\mathrm{b} + \tfrac{\delta L}{C} i_\mathrm{c}. \tag{4.14}$$

Thus, (4.12a) multiplied by L, (4.14), the voltage dynamics in (4.8a) and the control output (4.12b) together give

$$\begin{aligned}
L\dot{\varepsilon} &= L\kappa_\varepsilon(v - v^\star), && (4.15a)\\
L\dot{i}_\mathrm{z} &= -(R+\kappa_\mathrm{i})i - v - \kappa_\mathrm{v}(v - v^\star) - \varepsilon \\
&= -(R+\kappa_\mathrm{i})i_\mathrm{z} - (\kappa_\mathrm{v} + 1 - \delta\kappa_\mathrm{i} - \delta R)v - \varepsilon + \kappa_\mathrm{v} v^\star, && (4.15b)\\
C_\mathrm{ef}\dot{v} &= i + Y + i_\mathrm{int} + i_\mathrm{ex} = i_\mathrm{z} + (Y - \delta)v + i_\mathrm{int} + i_\mathrm{ex}. && (4.15c)
\end{aligned}$$

Shifting (4.15) w.r.t. the equilibrium $\hat{\varepsilon}$, $\hat{i}_\mathrm{z}$, $\hat{v} = v^\star$, $\hat{i}_\mathrm{int}$, and $\hat{i}_\mathrm{ex}$ yields the closed-loop bus dynamics

$$\underbrace{\begin{bmatrix} L\dot{\tilde{\varepsilon}} \\ L\dot{\tilde{i}}_\mathrm{z} \\ C_\mathrm{ef}\dot{\tilde{v}} \end{bmatrix}}_{\boldsymbol{Q}_\mathrm{s}\dot{\tilde{\boldsymbol{x}}}_\mathrm{s}} = \underbrace{\begin{bmatrix} 0 & 0 & \kappa_\varepsilon^\diamond \\ -1 & -\kappa_\mathrm{i}^\diamond & -\kappa_\mathrm{v}^\diamond \\ 0 & 1 & Y-\delta \end{bmatrix}}_{\boldsymbol{A}_\mathrm{s}} \underbrace{\begin{bmatrix} \tilde{\varepsilon} \\ \tilde{i}_\mathrm{z} \\ \tilde{v} \end{bmatrix}}_{\tilde{\boldsymbol{x}}_\mathrm{s}} + \underbrace{\begin{bmatrix} 0 \\ 0 \\ \tilde{i}_\mathrm{int} + \tilde{i}_\mathrm{ex} \end{bmatrix}}_{\boldsymbol{b}_\mathrm{s}\tilde{u}_\mathrm{s}}, \quad \tilde{y}_\mathrm{s} = \underbrace{\begin{bmatrix} 0 & 0 & 1 \end{bmatrix}}_{\boldsymbol{b}_\mathrm{s}^T} \tilde{\boldsymbol{x}}_\mathrm{s} = \tilde{v}, \tag{4.16}$$

$$\kappa_\mathrm{i}^\diamond := \kappa_\mathrm{i} + R, \qquad \kappa_\mathrm{v}^\diamond := \kappa_\mathrm{v} + 1 - \delta\kappa_\mathrm{i}^\diamond, \qquad \kappa_\varepsilon^\diamond := L\kappa_\varepsilon. \tag{4.17}$$

Next, the closed-loop bus dynamics shifted w.r.t. the equilibrium is given and the location of the equilibrium for a constant input is established.

Proposition 4.1 (Closed-Loop Voltage Setting Bus Equilibrium)
For a constant input $\tilde{u}_\mathrm{s} = 0$, *(4.16) has the equilibrium* $\tilde{\boldsymbol{x}}_\mathrm{s} = \boldsymbol{0}$.

Proof:
Since $\boldsymbol{A}_\mathrm{s}$ in (4.16) has full rank (due to $\kappa_\varepsilon^\diamond > 0$), the shifted voltage setting bus with $\tilde{u}_\mathrm{s} = 0$ has an equilibrium $\tilde{\boldsymbol{x}}_\mathrm{s} = \boldsymbol{0}$ with $\hat{v} = v^\star$, $\hat{i}_\mathrm{z} = -(Y-\delta)v^\star - \hat{u}_\mathrm{s}$, and $\hat{\varepsilon} = -\kappa_\mathrm{v}^\diamond v^\star - \kappa_\mathrm{i}^\diamond \hat{i}_\mathrm{z}$. ■

The closed-loop dynamics (4.16) written using the variable i_z in (4.13) demonstrate how the control gain δ injects damping onto the dynamics of the bus voltage v. Furthermore, the input-output port $(\tilde{i}_{\text{int}} + \tilde{i}_{\text{ex}}, \tilde{v})$ is unaffected by the variable transformation in (4.13). Proposition 4.1 also satisfies part of Objective 4.1 for the voltage setting buses by ensuring that $v = \hat{v}$ is an equilibrium of the closed loop for a constant input $\tilde{u}_{\text{s}} = 0$. Next, the OS-EIP of the shifted, closed-loop voltage setting bus (4.16) is investigated.

Theorem 4.2 (OS-EIP of a Voltage Setting Bus)
Consider the shifted closed-loop voltage setting bus (4.16). *If there is a* $\rho > 0$ *such that*

$$\delta > Y + \rho + \max\left(0, \frac{\kappa_{\varepsilon}^{\diamond} - \kappa_{\text{i}}^{\diamond}\kappa_{\text{v}}^{\diamond}}{(\kappa_{\text{i}}^{\diamond})^2}\right), \tag{4.18}$$

then there exists a $p_2 > 0$ *which ensures that* (4.16) *is OS-EIP w.r.t. the input-output pair* $(\tilde{u}_{\text{s}}, \tilde{y}_{\text{s}}) = (\tilde{i}_{\text{int}} + \tilde{i}_{\text{ex}}, \tilde{v})$ *with the storage function*

$$S_{\text{s}}(\tilde{\boldsymbol{x}}_{\text{s}}) = \tfrac{1}{2}\tilde{\boldsymbol{x}}_{\text{s}}^T \boldsymbol{Q}_{\text{s}} \boldsymbol{P}_{\text{s}} \tilde{\boldsymbol{x}}_{\text{s}}, \quad \boldsymbol{Q}_{\text{s}} = \operatorname{diag}(L, L, C_{\text{ef}}), \tag{4.19a}$$

$$\boldsymbol{P}_{\text{s}} = \begin{bmatrix} p_1 & p_2 & 0 \\ p_2 & p_3 & 0 \\ 0 & 0 & 1 \end{bmatrix}, \quad \begin{cases} p_1 := \dfrac{1}{\kappa_{\text{i}}^{\diamond}} p_2 + \dfrac{1}{\kappa_{\text{i}}^{\diamond}\kappa_{\varepsilon}^{\diamond}}, \\ p_3 := \kappa_{\text{i}}^{\diamond} p_2. \end{cases} \tag{4.19b}$$

Proof Sketch:
The proof centres on verifying that $\dot{S}_{\text{s}}(\tilde{\boldsymbol{x}}_{\text{s}}) \leq \tilde{y}_{\text{s}}\tilde{u}_{\text{s}} - \rho\tilde{y}_{\text{s}}^2$ and that S_{s} is positive definite in $\tilde{\boldsymbol{x}}_{\text{s}}$. Since $\boldsymbol{Q}_{\text{s}}\boldsymbol{P}_{\text{s}} = \boldsymbol{P}_{\text{s}}\boldsymbol{Q}_{\text{s}}$, the former condition reduces to showing that $-\boldsymbol{A}_{\text{s}}^T\boldsymbol{P}_{\text{s}} - \boldsymbol{P}_{\text{s}}\boldsymbol{A}_{\text{s}} - \rho\boldsymbol{b}_{\text{s}}\boldsymbol{b}_{\text{s}}^T \succcurlyeq 0$ while the latter condition is met if $\boldsymbol{P} \succ 0$. The first LMI is analysed using Sylvester's criterion for semi-definite matrices which requires the determinant of the matrix comprising the LMI to be zero while its trailing principal minors are positive. The elements of $\boldsymbol{P}_{\text{s}}$ are chosen such that the determinant of the matrix is zero. Next, the trailing zeros of the matrix comprising the LMI are investigated, yielding the condition in (4.18). Specifically, it is shown that there is always some $p_2 > 0$ which can be used alongside (4.19b) to verify the first LMI if the condition in (4.18) holds. Finally, the second condition $\boldsymbol{P} \succ 0$ is also evaluated using Sylvester's criterion and it is shown that the choices for the elements of $\boldsymbol{P}_{\text{s}}$ always fulfil this condition. The full proof can be found in Appendix A.1. ■

The condition in (4.18) allows control gains to be chosen such that the controlled voltage setting bus is OS-EIP w.r.t. the input currents $i_{\text{int}} + i_{\text{ex}}$ and the bus voltage $\tilde{v}$. Specifically, (4.18) requires that the damping injected through the capacitor current i_{c} must be strictly greater than the Y associated with the load (see Remark 4.2). Complementing the OS-EIP property of the closed-loop voltage setting bus, its EISO can be established.

Proposition 4.3 (EISO of a Voltage Setting Bus)
The controlled voltage setting bus (4.16) *is EISO.*

Proof:
Setting $\tilde{u}_\mathrm{s} \equiv \tilde{\imath}_\mathrm{int} + \tilde{\imath}_\mathrm{ex} \equiv 0$ and $\tilde{y}_\mathrm{s} \equiv \tilde{v} \equiv 0$ in (4.16) yields $C_\mathrm{ef}\dot{\tilde{v}} \equiv 0 \equiv \tilde{\imath}_\mathrm{z}$. This implies that $L\dot{\tilde{\imath}}_\mathrm{z} \equiv 0 \equiv -\tilde{\varepsilon}$, from which it follows that $(\tilde{\varepsilon}, \tilde{\imath}_\mathrm{z}, \tilde{v})^T = \tilde{\boldsymbol{x}}_\mathrm{s} = \boldsymbol{0}$. ■
Proposition 4.1, Theorem 4.2, and Proposition 4.3 together ensure that the sub-objective for the voltage setting buses in Objective 4.1 is met. Note that convergence to the steady state for the autonomous bus with $\tilde{u}_\mathrm{s} = 0$ is determined through Statement 3 of Proposition 3.1 and Remark 3.6.

Remark 4.6. *Consider condition* (4.18) *in Theorem 4.2 and recall that* $\kappa_\mathrm{v}^\diamond$ *in* (4.17) *is dependent on* δ. *Substituting* $\kappa_\mathrm{v}^\diamond$ *into* (4.18) *and simplifying thus yields*

$$\delta > Y + \rho \quad \wedge \quad \frac{R}{\kappa_\mathrm{i} + R}\delta > Y + \rho + \frac{\kappa_\varepsilon^\diamond - \kappa_\mathrm{i}^\diamond(\kappa_\mathrm{v} + 1)}{(\kappa_\mathrm{i}^\diamond)^2}. \tag{4.20}$$

Since δ *can always be increased to ensure that* (4.20) *holds, condition* (4.18) *in Theorem 4.2 is guaranteed to be feasible. However, a larger* δ *is associated with more aggressive control action.*

Remark 4.7. *Although condition* (4.18) *in Theorem 4.2 is dependent on the filter parameters* R, L, *and* C *as well as the load parameter* Y, *an OS-EIP that is robust against parameter changes can be ensured by verifying* (4.18) *for a range of* R, L, C *and* Y. *This can be achieved, for example, by adding safety factors to the choices of the control gains in* (4.12) *and employing interval arithmetic (see [MKC09, p. 12f.]).*

4.2.3 Voltage Following Controller

In the previous subsection, Theorem 4.2 demonstrates that the voltage setting controller in (4.12) can render a bus $j \in \mathcal{S}_m$, $m \in \mathcal{M}$ OS-EIP if sufficient steady-state power is available. However, for the voltage following buses $k \in \mathcal{F}_m$, $m \in \mathcal{M}$, the lack of steady-state power availability prevents an EIP result on the equivalent input-output port $(\tilde{\imath}_{\mathrm{int},k} + \tilde{\imath}_{\mathrm{ex},k}, \tilde{v}_k)$ in the presence of non-monotone loads. This is demonstrated in the following proposition.

Proposition 4.4 (Lack of EIP for Voltage Following Buses)
Consider a bus $k \in \mathcal{F}_m$ in a cluster $m \in \mathcal{M}$ with the dynamics (4.8) where no steady-state power is available, i.e., $\hat{i}_k = 0$. Let the bus have no P source, i.e., $i_{\mathrm{ps},k}(v_k) = 0$ and let it have a load function $i_{\mathrm{l},k}(v_k)$ that is not monotone. Then, such a bus is not EIP w.r.t. the input-output pair $(\tilde{i}_{\mathrm{int},k} + \tilde{i}_{\mathrm{ex},k}, \tilde{v}_k)$.

Proof:
Since the EIP property of a system must hold for all time, it is necessary that its steady-state representation must exhibit the same EIP property. Consider now the bus $k \in \mathcal{F}_m$ where no steady-state power is injected by the DGU, i.e., $\hat{i}_k = 0$. From (4.8a), the steady-state input-output relation is derived as

$$C_{\mathrm{ef},k}\dot{\hat{v}}_k = 0 = Y_k\hat{v}_k + \hat{i}_{\mathrm{int},k} + \hat{i}_{\mathrm{ex},k} \implies -Y_k\hat{v}_k = \hat{i}_{\mathrm{int},k} + \hat{i}_{\mathrm{ex},k}. \tag{4.21}$$

For EIP, the steady-state relation in (4.21) must be monotone (see Lemma 3.5, which is necessary and sufficient for a static function). However, since the load function $i_{\mathrm{l},k}(v_k)$ is not monotone, $Y_k > 0$ in (4.21) (see Remark 4.2). Thus, the voltage following bus with a non-monotone load cannot be EIP. ■

By Proposition 4.4, any controller which achieves the steady state $\hat{i}_k = 0$ cannot guarantee that the bus is EIP if the load at the bus is not monotone. Instead, a damping injection similar to that of the voltage setting controllers (4.12b) is combined with an integrator state that allows the converter voltage $v_{\mathrm{b},k}$ to slowly follow the natural bus voltage v_k. This allows the controlled voltage following bus to react to sudden changes using only a finite supply of local energy, since $\lim_{t\to\infty} i_k = 0$. This voltage following controller has the form

$$\dot{\varepsilon}_k = \psi_{\varepsilon,k}(\psi_{\mathrm{v},k}v_k - \varepsilon_k), \tag{4.22a}$$
$$v_{\mathrm{b},k} = -(\psi_{\mathrm{v},k} - 1)v_k - \psi_{\mathrm{i},k}i_k + \varepsilon_k - \tfrac{\delta_k L_k}{C_k} i_{\mathrm{c},k}, \quad k \in \mathcal{F}_m, \tag{4.22b}$$

where $\psi_{\mathrm{v},k}$, $\psi_{\mathrm{i},k}$, $\delta_k \in \mathbb{R}_{>0}$ are control gains which inject damping onto the bus voltage v_k, the filter current i_k, and the capacitor current $i_{\mathrm{c},k}$, respectively. As visualised in Figure 4.5, the integrator state ε_k represents the output of a low-pass filter with a time constant $\psi_{\varepsilon,k} \in \mathbb{R}_{>0}$. The low-pass filter ensures that the weighted bus voltage $\psi_{\mathrm{v},k}v_k$ is followed by ε_k. For simplicity, the subscript k is omitted for the rest of this subsection.

Applying the voltage following controller (4.22) to the bus dynamics (4.8) gives

$$\begin{bmatrix} \dot{\varepsilon} \\ L\dot{i} \\ C_{\mathrm{ef}}\dot{v} \end{bmatrix} = \begin{bmatrix} -\psi_\varepsilon & 0 & \psi_\varepsilon\psi_{\mathrm{v}} \\ 1 & -\psi_{\mathrm{i}}^\diamond & -\psi_{\mathrm{v}} \\ 0 & 1 & Y \end{bmatrix} \begin{bmatrix} \varepsilon \\ i \\ v \end{bmatrix} + \begin{bmatrix} 0 \\ -\frac{\delta L}{C} i_{\mathrm{c}} \\ i_{\mathrm{int}} + i_{\mathrm{ex}} \end{bmatrix}, \tag{4.23}$$

$$\psi_{\mathrm{i}}^\diamond := R + \psi_{\mathrm{i}}, \quad \psi_{\mathrm{v}}^\diamond := \psi_{\mathrm{v}} - \psi_{\mathrm{i}}^\diamond\delta. \tag{4.24}$$

Further applying the transformation in (4.13) and shifting w.r.t. the equilibrium $\hat{\varepsilon} = \psi_\mathrm{v}\hat{v}$, $\hat{i}_\mathrm{z}$, $\hat{v}$, $\hat{i}_\mathrm{int}$ and $\hat{i}_\mathrm{ex}$ yields the error dynamics of the controlled bus, i.e.,

$$\begin{bmatrix} \dot{\tilde{\varepsilon}} \\ L\dot{\tilde{i}}_\mathrm{z} \\ C_\mathrm{ef}\dot{\tilde{v}} \end{bmatrix} = \begin{bmatrix} -\psi_\varepsilon & 0 & \psi_\varepsilon\psi_\mathrm{v} \\ 1 & -\psi_\mathrm{i}^\diamond & -\psi_\mathrm{v}^\diamond \\ 0 & 1 & Y-\delta \end{bmatrix} \begin{bmatrix} \tilde{\varepsilon} \\ \tilde{i}_\mathrm{z} \\ \tilde{v} \end{bmatrix} + \begin{bmatrix} 0 \\ 0 \\ \tilde{i}_\mathrm{int} + \tilde{i}_\mathrm{ex} \end{bmatrix}. \tag{4.25}$$

Using the controlled bus dynamics, the requirement in Objective 4.1 is verified in the following proposition.

Proposition 4.5 (Closed-Loop Voltage Following Bus Equilibrium)
A voltage following bus (4.8) controlled with (4.22) has an equilibrium characterised by $\hat{i} = 0$.

Proof:
From the steady state of (4.23), it can be seen that $\dot{\varepsilon} = 0 \implies \hat{\varepsilon} = \psi_\mathrm{v}\hat{v}$. Moreover, considering the capacitor equation in steady state, $\hat{i}_\mathrm{c} = C\dot{\hat{v}} = 0$, the second equation in (4.23) in steady state gives

$$L\dot{\hat{i}} = 0 = \hat{\varepsilon} - \psi_\mathrm{i}^\diamond\hat{i} - \psi_\mathrm{v}\hat{v} - \delta L\dot{\hat{v}} \implies 0 = -\psi_\mathrm{i}^\diamond\hat{i}, \tag{4.26}$$

thus enforcing the steady state $\hat{i} = 0$. ■

Proposition 4.5 thus verifies that the DGU does not inject power in steady state. Although Proposition 4.4 demonstrates a lack of EIP for the controlled voltage following bus (4.25) in the presence of non-monotone loads, the controlled bus nevertheless injects transient damping which can ensure stability when the voltage following bus is considered together with other neighbouring nodes, e.g., in a cluster. This is demonstrated in the following example, where the voltage following controller ensures a stable steady state for a non-monotone P load subjected to a voltage disturbance. Note that the setup in the example becomes unstable if the voltage following controller is deactivated.

Example 4.1:
Consider the setup in Figure 4.5, where the bus is equipped with the voltage following controller (4.22) and the bus is connected via a single transmission line to an ideal voltage source. Let $i_\mathrm{ps} = 0$ for the bus and let load $i_\mathrm{l}(v)$ consist of a $120\,\mathrm{kW}$ P load. The resulting closed-loop system is simulated and the voltage of the ideal source is stepped. The resulting simulated trajectories of the bus voltage v and the line current i_t are shown in Figure 4.6.

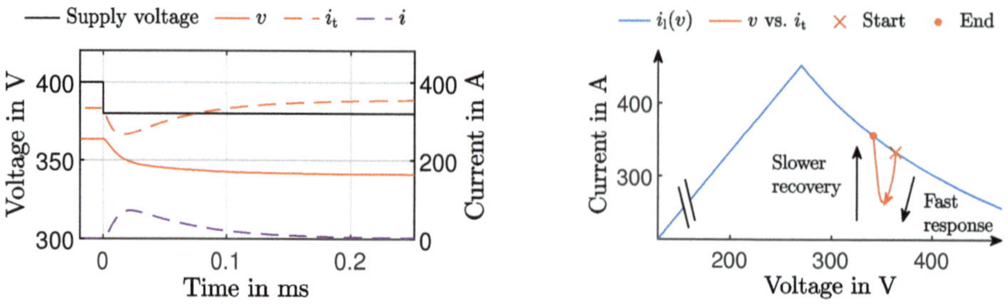

(a) Current and voltage states versus time.

(b) Voltage versus current.

Figure 4.6: Simulated filter current i, bus voltage v, and line current i_t for a controlled voltage following bus connected to an ideal voltage source via a line. (a) shows the states over time and (b) the bus voltage against the line current.

The static P load requires more current when the bus voltage decreases (see the static load curve in Figure 4.6b). During the initial transient reaction, an excess current is supplied by the controller through the filter current i. This causes the initial transient voltage-current curve to move along a path with $di_t/dv > 0$, which corresponds to an initial monotone reaction for the bus as seen through the port (i_t, v). As the system tends towards the equilibrium with $\hat{i} = 0$, the excess injected current is decreased and the load is again supplied fully by the current delivered by the line i_t.

Remark 4.8. *For both the voltage setting* (4.12) *and voltage following* (4.22) *cases, the controllers exhibit a structure similar to other approaches in the literature (see the PI structures in [Str+20b; Nah+20; FCS23]). However, a significant difference comes through the inclusion of the filter capacitor current $i_{c,n}$. Notice that*

$$C_n \dot{v}_n = i_{c,n} = i_n - i_{out,n}, \quad n \in \mathcal{N}_{el}, \tag{4.27}$$

where $i_{out,n}$ is the filter output current (see Figure 4.5). Thus, using $i_{c,n}$ has the same effect as a voltage derivative in the control, leading to a proportional-integral-derivative (PID). This is similar to the controller in [CKS23], but replaces a numerical derivation of the measured bus voltage with a current measurement. Moreover, it can be seen from (4.27) *that the controllers may be adapted to use a measurement of $i_{out,n}$ instead of $i_{c,n}$ without changing the results.*

4.3 Passivity and Stability Analysis

Using the DC network component models in Section 4.1 and by applying the decentralised bus controllers in Section 4.2, the stability of an entire DC network is investigated in this section. While it is established in Theorem 4.2 that voltage setting buses with steady-state power can have useful EIP properties, this is not the case for voltage following buses where no steady-state power is available (see Proposition 4.4). As hinted at in Example 4.1, however, AS of the desired equilibrium in Assumption 4.1 can still be achieved when considering more than one bus.

In this section, the AS of any equilibrium of an entire DC network is thus investigated by deriving OS-EIP properties for the network clusters, as per Objective 4.1. This section commences with **Section 4.3.1**, where an EIP property for the transmission lines is established. Next, in **Section 4.3.2**, an LMI is presented by which the OS-EIP of an entire cluster can be verified. This result is then used in **Section 4.3.3**, where the AS of any equilibrium of the DC network is established. Finally, in **Section 4.3.4**, the cluster OS-EIP robustness is investigated w.r.t. to changes in the network parameters and topology.

4.3.1 EIP of Transmission Lines

Recalling Assumption 4.1, the line dynamics (4.7) shifted w.r.t. the equilibrium are

$$L_{\mathrm{t},o}\dot{\tilde{i}}_{\mathrm{t},o} = -R_{\mathrm{t},o}\tilde{i}_{\mathrm{t},o} + \tilde{v}_{n_{o,+}} - \tilde{v}_{n_{o,-}}, \quad o \in \mathcal{E}_{\mathrm{el}}. \tag{4.28}$$

These shifted dynamics form the basis for an EIP investigation.

Proposition 4.6 (Transmission Line OS-EIP and EISO)
The shifted linear transmission line $o \in \mathcal{E}_{\mathrm{el}}$ (4.28) with the storage function $S_{\mathrm{t},o}(\tilde{i}_{\mathrm{t},o}) = \frac{1}{2}L_{\mathrm{t},o}\tilde{i}_{\mathrm{t},o}^2$ is OS-EIP and EISO w.r.t. the input-output port $(\tilde{v}_{n_{o,+}} - \tilde{v}_{n_{o,-}}, \tilde{i}_{\mathrm{t},o})$ with $\rho = R_{\mathrm{t},o}$.

Proof:
The storage function $S_{\mathrm{t},o}$ is positive definite in $\tilde{i}_{\mathrm{t},o}$. Furthermore, computing the time derivative of $S_{\mathrm{t},o}$ yields

$$\dot{S}_{\mathrm{t},o} = \tilde{i}_{\mathrm{t},o}(\tilde{v}_{n_{o,+}} - \tilde{v}_{n_{o,-}}) - R_{\mathrm{t},o}\tilde{i}_{\mathrm{t},o}^2, \tag{4.29}$$

which is equal to an OS-EIP supply rate with the input $\tilde{v}_{n_{o,+}} - \tilde{v}_{n_{o,-}}$, output $\tilde{i}_{\mathrm{t},o}$ and $\rho = R_{\mathrm{t},o}$ (see Definition 3.4). EISO follows since the output $\tilde{i}_{\mathrm{t},o}$ is equal to the state variable in (4.28). ∎

4.3.2 EIP of Clusters

In this subsection, the OS-EIP and the EISD for a cluster $m \in \mathcal{M}$ with the dynamics in (4.10) is established. Let each voltage setting bus in $\mathcal{S}_m$ be equipped with the voltage setting controller (4.12) and each voltage following bus in $\mathcal{F}_m$ be equipped with the voltage following controller (4.22). Shifting the resulting cluster dynamics w.r.t. to the equilibrium in Assumption 4.1 gives the controlled cluster dynamics

$$\underbrace{\begin{bmatrix} \boldsymbol{L}_{\mathrm{s},m}\dot{\tilde{\boldsymbol{\varepsilon}}}_{\mathrm{s},m} \\ \boldsymbol{L}_{\mathrm{s},m}\dot{\tilde{\boldsymbol{i}}}_{\mathrm{s,z},m} \\ \boldsymbol{C}_{\mathrm{s},m}\dot{\tilde{\boldsymbol{v}}}_{\mathrm{s},m} \\ \dot{\tilde{\boldsymbol{\varepsilon}}}_{\mathrm{f},m} \\ \boldsymbol{L}_{\mathrm{f},m}\dot{\tilde{\boldsymbol{i}}}_{\mathrm{f,z},m} \\ \boldsymbol{C}_{\mathrm{f},m}\dot{\tilde{\boldsymbol{v}}}_{\mathrm{f},m} \\ \boldsymbol{L}_{\mathrm{t},m}\dot{\tilde{\boldsymbol{i}}}_{\mathrm{t},m} \end{bmatrix}}_{\boldsymbol{Q}_{\mathrm{cl},m}\dot{\tilde{\boldsymbol{x}}}_{\mathrm{cl},m}} = \underbrace{\begin{bmatrix} \boldsymbol{0} & \boldsymbol{0} & \boldsymbol{\kappa}^{\diamond}_{\varepsilon,m} & \boldsymbol{0} & \boldsymbol{0} & \boldsymbol{0} & \boldsymbol{0} \\ -\boldsymbol{I}_{|\mathcal{S}_m|} & -\boldsymbol{\kappa}^{\diamond}_{\mathrm{i},m} & -\boldsymbol{\kappa}^{\diamond}_{\mathrm{v},m} & \boldsymbol{0} & \boldsymbol{0} & \boldsymbol{0} & \boldsymbol{0} \\ \boldsymbol{0} & \boldsymbol{I}_{|\mathcal{S}_m|} & \boldsymbol{Y}_{\mathrm{s},m}-\boldsymbol{\delta}_{\mathrm{s},m} & \boldsymbol{0} & \boldsymbol{0} & \boldsymbol{0} & \boldsymbol{E}_{\mathrm{s},m} \\ \boldsymbol{0} & \boldsymbol{0} & \boldsymbol{0} & -\boldsymbol{\psi}_{\varepsilon,m} & \boldsymbol{0} & \boldsymbol{\psi}_{\varepsilon,m}\boldsymbol{\psi}_{\mathrm{v},m} & \boldsymbol{0} \\ \boldsymbol{0} & \boldsymbol{0} & \boldsymbol{0} & \boldsymbol{I}_{|\mathcal{F}_m|} & -\boldsymbol{\psi}^{\diamond}_{\mathrm{i},m} & -\boldsymbol{\psi}^{\diamond}_{\mathrm{v},m} & \boldsymbol{0} \\ \boldsymbol{0} & \boldsymbol{0} & \boldsymbol{0} & \boldsymbol{0} & \boldsymbol{I}_{|\mathcal{F}_m|} & \boldsymbol{Y}_{\mathrm{f}}-\boldsymbol{\delta}_m & \boldsymbol{E}_{\mathrm{f},m} \\ \boldsymbol{0} & \boldsymbol{0} & -\boldsymbol{E}^T_{\mathrm{s},m} & \boldsymbol{0} & \boldsymbol{0} & -\boldsymbol{E}^T_{\mathrm{f},m} & -\boldsymbol{R}_{\mathrm{t},m} \end{bmatrix}}_{\boldsymbol{A}_{\mathrm{cl},m}} \underbrace{\begin{bmatrix} \tilde{\boldsymbol{\varepsilon}}_{\mathrm{s},m} \\ \tilde{\boldsymbol{i}}_{\mathrm{s,z},m} \\ \tilde{\boldsymbol{v}}_{\mathrm{s},m} \\ \tilde{\boldsymbol{\varepsilon}}_{\mathrm{f},m} \\ \tilde{\boldsymbol{i}}_{\mathrm{f,z},m} \\ \tilde{\boldsymbol{v}}_{\mathrm{f},m} \\ \tilde{\boldsymbol{i}}_{\mathrm{t},m} \end{bmatrix}}_{\tilde{\boldsymbol{x}}_{\mathrm{cl},m}} \dots$$

$$+ \boldsymbol{B}_{\mathrm{cl},m} \underbrace{\begin{bmatrix} \tilde{\boldsymbol{i}}_{\mathrm{ex,s},m} \\ \tilde{\boldsymbol{i}}_{\mathrm{ex,f},m} \end{bmatrix}}_{\tilde{\boldsymbol{u}}_{\mathrm{cl},m}}, \tag{4.30a}$$

$$\tilde{\boldsymbol{y}}_{\mathrm{cl},m} = \underbrace{\begin{bmatrix} \boldsymbol{0} & \boldsymbol{0} & \boldsymbol{I}_{|\mathcal{S}_m|} & \boldsymbol{0} & \boldsymbol{0} & \boldsymbol{0} & \boldsymbol{0} \\ \boldsymbol{0} & \boldsymbol{0} & \boldsymbol{0} & \boldsymbol{0} & \boldsymbol{0} & \boldsymbol{I}_{|\mathcal{F}_m|} & \boldsymbol{0} \end{bmatrix}}_{\boldsymbol{B}^T_{\mathrm{cl},m}} \tilde{\boldsymbol{x}}_{\mathrm{cl},m} = \begin{bmatrix} \tilde{\boldsymbol{v}}_{\mathrm{s},m} \\ \tilde{\boldsymbol{v}}_{\mathrm{f},m} \end{bmatrix}, \quad m \in \mathcal{M}, \tag{4.30b}$$

where $\tilde{\boldsymbol{x}}_{\mathrm{cl},m}$ consists of the states of the voltage setting buses (4.16) for $j \in \mathcal{S}_m$, ($\tilde{\boldsymbol{\varepsilon}}_{\mathrm{s},m}$, $\tilde{\boldsymbol{i}}_{\mathrm{s,z},m}$, $\tilde{\boldsymbol{v}}_{\mathrm{s},m}$); the states of the voltage following buses (4.25) for $k \in \mathcal{F}_m$, ($\tilde{\boldsymbol{\varepsilon}}_{\mathrm{f},m}$, $\tilde{\boldsymbol{i}}_{\mathrm{f,z},m}$, $\tilde{\boldsymbol{v}}_{\mathrm{f},m}$); and the currents $\tilde{\boldsymbol{i}}_{\mathrm{t},m}$ from the transmission lines (4.28) for $o \in \mathcal{E}_m$. The cluster input $\tilde{\boldsymbol{u}}_{\mathrm{cl},m}$ comprises the stacked external currents of the buses. The bus and line parameters are stacked diagonally to obtain $\boldsymbol{L}_{\mathrm{s},m}$, $\boldsymbol{C}_{\mathrm{s},m}$, $\boldsymbol{\kappa}^{\diamond}_{\varepsilon,m}$, $\boldsymbol{\kappa}^{\diamond}_{\mathrm{i},m}$, $\boldsymbol{\kappa}^{\diamond}_{\mathrm{v},m}$, $\boldsymbol{Y}_{\mathrm{s},m}$, and $\boldsymbol{\delta}_{\mathrm{s},m}$ for the buses $j \in \mathcal{S}_m$ from (4.16); $\boldsymbol{L}_{\mathrm{f},m}$, $\boldsymbol{C}_{\mathrm{f},m}$, $\boldsymbol{\psi}_{\varepsilon,m}$, $\boldsymbol{\psi}^{\diamond}_{\mathrm{i},m}$, $\boldsymbol{\psi}_{\mathrm{v},m}$, $\boldsymbol{Y}_{\mathrm{f},m}$, and $\boldsymbol{\delta}_{\mathrm{f},m}$ for the buses $k \in \mathcal{F}_m$ from (4.25); and $\boldsymbol{L}_{\mathrm{t},m}$ and $\boldsymbol{R}_{\mathrm{t},m}$ for the lines $o \in \mathcal{E}_m$ from (4.28).

Theorem 4.7 (Cluster OS-EIP)

The shifted cluster dynamics (4.30) are OS-EIP w.r.t. the input-output pair ($\tilde{\boldsymbol{u}}_{\mathrm{cl},m}$, $\tilde{\boldsymbol{y}}_{\mathrm{cl},m}$) and with the storage function

$$S_{\mathrm{cl},m}(\tilde{\boldsymbol{x}}_{\mathrm{cl},m}) = \frac{1}{2}\tilde{\boldsymbol{x}}^T_{\mathrm{cl},m}\boldsymbol{Q}_{\mathrm{cl},m}\boldsymbol{P}_{\mathrm{cl},m}\boldsymbol{Q}_{\mathrm{cl},m}\tilde{\boldsymbol{x}}_{\mathrm{cl},m}, \tag{4.31}$$

if there is a matrix $\boldsymbol{P}_{\mathrm{cl},m} = \boldsymbol{P}^T_{\mathrm{cl},m} \succ 0$ and a $\rho > 0$ such that

$$\boldsymbol{Q}_{\mathrm{cl},m}\boldsymbol{P}_{\mathrm{cl},m}\boldsymbol{A}_{\mathrm{cl},m} + \boldsymbol{A}^T_{\mathrm{cl},m}\boldsymbol{P}_{\mathrm{cl},m}\boldsymbol{Q}_{\mathrm{cl},m} + 2\rho\boldsymbol{B}_{\mathrm{cl},m}\boldsymbol{B}^T_{\mathrm{cl},m} \preccurlyeq 0, \tag{4.32a}$$

$$\boldsymbol{Q}_{\mathrm{cl},m}\boldsymbol{P}_{\mathrm{cl},m}\boldsymbol{B}_{\mathrm{cl},m} = \boldsymbol{B}_{\mathrm{cl},m}. \tag{4.32b}$$

Proof:
Since $\boldsymbol{Q}_{\text{cl},m} \succ 0$ from (4.30a), $S_{\text{cl},m}$ is positive definite in $\tilde{\boldsymbol{x}}_{\text{cl},m}$ if $\boldsymbol{P}_{\text{cl},m} \succ 0$ [HJ12, Observation 7.1.8]. Calculating the OS-EIP inequality $\dot{S}_{\text{cl},m} \leq \tilde{\boldsymbol{y}}_{\text{cl},m}^T \tilde{\boldsymbol{u}}_{\text{cl},m} - \rho \tilde{\boldsymbol{y}}_{\text{cl},m}^T \tilde{\boldsymbol{y}}_{\text{cl},m}$ for the controlled cluster (4.30) and substituting in $\tilde{\boldsymbol{y}}_{\text{cl},m} = \boldsymbol{B}_{\text{cl},m}^T \tilde{\boldsymbol{x}}_{\text{cl},m}$ leads to

$$\frac{1}{2} \tilde{\boldsymbol{x}}_{\text{cl},m}^T \left(\boldsymbol{Q}_{\text{cl},m} \boldsymbol{P}_{\text{cl},m} \boldsymbol{A}_{\text{cl},m} + \boldsymbol{A}_{\text{cl},m}^T \boldsymbol{P}_{\text{cl},m} \boldsymbol{Q}_{\text{cl},m} \right) \tilde{\boldsymbol{x}}_{\text{cl},m} + \tilde{\boldsymbol{x}}_{\text{cl},m}^T \boldsymbol{Q}_{\text{cl},m} \boldsymbol{P}_{\text{cl},m} \boldsymbol{B}_{\text{cl},m} \tilde{\boldsymbol{u}}_{\text{cl},m} \dots$$
$$\leq \tilde{\boldsymbol{x}}_{\text{cl},m}^T \boldsymbol{B}_{\text{cl},m} \tilde{\boldsymbol{u}}_{\text{cl},m} - \rho \tilde{\boldsymbol{x}}_{\text{cl},m}^T \boldsymbol{B}_{\text{cl},m} \boldsymbol{B}_{\text{cl},m}^T \tilde{\boldsymbol{x}}_{\text{cl},m}, \quad (4.33)$$

from which (4.32a) follows if (4.32b) holds. ■

Remark 4.9. *The restriction in* (4.32b) *ensures an EIP result by constraining the form of the storage function in* (4.31)*. Specifically, with the matrix* $\boldsymbol{B}_{\text{cl},m}$ *in* (4.30)*, the quadratic storage function* $S_{\text{cl},m}$ *may not contain any terms in which a bus voltage* $\tilde{v}_n$ *is multiplied by any other state except by itself.*

Using Theorem 4.7, the OS-EIP of a cluster can be verified by solving an LMI. Once OS-EIP for the cluster is established, connections to other clusters may be made at any bus in the cluster, since each bus voltage and the corresponding external current are present in $\tilde{\boldsymbol{y}}_{\text{cl},m}$ and $\tilde{\boldsymbol{u}}_{\text{cl},m}$, respectively. A successful verification of the LMI (4.32) also demonstrates that the interconnection of voltage setting and voltage following buses can ensure that the latter obtains an OS-EIP property in the cluster setup. With the aim of interconnecting the OS-EIP clusters to obtain an AS result for any equilibrium of the DC network, the following proposition investigates the equilibrium-independent detectability of the system.

Proposition 4.8 (Cluster EISD)
The cluster dynamics (4.30) *are EISD.*

Proof:
Set $\tilde{\boldsymbol{u}}_{\text{cl},m} \equiv \boldsymbol{0}$ and $\tilde{\boldsymbol{y}}_{\text{cl},m} \equiv \boldsymbol{0}$ in (4.30) to obtain

$$\boldsymbol{L}_{\text{s},m} \dot{\tilde{\boldsymbol{\varepsilon}}}_{\text{s},m} = \boldsymbol{0}, \quad (4.34a)$$

$$\begin{bmatrix} \boldsymbol{L}_{\text{s},m} \dot{\tilde{\boldsymbol{i}}}_{\text{s,z},m} \\ \dot{\tilde{\boldsymbol{\varepsilon}}}_{\text{f},m} \\ \boldsymbol{L}_{\text{f},m} \dot{\tilde{\boldsymbol{i}}}_{\text{f,z},m} \\ \boldsymbol{L}_{\text{t},m} \dot{\tilde{\boldsymbol{i}}}_{\text{t},m} \end{bmatrix} = \begin{bmatrix} -\boldsymbol{\psi}_{\text{i},m}^{\diamond} & \boldsymbol{0} & \boldsymbol{0} & \boldsymbol{0} \\ \boldsymbol{0} & -\boldsymbol{\psi}_{\varepsilon,m} & \boldsymbol{0} & \boldsymbol{0} \\ \boldsymbol{0} & \boldsymbol{I}_{|\mathcal{F}_m|} & -\boldsymbol{\psi}_{\text{i},m}^{\diamond} & \boldsymbol{0} \\ \boldsymbol{0} & \boldsymbol{0} & \boldsymbol{0} & -\boldsymbol{R}_{\text{t},m} \end{bmatrix} \begin{bmatrix} \tilde{\boldsymbol{i}}_{\text{s,z},m} \\ \tilde{\boldsymbol{\varepsilon}}_{\text{f},m} \\ \tilde{\boldsymbol{i}}_{\text{f,z},m} \\ \tilde{\boldsymbol{i}}_{\text{t},m} \end{bmatrix} + \begin{bmatrix} -\tilde{\boldsymbol{\varepsilon}}_{\text{s},m} \\ \boldsymbol{0} \\ \boldsymbol{0} \\ \boldsymbol{0} \end{bmatrix}. \quad (4.34b)$$

From (4.34a), $\tilde{\boldsymbol{\varepsilon}}_{\text{s},m}$ is in steady state and since the states are shifted to the equilibrium, $\tilde{\boldsymbol{\varepsilon}}_{\text{s},m} = \boldsymbol{0}$. Since $\tilde{\boldsymbol{\varepsilon}}_{\text{s},m} = \boldsymbol{0}$, the remaining dynamics in (4.34b) represent a linear autonomous system with strictly negative eigenvalues. Thus, the shifted cluster (4.30) is EISD. ■

Remark 4.10. *Consider the LMI in* (4.32) *for the cluster* (4.30). *While it is hard to predict the LMI feasibility, certain parameters are observed to play a pronounced role. Since* $\boldsymbol{Y}_{\mathrm{s},m} - \boldsymbol{\delta}_{\mathrm{s},m}$ *and* $\boldsymbol{Y}_{\mathrm{f},m} - \boldsymbol{\delta}_{\mathrm{f},m}$ *directly act on the voltage error states, they directly affect the feasibility of obtaining an OS-EIP result (see Equation 4.18). Specifically, larger element values in* $\boldsymbol{\delta}_{\mathrm{s},m}$ *and* $\boldsymbol{\delta}_{\mathrm{f},m}$ *along with smaller element values in* $\boldsymbol{Y}_{\mathrm{s},m}$ *and* $\boldsymbol{Y}_{\mathrm{f},m}$ *improve the LMI feasibility. In the interconnected cluster, the transmission lines couple the OS-EIP buses in* $\mathcal{S}_m$ *and the non-EIP buses in* $\mathcal{F}_m$. *Increasing the strength of coupling, e.g., by*

- *increasing the number of lines in* $\mathcal{E}_m$*;*
- *decreasing the line resistances in* $\boldsymbol{R}_{\mathrm{t},m}$*;*
- *decreasing the electrical distance*[3] *between a voltage following bus in* $\mathcal{F}_m$ *and its closest voltage setting bus neighbour in* $\mathcal{S}_m$*;*

will thus also improve the feasibility of the LMI. The potential robustness properties arising from these last three points are considered more closely in Section 4.3.4.

Remark 4.11. *The OS-EIP result in Theorem 4.7 can be made robust against parameter uncertainties by considering a range of possible parameter values and extending the LMI to a robust optimisation problem (see [Ama06, p. 32]). Such an approach allows the cluster to be robustly OS-EIP in the presence of uncertain bus, load, line and control parameters.*

Remark 4.12. *The LMI in Theorem 4.7 is of the order* $3|\mathcal{N}_m| + |\mathcal{E}_m|$*, and thus quickly increases in size for larger clusters. A computationally efficient clustering for Theorem 4.7 will thus seek to maximise the number of clusters in the DC network by restricting the number of voltage setting buses to one per cluster, i.e.,* $|\mathcal{S}_m| = 1$. *The voltage following buses can then be added to the respective clusters to which they are closest in an electrical sense.*

Remark 4.13. *If a cluster consists of a single voltage setting bus, i.e.,* $|\mathcal{S}_m| = 1$, $|\mathcal{F}_m| = 0$, $|\mathcal{E}_m| = 0$, *its OS-EIP can be determined from the control design requirements in Theorem 4.2 instead of verifying the LMI in Theorem 4.7.*

4.3.3 GAS of DC Network Equilibria

Using the established OS-EIP properties of the lines (Proposition 4.6) and the clusters (Theorem 4.7), the AS of any DC network equilibrium can be investigated using the methods in Chapter 3. The DC network is constructed by stacking the clusters, the

[3] The electrical distance between two nodes can be computed (see [DB13, Eq. (3)]) using the Laplacian matrix derived from the graph edges weighted by the line conductances. Similarly, the electrical distance between any two nodes is characterised by the algebraic connectivity of the Laplacian (see Section 3.1).

monotone parts of the load functions, the P source functions, and the lines interconnecting the clusters, i.e.,

$$\tilde{\boldsymbol{u}}_{\text{cl}} = \text{stack}_{m\in\mathcal{M}}(\tilde{\boldsymbol{u}}_{\text{cl},m}), \quad \tilde{\boldsymbol{y}}_{\text{cl}} = \text{stack}_{m\in\mathcal{M}}(\tilde{\boldsymbol{y}}_{\text{cl},m}), \quad \tilde{\boldsymbol{x}}_{\text{cl}} = \text{stack}_{m\in\mathcal{M}}(\tilde{\boldsymbol{x}}_{\text{cl},m}), \tag{4.35a}$$

$$\tilde{\boldsymbol{u}}_{\text{l}}^{\text{m}} = \text{stack}_{n\in\mathcal{N}_{\text{el}}}(\tilde{\boldsymbol{u}}_{\text{l},n}^{\text{m}}), \quad \tilde{\boldsymbol{y}}_{\text{l}}^{\text{m}} = \text{stack}_{n\in\mathcal{N}_{\text{el}}}(\tilde{\boldsymbol{y}}_{\text{l},n}^{\text{m}}), \tag{4.35b}$$

$$\tilde{\boldsymbol{u}}_{\text{ps}} = \text{stack}_{n\in\mathcal{N}_{\text{el}}}(\tilde{\boldsymbol{u}}_{\text{ps},n}), \quad \tilde{\boldsymbol{y}}_{\text{ps}} = \text{stack}_{n\in\mathcal{N}_{\text{el}}}(\tilde{\boldsymbol{y}}_{\text{ps},n}), \tag{4.35c}$$

$$\tilde{\boldsymbol{u}}_{\mathcal{T}} = \text{stack}_{o\in\mathcal{T}}(\tilde{v}_{n_{o,+}} - \tilde{v}_{n_{o,-}}), \quad \tilde{\boldsymbol{y}}_{\text{cl}} = \tilde{\boldsymbol{i}}_{\mathcal{T}} = \text{stack}_{o\in\mathcal{T}}(\tilde{i}_{\mathcal{T},o}). \tag{4.35d}$$

Recalling Assumption 4.1, the stability of the assumed equilibrium of the DC network is now investigated, i.e.,

$$\left\{\tilde{\boldsymbol{x}}_{\text{cl}} = \mathbf{0}, \tilde{\boldsymbol{i}}_{\mathcal{T}} = \mathbf{0}\right\}. \tag{4.36}$$

The clusters, monotone load functions, P source functions, and the lines in $\mathcal{T}$ are interconnected as shown in Figure 4.7, where $\boldsymbol{E}_{\mathcal{T}}$ denotes the incidence matrix (3.1) describing the interconnection of the DC network buses $\mathcal{N}_{\text{el}}$ via the lines in $\mathcal{T}$.

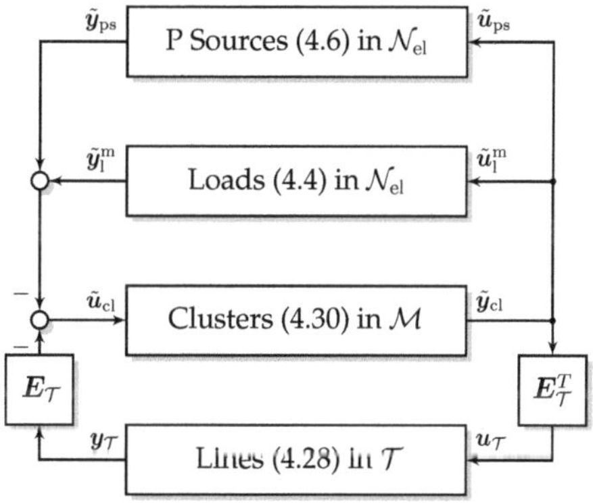

Figure 4.7: Skew-symmetric interconnection of the clusters with the external lines and the monotone load and P source functions using the input and output variables defined in (4.35).

Theorem 4.9 (GAS of DC Network Equilibria)
Consider a DC network of $|\mathcal{M}|$ clusters (4.30) weakly interconnected via $|\mathcal{T}|$ lines (4.28). Each bus $n \in \mathcal{N}_{\mathrm{el}}$ in the DC network is optionally equipped with a load (4.4) and a P source (4.6). Let each cluster and each line be OS-EIP and EISD, let the static loads and P sources be EIP, and let the components be interconnected as in Figure 4.7. Then, the resulting equilibrium of the DC network is GAS with the radially unbounded storage function

$$S_{\mathrm{el}}(\tilde{\boldsymbol{x}}_{\mathrm{cl}}, \tilde{\boldsymbol{i}}_{\mathcal{T}}) = \sum_{m \in \mathcal{M}} S_{\mathrm{cl},m}(\tilde{\boldsymbol{x}}_{\mathrm{cl},m}) + \sum_{o \in \mathcal{T}} S_{\mathrm{t},o}(\tilde{i}_{\mathcal{T},o}). \tag{4.37}$$

Proof:
The linear interconnection in Figure 4.7 can be expressed as

$$\begin{bmatrix} \tilde{\boldsymbol{u}}_{\mathrm{cl}} \\ \tilde{\boldsymbol{u}}_{\mathcal{T}} \\ \tilde{\boldsymbol{u}}_{\mathrm{l}}^{\mathrm{m}} \\ \tilde{\boldsymbol{u}}_{\mathrm{ps}} \end{bmatrix} = \begin{bmatrix} \boldsymbol{0} & -\boldsymbol{E}_{\mathcal{T}} & -\boldsymbol{I}_{|\mathcal{N}_{\mathrm{el}}|} & -\boldsymbol{I}_{|\mathcal{N}_{\mathrm{el}}|} \\ \boldsymbol{E}_{\mathcal{T}}^{T} & \boldsymbol{0} & \boldsymbol{0} & \boldsymbol{0} \\ \boldsymbol{I}_{|\mathcal{N}_{\mathrm{el}}|} & \boldsymbol{0} & \boldsymbol{0} & \boldsymbol{0} \\ \boldsymbol{I}_{|\mathcal{N}_{\mathrm{el}}|} & \boldsymbol{0} & \boldsymbol{0} & \boldsymbol{0} \end{bmatrix} \begin{bmatrix} \tilde{\boldsymbol{y}}_{\mathrm{cl}} \\ \tilde{\boldsymbol{y}}_{\mathcal{T}} \\ \tilde{\boldsymbol{y}}_{\mathrm{l}}^{\mathrm{m}} \\ \tilde{\boldsymbol{y}}_{\mathrm{ps}} \end{bmatrix}. \tag{4.38}$$

Recalling the subsystem partitioning in Theorem 3.3, the clusters and lines are included in the set $\mathcal{J}_{-,\mathrm{EIID}}$. Furthermore, since the EIP load and P source functions are static, they are included in the set $\mathcal{J}_{0,\mathrm{GAS}}$, whereas $\mathcal{J}_{-} = \emptyset = \mathcal{J}_{0}$. Thus, the interconnection in (4.38) results in the skew-symmetric partitioned interconnection from (3.10a)

$$\begin{bmatrix} \tilde{\boldsymbol{u}}_{-,\mathrm{EIID}} \\ \tilde{\boldsymbol{u}}_{0,\mathrm{GAS}} \end{bmatrix} = \begin{bmatrix} \boldsymbol{H}_{22} & \boldsymbol{H}_{24} \\ \boldsymbol{H}_{42} & \boldsymbol{H}_{44} \end{bmatrix} \begin{bmatrix} \tilde{\boldsymbol{y}}_{-,\mathrm{EIID}} \\ \tilde{\boldsymbol{y}}_{0,\mathrm{GAS}} \end{bmatrix}, \quad \boldsymbol{H}_{44} = \boldsymbol{0}, \tag{4.39a}$$

$$\boldsymbol{H}_{22} = \begin{bmatrix} \boldsymbol{0} & -\boldsymbol{E}_{\mathcal{T}} \\ \boldsymbol{E}_{\mathcal{T}}^{T} & \boldsymbol{0} \end{bmatrix}, \quad \boldsymbol{H}_{24} = \begin{bmatrix} -\boldsymbol{I}_{|\mathcal{N}_{\mathrm{el}}|} & -\boldsymbol{I}_{|\mathcal{N}_{\mathrm{el}}|} \\ \boldsymbol{0} & \boldsymbol{0} \end{bmatrix} = -\boldsymbol{H}_{42}^{T}. \tag{4.39b}$$

Therefore, AS follows from Theorem 3.3 and Corollary 3.4, whereas GAS follows from Proposition 3.1. ∎

By applying Theorem 4.9, the GAS of the equilibria of a DC network comprising clusters that are OS-EIP and EISD (see Theorem 4.7 and Proposition 4.8) and lines that are OS-EIP and EISO (see Proposition 4.6) can be assured. Note that only the cluster OS-EIP must actively be verified. All other steps leading to GAS follow automatically. Theorem 4.9 thus answers RQ 1.2 with a stability result that is robust against any changes to the lines in $\mathcal{T}$ as well as changes to the clusters as per Remark 4.11.

Remark 4.14. *Consider the subsystem interconnection in Figure 4.7 and let a static load at any bus $n \in \mathcal{N}_{\mathrm{el}}$ in the network be replaced by a dynamic load (optionally with direct feedthrough). The inclusion of such dynamic loads will not affect the stability of the DC network, if each equilibrium of the loads is GAS for a constant bus voltage $\tilde{v}_n = 0$, and if the loads are IF-EIP with a $\nu > -Y_n$ for the load parameter Y_n included in the cluster dynamics (see Remark 4.2). Under these conditions, Theorem 4.9 can be applied without modification.*

4.3.4 Robust Cluster OS-EIP

The biggest disadvantage of the GAS result in Theorem 4.9 is the LMI-based OS-EIP verification of the clusters in Theorem 4.7. While the verification can be made robust against parameter uncertainties as discussed in Remark 4.11, a re-evaluation of the LMI is required if the topology of the cluster changes. To circumvent the need for this re-evaluation and to extend the stability guarantee required for RQ 1.2, the robustness of the OS-EIP property of a cluster against certain parameter and network topology changes is investigated in this section. The robustness property is evaluated by exploiting the timescale difference between the slow bus dynamics and the fast transmission line dynamics via singular perturbation theory (see [KKO99], [Kha02, Chapter 11]).

Singularly Perturbed Clusters

Although singular perturbation theory generally investigates the stability of an equilibrium by considering the stability of the slow and the fast system dynamics, these methods have also been extended to passivity. Indeed, as shown in [CGW99], the passivity of a linear singularly perturbed system is equivalent to the passivity of its slow and fast dynamics. To enable the use of singular perturbation theory for the OS-EIP clusters, the following assumption is made.

Assumption 4.2 (Fast Transmission Line Dynamics)
In a cluster (4.30) *with controlled voltage setting and voltage following buses interconnected by transmission lines, the bus dynamics are sufficiently slow compared to the dynamics of the transmission lines to allow a singular perturbation analysis. Specifically, it is assumed that* $L_{t,o} \ll L_n, C_{ef,n}$ *for each line* $o \in \mathcal{E}_m$ *and each bus* $n \in \mathcal{N}_m$.

Remark 4.15. *The validity of Assumption 4.2 has been shown in [Tuc+16], where a singular perturbation analysis is made for a DC network. Furthermore, Assumption 4.2 is similar to assuming that the transmission lines are purely static (as done in [SSK18; TRFT18; Fan+20]).*

If Assumption 4.2 holds, the controlled cluster dynamics (4.30) can be divided into the fast dynamics comprising the transmission line dynamics

$$\boldsymbol{L}_{t,m}\dot{\tilde{\boldsymbol{i}}}_{t,m} = -\boldsymbol{R}_{t,m}\tilde{\boldsymbol{i}}_{t,m} - \boldsymbol{E}_{s,m}^{T}\tilde{\boldsymbol{v}}_{s,m} - \boldsymbol{E}_{f,m}^{T}\tilde{\boldsymbol{v}}_{f,m}, \quad \boldsymbol{v}_{s,m}, \boldsymbol{v}_{f,m} \text{ const.}, m \in \mathcal{M}, \tag{4.40}$$

and the slow dynamics consisting of the bus dynamics with $\dot{\tilde{\boldsymbol{i}}}_{\mathrm{t},m} = \mathbf{0}$

$$\underbrace{\begin{bmatrix} \boldsymbol{L}_{\mathrm{s},m}\dot{\tilde{\boldsymbol{\varepsilon}}}_{\mathrm{s},m} \\ \boldsymbol{L}_{\mathrm{s},m}\dot{\tilde{\boldsymbol{i}}}_{\mathrm{s,z},m} \\ \boldsymbol{C}_{\mathrm{s},m}\dot{\tilde{\boldsymbol{v}}}_{\mathrm{s},m} \\ \dot{\tilde{\boldsymbol{\varepsilon}}}_{\mathrm{f},m} \\ \boldsymbol{L}_{\mathrm{f},m}\dot{\tilde{\boldsymbol{i}}}_{\mathrm{f,z},m} \\ \boldsymbol{C}_{\mathrm{f},m}\dot{\tilde{\boldsymbol{v}}}_{\mathrm{f},m} \end{bmatrix}}_{\boldsymbol{Q}_{\mathrm{r},m}\dot{\tilde{\boldsymbol{x}}}_{\mathrm{r},m}} = \underbrace{\begin{bmatrix} \mathbf{0} & \mathbf{0} & \boldsymbol{\kappa}^{\diamond}_{\varepsilon,m} & \mathbf{0} & \mathbf{0} & \mathbf{0} \\ -\boldsymbol{I}_{|\mathcal{S}_m|} & -\boldsymbol{\kappa}^{\diamond}_{\mathrm{i},m} & -\boldsymbol{\kappa}^{\diamond}_{\mathrm{v},m} & \mathbf{0} & \mathbf{0} & \mathbf{0} \\ \mathbf{0} & \boldsymbol{I}_{|\mathcal{S}_m|} & \mathbf{Y}_{\mathrm{s},m} - \boldsymbol{\delta}_{\mathrm{s},m} - \boldsymbol{\mathcal{L}}_{11,m} & \mathbf{0} & \mathbf{0} & -\boldsymbol{\mathcal{L}}_{12,m} \\ \mathbf{0} & \mathbf{0} & \mathbf{0} & -\boldsymbol{\psi}_{\varepsilon,m} & \mathbf{0} & \boldsymbol{\psi}_{\varepsilon,m}\boldsymbol{\psi}_{\mathrm{v},m} \\ \mathbf{0} & \mathbf{0} & \mathbf{0} & \boldsymbol{I}_{|\mathcal{F}_m|} & -\boldsymbol{\psi}^{\diamond}_{\mathrm{i},m} & -\boldsymbol{\psi}^{\diamond}_{\mathrm{v},m} \\ \mathbf{0} & \mathbf{0} & -\boldsymbol{\mathcal{L}}^T_{12,m} & \mathbf{0} & \boldsymbol{I}_{|\mathcal{F}_m|} & \mathbf{Y}_{\mathrm{f},m} - \boldsymbol{\delta}_{\mathrm{f},m} - \boldsymbol{\mathcal{L}}_{22,m} \end{bmatrix}}_{\boldsymbol{A}_{\mathrm{r},m}} \underbrace{\begin{bmatrix} \tilde{\boldsymbol{\varepsilon}}_{\mathrm{s},m} \\ \tilde{\boldsymbol{i}}_{\mathrm{s,z},m} \\ \tilde{\boldsymbol{v}}_{\mathrm{s},m} \\ \tilde{\boldsymbol{\varepsilon}}_{\mathrm{f},m} \\ \tilde{\boldsymbol{i}}_{\mathrm{f,z},m} \\ \tilde{\boldsymbol{v}}_{\mathrm{f},m} \end{bmatrix}}_{\tilde{\boldsymbol{x}}_{\mathrm{r},m}} \cdots$$

$$+\boldsymbol{B}_{\mathrm{r},m}\tilde{\boldsymbol{u}}_{\mathrm{cl},m}, \tag{4.41a}$$

$$\tilde{\boldsymbol{y}}_{\mathrm{cl},m} = \underbrace{\begin{bmatrix} \mathbf{0} & \mathbf{0} & \boldsymbol{I}_{|\mathcal{S}_m|} & \mathbf{0} & \mathbf{0} & \mathbf{0} \\ \mathbf{0} & \mathbf{0} & \mathbf{0} & \mathbf{0} & \mathbf{0} & \boldsymbol{I}_{|\mathcal{F}_m|} \end{bmatrix}}_{\boldsymbol{B}^T_{\mathrm{r},m}} \tilde{\boldsymbol{x}}_{\mathrm{r},m}, \quad m \in \mathcal{M}. \tag{4.41b}$$

The graph Laplacian is constructed from the cluster incidence matrix $\boldsymbol{E}_m$ (4.11) weighted by the conductances of the lines $\boldsymbol{R}^{-1}_{\mathrm{t},m}$, i.e.,

$$\begin{bmatrix} \boldsymbol{\mathcal{L}}_{11,m} & \boldsymbol{\mathcal{L}}_{12,m} \\ \boldsymbol{\mathcal{L}}^T_{12,m} & \boldsymbol{\mathcal{L}}_{22,m} \end{bmatrix} := \boldsymbol{\mathcal{L}}_m = \boldsymbol{E}_m \boldsymbol{R}^{-1}_{\mathrm{t},m} \boldsymbol{E}^T_m, \qquad \boldsymbol{\mathcal{L}}_m = \boldsymbol{\mathcal{L}}^T_m. \tag{4.42}$$

Notice from (4.41) that the input $\tilde{\boldsymbol{u}}_{\mathrm{cl},m}$ and output $\tilde{\boldsymbol{y}}_{\mathrm{cl},m}$ are unchanged from the full cluster dynamics in (4.30). Furthermore, the Laplacian may be extracted from the system matrix in (4.41) as follows

$$\boldsymbol{A}_{\mathrm{r},m} = \boldsymbol{A}_{\mathrm{base},m} - \boldsymbol{B}_{\mathrm{r},m}\boldsymbol{\mathcal{L}}_m\boldsymbol{B}^T_{\mathrm{r},m}. \tag{4.43}$$

Remark 4.16. *The fast system* (4.40) *and the slow system* (4.41) *are both shifted w.r.t. their assumed equilibria. Thus,* $\tilde{\boldsymbol{x}}_{\mathrm{r},m} = \mathbf{0}$ *holds in the fast system, and* $\tilde{\boldsymbol{i}}_{\mathrm{t},m} = \mathbf{0}$ *in the slow system.*

Using the separated fast dynamics (4.40) and slow dynamics (4.41), the OS-EIP of a singularly perturbed cluster can now be investigated.

Proposition 4.10 (OS-EIP Singularly Perturbed Cluster)
Consider a cluster $m \in \mathcal{M}$ with the dynamics (4.30) for which Assumption 4.2 holds. Such a cluster with a storage function for the reduced dynamics

$$S_{\mathrm{r},m}(\tilde{\boldsymbol{x}}_{\mathrm{r},m}) = \frac{1}{2}\tilde{\boldsymbol{x}}_{\mathrm{r},m}^T \boldsymbol{Q}_{\mathrm{r},m}\boldsymbol{P}_{\mathrm{r},m}\boldsymbol{Q}_{\mathrm{r},m}\tilde{\boldsymbol{x}}_{\mathrm{r},m} \tag{4.44}$$

is OS-EIP w.r.t. the input-output pair $(\tilde{\boldsymbol{u}}_{\mathrm{cl},m}, \tilde{\boldsymbol{y}}_{\mathrm{cl},m})$, if there is a $\rho > 0$ and a matrix $\boldsymbol{P}_{\mathrm{r},m} = \boldsymbol{P}_{\mathrm{r},m}^T \succ 0$ such that

$$\boldsymbol{Q}_{\mathrm{r},m}\boldsymbol{P}_{\mathrm{r},m}\boldsymbol{A}_{\mathrm{r},m} + \boldsymbol{A}_{\mathrm{r},m}^T\boldsymbol{P}_{\mathrm{r},m}\boldsymbol{Q}_{\mathrm{r},m} + 2\rho\boldsymbol{B}_{\mathrm{r},m}\boldsymbol{B}_{\mathrm{r},m}^T \preccurlyeq 0, \quad \boldsymbol{Q}_{\mathrm{r},m}\boldsymbol{P}_{\mathrm{r},m}\boldsymbol{B}_{\mathrm{r},m} = \boldsymbol{B}_{\mathrm{r},m}. \tag{4.45}$$

Proof:
Consider the fast dynamics of the lines (4.40), where the slow system is constant, i.e., where $\dot{\tilde{\boldsymbol{x}}}_{\mathrm{r},m} = \boldsymbol{0}$ and thus $\tilde{\boldsymbol{x}}_{\mathrm{r},m} = \boldsymbol{0}$ through a shift to the equilibrium (see Remark 4.16). In this case, the storage function for the cluster in (4.31) reduces to

$$S_{\mathrm{cl},m}(\tilde{\boldsymbol{x}}_{\mathrm{r},m} = \boldsymbol{0}, \tilde{\boldsymbol{i}}_{\mathrm{t},m}) = \frac{1}{2}\tilde{\boldsymbol{i}}_{\mathrm{t},m}^T\boldsymbol{L}_{\mathrm{t},m}\boldsymbol{P}_{\mathrm{t},m}\boldsymbol{L}_{\mathrm{t},m}\tilde{\boldsymbol{i}}_{\mathrm{t},m}. \tag{4.46}$$

From $\tilde{\boldsymbol{x}}_{\mathrm{r},m} = \boldsymbol{0}$, it follows that $\tilde{\boldsymbol{y}}_{\mathrm{cl},m} = \boldsymbol{0}$ and thus, testing the fast dynamics for OS-EIP on the port $(\tilde{\boldsymbol{u}}_{\mathrm{cl},m}, \tilde{\boldsymbol{y}}_{\mathrm{cl},m})$ using (4.46) results in

$$\dot{S}_{\mathrm{cl},m}(\tilde{\boldsymbol{x}}_{\mathrm{r},m} = \boldsymbol{0}, \tilde{\boldsymbol{i}}_{\mathrm{t},m}) \leq \tilde{\boldsymbol{y}}_{\mathrm{cl},m}^T\tilde{\boldsymbol{u}}_{\mathrm{cl},m} - \rho\tilde{\boldsymbol{y}}_{\mathrm{cl},m}^T\tilde{\boldsymbol{y}}_{\mathrm{cl},m} = 0. \tag{4.47}$$

This holds for the fast dynamics (4.40) through Lemma 3.6, since $-\nabla\boldsymbol{f}(\tilde{\boldsymbol{i}}_{\mathrm{t},m}) = \boldsymbol{R}_{\mathrm{t},m} \succ 0$. Furthermore, for the slow bus dynamics where $\dot{\tilde{\boldsymbol{i}}}_{\mathrm{t},m} = \boldsymbol{0}$ and thus $\tilde{\boldsymbol{i}}_{\mathrm{t},m} = \boldsymbol{0}$ through an appropriate shift (see Remark 4.16), the cluster storage function (4.31) reduces to $S_{\mathrm{r},m}$ in (4.44). OS-EIP for the slow system dynamics is then verified by successfully solving the LMI in (4.45) which is derived in the same way as in the proof of Theorem 4.7. Since the fast dynamics are automatically OS-EIP, determining OS-EIP for the slow dynamics through (4.45) is equivalent to determining the OS-EIP of the full cluster dynamics [CGW99, Theorem 1]. ∎

Remark 4.17. *An immediate benefit of making Assumption 4.2 is the reduction of the LMI used to verify the cluster OS-EIP from an order of $3|\mathcal{N}_m| + |\mathcal{E}_m|$ in Theorem 4.7 to an order of $3|\mathcal{N}_m|$ in Proposition 4.10.*

Remark 4.18. *Assumption 4.2, which allows a singular perturbation analysis, requires a sufficiently large capacitance at the buses in the network and might therefore not hold if no DGU is present at a bus. In such a case, the dynamics of buses without DGUs also form part of the fast system and a strictly monotone incremental load impedance would be necessary to ensure the desired OS-EIP result. Expanding on this line of thought, a dynamic load which exhibits a strictly monotone incremental impedance for its fast dynamics while possibly having non-monotone behaviour on a slower timescale will be compatible with the singular perturbation analysis in this subsection. Indeed, such a load would show the same behaviour as in Figure 4.6.*

Passivity Preserving Cluster Changes

Through the appearance of the Laplacian matrix in the slow dynamics (4.43) of a cluster, new pathways are opened for comparing clusters with different topologies. Indeed, the robustness of the OS-EIP property of a cluster can be proven for certain topology changes, as shown in the following proposition.

Theorem 4.11 (Cluster Laplacian Comparison)
Consider a cluster m_1 with the dynamics (4.30) which is OS-EIP w.r.t. the port $(\tilde{\boldsymbol{u}}_{\mathrm{cl},m_1}, \tilde{\boldsymbol{y}}_{\mathrm{cl},m_1})$. Consider now another cluster m_2 with the same buses[4] as cluster m_1 and thus with the same port, i.e., $\mathcal{N}_{m_2} = \mathcal{N}_{m_1}$ and $(\tilde{\boldsymbol{u}}_{\mathrm{cl},m_1}, \tilde{\boldsymbol{y}}_{\mathrm{cl},m_1}) = (\tilde{\boldsymbol{u}}_{\mathrm{cl},m_2}, \tilde{\boldsymbol{y}}_{\mathrm{cl},m_2})$, but with different lines connecting the buses. Let Assumption 4.2 hold for both clusters. Then, cluster m_2 is OS-EIP w.r.t. the port $(\tilde{\boldsymbol{u}}_{\mathrm{cl},m_2}, \tilde{\boldsymbol{y}}_{\mathrm{cl},m_2})$ if

$$\boldsymbol{\mathcal{L}}_{m_1} - \boldsymbol{\mathcal{L}}_{m_2} \preccurlyeq 0, \tag{4.48}$$

where $\boldsymbol{\mathcal{L}}_{m_1}$ and $\boldsymbol{\mathcal{L}}_{m_2}$ are the respective Laplacian matrices (4.42).

Proof:
For simplicity in this proof, the cluster subscript m is omitted for matrices which are the same for both clusters m_1 and m_2. Starting from the LMI in (4.45), which is verified for cluster m_1, substitute $\boldsymbol{A}_{\mathrm{r},m_1}$ in (4.43) to find

$$\boldsymbol{Q}_{\mathrm{r}}\boldsymbol{P}_{\mathrm{r}}\boldsymbol{A}_{\mathrm{base}} + \boldsymbol{A}_{\mathrm{base}}^T\boldsymbol{P}_{\mathrm{r}}\boldsymbol{Q}_{\mathrm{r}} - \boldsymbol{Q}_{\mathrm{r}}\boldsymbol{P}_{\mathrm{r}}\boldsymbol{B}_{\mathrm{r}}\boldsymbol{\mathcal{L}}_{m_1}\boldsymbol{B}_{\mathrm{r}}^T - \boldsymbol{B}_{\mathrm{r}}\boldsymbol{\mathcal{L}}_{m_1}^T\boldsymbol{B}_{\mathrm{r}}^T\boldsymbol{P}_{\mathrm{r}}\boldsymbol{Q}_{\mathrm{r}} + 2\rho\boldsymbol{B}_{\mathrm{r}}\boldsymbol{B}_{\mathrm{r}}^T \preccurlyeq 0. \tag{4.49}$$

Further substituting the equality constraint in (4.45) into (4.49) leads to

$$\boldsymbol{Q}_{\mathrm{r}}\boldsymbol{P}_{\mathrm{r}}\boldsymbol{A}_{\mathrm{base}} + \boldsymbol{A}_{\mathrm{base}}^T\boldsymbol{P}_{\mathrm{r}}\boldsymbol{Q}_{\mathrm{r}} - \boldsymbol{B}_{\mathrm{r}}\boldsymbol{\mathcal{L}}_{m_1}\boldsymbol{B}_{\mathrm{r}}^T - \boldsymbol{B}_{\mathrm{r}}\boldsymbol{\mathcal{L}}_{m_1}^T\boldsymbol{B}_{\mathrm{r}}^T + 2\rho\boldsymbol{B}_{\mathrm{r}}\boldsymbol{B}_{\mathrm{r}}^T \preccurlyeq 0 \tag{4.50a}$$

$$\boldsymbol{Q}_{\mathrm{r}}\boldsymbol{P}_{\mathrm{r}}\boldsymbol{A}_{\mathrm{base}} + \boldsymbol{A}_{\mathrm{base}}^T\boldsymbol{P}_{\mathrm{r}}\boldsymbol{Q}_{\mathrm{r}} - 2\boldsymbol{B}_{\mathrm{r}}\boldsymbol{\mathcal{L}}_{m_1}\boldsymbol{B}_{\mathrm{r}}^T + 2\rho\boldsymbol{B}_{\mathrm{r}}\boldsymbol{B}_{\mathrm{r}}^T \preccurlyeq 0, \tag{4.50b}$$

where $\boldsymbol{B}_{\mathrm{r}}\boldsymbol{\mathcal{L}}\boldsymbol{B}_{\mathrm{r}}^T$ is symmetric since $\boldsymbol{\mathcal{L}}$ is symmetric. Since cluster m_1 is OS-EIP and thus (4.50b) is already verified, the OS-EIP of cluster m_2 with the same buses, i.e., with $\boldsymbol{A}_{\mathrm{base},m_1} = \boldsymbol{A}_{\mathrm{base},m_2}$, can be assured if

$$\begin{aligned}\boldsymbol{Q}_{\mathrm{r}}\boldsymbol{P}_{\mathrm{r}}\boldsymbol{A}_{\mathrm{base}} + \boldsymbol{A}_{\mathrm{base}}^T\boldsymbol{P}_{\mathrm{r}}\boldsymbol{Q}_{\mathrm{r}} - 2\boldsymbol{B}_{\mathrm{r}}\boldsymbol{\mathcal{L}}_{m_2}\boldsymbol{B}_{\mathrm{r}}^T + 2\rho\boldsymbol{B}_{\mathrm{r}}\boldsymbol{B}_{\mathrm{r}}^T \preccurlyeq \\ \boldsymbol{Q}_{\mathrm{r}}\boldsymbol{P}_{\mathrm{r}}\boldsymbol{A}_{\mathrm{base}} + \boldsymbol{A}_{\mathrm{base}}^T\boldsymbol{P}_{\mathrm{r}}\boldsymbol{Q}_{\mathrm{r}} - 2\boldsymbol{B}_{\mathrm{r}}\boldsymbol{\mathcal{L}}_{m_1}\boldsymbol{B}_{\mathrm{r}}^T + 2\rho\boldsymbol{B}_{\mathrm{r}}\boldsymbol{B}_{\mathrm{r}}^T \preccurlyeq 0,\end{aligned} \tag{4.51}$$

which simplifies to $\boldsymbol{B}_{\mathrm{r}}\boldsymbol{\mathcal{L}}_{m_1}\boldsymbol{B}_{\mathrm{r}}^T \preccurlyeq \boldsymbol{B}_{\mathrm{r}}\boldsymbol{\mathcal{L}}_{m_2}\boldsymbol{B}_{\mathrm{r}}^T$ and finally results in (4.48). ■

If a cluster is known to be OS-EIP, this property is automatically retained when the lines undergo changes as long as the new Laplacian from the changed set of lines fulfils (4.48). Note that this includes changes in line parameters as well as the removal

[4] The buses in clusters m_1 and m_2 have the identical control and system parameters.

and/or the addition of lines in the cluster. Evaluating the retention of OS-EIP via Theorem 4.11 also only involves computing the eigenvalues of an $|\mathcal{N}_m| \times |\mathcal{N}_m|$ matrix as opposed to solving the order $3|\mathcal{N}_m|$ LMI in Proposition 4.10, thus lowering the numerical complexity. Furthermore, the following two corollaries result directly from the application of Theorem 4.11.

Corollary 4.12 (Cluster Line Resistance Comparison)

Consider a cluster m_1 (4.30) which is OS-EIP w.r.t. the port $(\tilde{\boldsymbol{u}}_{\mathrm{cl},m_1}, \tilde{\boldsymbol{y}}_{\mathrm{cl},m_1})$. Let another cluster m_2 have the same buses and lines as cluster m_1, but let the line resistances of the transmission lines in cluster m_2 be different. Let Assumption 4.2 hold for both clusters. Then, cluster m_2 is OS-EIP w.r.t. the port $(\tilde{\boldsymbol{u}}_{\mathrm{cl},m_2}, \tilde{\boldsymbol{y}}_{\mathrm{cl},m_2})$ if

$$0 < R_{\mathrm{t},m_2,o} \leq R_{\mathrm{t},m_1,o}, \quad \forall o \in \mathcal{E}_{m_1}. \tag{4.52}$$

Proof:

Since the two clusters have the same lines, their incidence matrices are equal, i.e., $\boldsymbol{E}_{m_1} = \boldsymbol{E}_{m_2}$. Thus, substituting (4.42) into (4.48) yields

$$\boldsymbol{E}_{m_1}\boldsymbol{R}_{\mathrm{t},m_1}^{-1}\boldsymbol{E}_{m_1}^T - \boldsymbol{E}_{m_1}\boldsymbol{R}_{\mathrm{t},m_2}^{-1}\boldsymbol{E}_{m_1}^T \preccurlyeq 0 \implies \boldsymbol{R}_{\mathrm{t},m_1}^{-1} - \boldsymbol{R}_{\mathrm{t},m_2}^{-1} \preccurlyeq 0, \tag{4.53}$$

where $\boldsymbol{R}_{\mathrm{t},m_1}$ and $\boldsymbol{R}_{\mathrm{t},m_2}$ are positive diagonal matrices. This directly results in the matrix element comparison in (4.52). ■

Corollary 4.13 (Cluster Line Addition)

Consider a cluster m_1 (4.30) which is OS-EIP w.r.t. the port $(\tilde{\boldsymbol{u}}_{\mathrm{cl},m_1}, \tilde{\boldsymbol{y}}_{\mathrm{cl},m_1})$. Let cluster m_2 be obtained by duplicating cluster m_1 and adding additional lines such that $\mathcal{E}_{m_2} \setminus \mathcal{E}_{m_1} = \mathcal{E}_{\mathrm{new}}$. Let Assumption 4.2 hold for both clusters. Then, cluster m_2 is OS-EIP w.r.t. the port $(\tilde{\boldsymbol{u}}_{\mathrm{cl},m_2}, \tilde{\boldsymbol{y}}_{\mathrm{cl},m_2})$.

Proof:

The incidence matrix of the modified cluster is found by adding the columns associated with the lines in $\mathcal{E}_{\mathrm{new}}$ to the incidence matrix of cluster m_1, i.e., $\boldsymbol{E}_{m_2} = [\boldsymbol{E}_{m_1}, \boldsymbol{E}_{\mathrm{new}}]$. The line resistances associated with $\boldsymbol{E}_{m_2}$ are thus $\boldsymbol{R}_{\mathrm{t},m_2} = \mathrm{diag}[\boldsymbol{R}_{\mathrm{t},m_1}, \boldsymbol{R}_{\mathrm{t,new}}]$, where $\boldsymbol{R}_{\mathrm{t,new}}$ are the resistances associated with the lines in $\mathcal{E}_{\mathrm{new}}$. The LMI in (4.48) thus becomes

$$\boldsymbol{E}_{m_1}\boldsymbol{R}_{\mathrm{t},m_1}^{-1}\boldsymbol{E}_{m_1}^T - \begin{bmatrix}\boldsymbol{E}_{m_1} & \boldsymbol{E}_{\mathrm{new}}\end{bmatrix}\begin{bmatrix}\boldsymbol{R}_{\mathrm{t},m_1}^{-1} & \boldsymbol{0} \\ \boldsymbol{0} & \boldsymbol{R}_{\mathrm{t,new}}^{-1}\end{bmatrix}\begin{bmatrix}\boldsymbol{E}_{m_1}^T \\ \boldsymbol{E}_{\mathrm{new}}^T\end{bmatrix} \preccurlyeq 0 \tag{4.54a}$$

$$\boldsymbol{E}_{m_1}\boldsymbol{R}_{\mathrm{t},m_1}^{-1}\boldsymbol{E}_{m_1}^T - \boldsymbol{E}_{m_1}\boldsymbol{R}_{\mathrm{t},m_1}^{-1}\boldsymbol{E}_{m_1}^T - \boldsymbol{E}_{\mathrm{new}}\boldsymbol{R}_{\mathrm{t,new}}^{-1}\boldsymbol{E}_{\mathrm{new}}^T \preccurlyeq 0 \tag{4.54b}$$

$$-\boldsymbol{E}_{\mathrm{new}}\boldsymbol{R}_{\mathrm{t,new}}^{-1}\boldsymbol{E}_{\mathrm{new}}^T \preccurlyeq 0, \tag{4.54c}$$

which always holds for positive line resistances. ■

Corollary 4.12 and Corollary 4.13 show that the OS-EIP property of a cluster is inherently robust against decreasing line resistances or the addition of new lines to the cluster. These corollaries also demonstrate some of the points made on the feasibility of the LMI verifying the OS-EIP of a cluster in Remark 4.10. Specifically, robustness against factors such as line wear, line ageing or line loading can thus be ensured by verifying the OS-EIP of a cluster for line resistances increased by a chosen safety factor.

Remark 4.19. *The OS-EIP retention when decreasing the line resistances or adding lines in Corollary 4.12 and Corollary 4.13 does not contradict Braess' paradox for electrical networks (see [Sch+22]). Whereas the former deals with a passivity and ultimately a stability result, Braess' paradox describes the phenomenon where adding new lines in the network might change the effective resistance of the network in such a way that certain lines might be overloaded after the addition, i.e., it describes operational (steady-state) limits.*

Whereas the previous two corollaries simplify the LMI in Theorem 4.11 for resistance changes or line additions, an additional interesting comparison can be made when considering the Laplacian of a complete graph. Notably, in such an all-to-all topology in which the weights of the graph are identical, all the non-zero eigenvalues of the Laplacian are equal. This allows the Laplacian comparison in (4.48) to simplify to a comparison of the algebraic connectivity (see Section 3.1) if one of the clusters has a complete graph topology.

Proposition 4.14 (Cluster Algebraic Connectivity Comparison)
Consider a cluster m_1 (4.30) with a complete graph topology and identical line resistances which is OS-EIP w.r.t. the port $(\tilde{\boldsymbol{u}}_{\mathrm{cl},m_1}, \tilde{\boldsymbol{y}}_{\mathrm{cl},m_1})$. Let another cluster m_2 have the same buses as cluster m_1, but with different set of lines, i.e., $\mathcal{N}_{m_1} = \mathcal{N}_{m_2}$ and $\mathcal{E}_{m_1} \neq \mathcal{E}_{m_2}$. Let Assumption 4.2 hold for both clusters. Then, cluster m_2 is OS-EIP w.r.t. the port $(\tilde{\boldsymbol{u}}_{\mathrm{cl},m_2}, \tilde{\boldsymbol{y}}_{\mathrm{cl},m_2})$ if

$$\lambda_2(\boldsymbol{\mathcal{L}}_{m_1}) \leq \lambda_2(\boldsymbol{\mathcal{L}}_{m_2}), \tag{4.55}$$

where $\lambda_2(\boldsymbol{\mathcal{L}}_m)$ is the algebraic connectivity of cluster m.

Proof:
Substitute the Laplacian decomposition in (3.3) into (4.48) of Theorem 4.11 to obtain

$$\boldsymbol{\Xi}_{m_1}\boldsymbol{\Lambda}_{m_1}\boldsymbol{\Xi}_{m_1}^{-1} - \boldsymbol{\Xi}_{m_2}\boldsymbol{\Lambda}_{m_2}\boldsymbol{\Xi}_{m_2}^{-1} \preccurlyeq 0. \tag{4.56}$$

Due to the $|\mathcal{N}_m| - 1$ eigenvalue multiplicity of a complete graph [BH12, p. 8], the image of the Laplacian $\boldsymbol{\Xi}_{\mathrm{Im},m_1}$ consists of any set of $|\mathcal{N}_m| - 1$ linearly independent vectors such that $\boldsymbol{\Xi}_{\mathrm{Im},m_1} \perp \mathbb{1}_{|\mathcal{N}_m|}$. This includes the choice $\boldsymbol{\Xi}_{\mathrm{Im},m_1} := \boldsymbol{\Xi}_{\mathrm{Im},m_2}$, by which (4.56) is simplified to a comparison of the diagonal eigenvalue matrix, i.e.,

$$\lambda_n(\boldsymbol{\mathcal{L}}_{m_1}) \leq \lambda_n(\boldsymbol{\mathcal{L}}_{m_2}), \quad n = 1, \dots, |\mathcal{N}_m|. \tag{4.57}$$

The eigenvalue multiplicity of the complete graph topology in cluster m_1 then reduces the eigenvalue comparison (4.57) to the algebraic connectivity comparison in (4.55). ■

By constructing a fictitious cluster with a complete graph and verifying its OS-EIP using Proposition 4.10, the OS-EIP of other clusters with the same buses but with other topologies can easily be verified by comparing a single eigenvalue using Proposition 4.14. Furthermore, Proposition 4.14 demonstrates the last of the points in Remark 4.10 since $\lambda_2(\boldsymbol{\mathcal{L}})$ for the Laplacian weighted by the line conductance is a measure for the electrical distances between the buses in a cluster.

Remark 4.20. *The algebraic connectivity for a few lines with low resistances can be greater than that of a complete graph with high line resistances. When verifying the OS-EIP of a cluster with a complete graph using Proposition 4.10, it is therefore desirable to set the line resistance as high as possible such that the algebraic connectivity $\lambda_2(\boldsymbol{\mathcal{L}})$ for the complete graph will be as low as possible. Note that the minimum $\lambda_2(\boldsymbol{\mathcal{L}})$ which yields a robustly OS-EIP cluster comprising a given number of nodes can be computed by maximising $\boldsymbol{R}_{\mathrm{t},m}$ subject to the LMI* (4.45).

The results of this subsection detailing conditions for the OS-EIP robustness of a cluster under line changes are visualised in Figure 4.8 and a full summary for bus and line changes is provided in Table 4.1. Note that the OS-EIP robustness of a cluster in which buses are added or removed can be ensured if the respective configurations before and after the bus changes exhibit an OS-EIP property. Such a verification, however, does not follow automatically, but requires separate OS-EIP verifications for both the before and the after states.

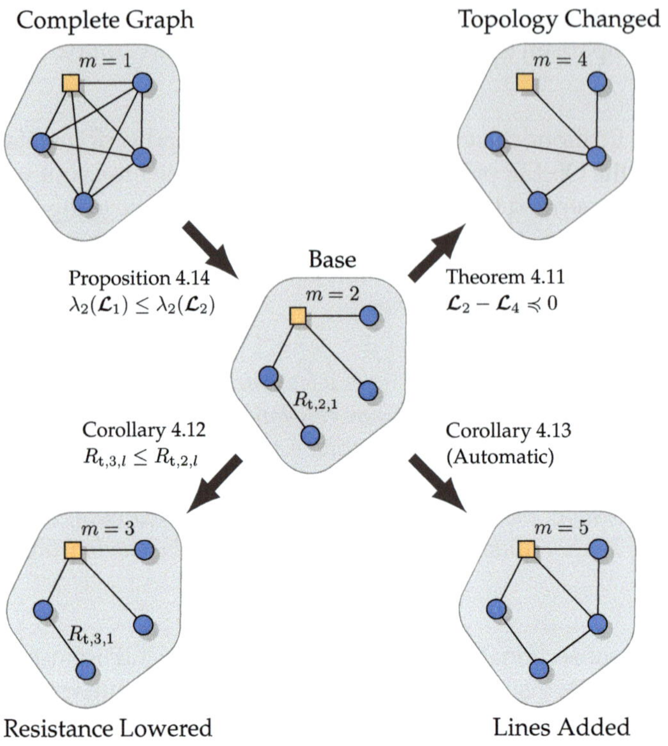

Figure 4.8: Overview and examples of the OS-EIP-preserving results when changing the lines within a cluster.

Table 4.1: Cluster OS-EIP Robustness

Modification		OS-EIP Robustness Guarantee
Line changes	Resistance	Inherent if decreasing (Corollary 4.12), Laplacian comparison (Theorem 4.11) or complete graph comparison (Proposition 4.14) otherwise.
	Inductance	Inherent if Assumption 4.2 still holds, robust LMI check otherwise (Remark 4.11 with Proposition 4.10).
	Topology	Inherent if lines added (Corollary 4.13), Laplacian comparison (Theorem 4.11) or complete graph comparison (Proposition 4.14) otherwise.
Bus changes	Load parameter Y_n	Inherent if decreasing (see Remark 4.2 and Remark 4.10), robust LMI check otherwise (Remark 4.11 with Theorem 4.7).
	DGU or control parameter	Robust LMI check (Remark 4.11 with Theorem 4.7).
	Bus addition or removal	Verify OS-EIP before and after change using a complete graph comparison Proposition 4.14, a Laplacian comparison Theorem 4.7 or an LMI check Proposition 4.10.

4.4 Simulation

In this section, a simulation is used to illustrate the GAS of the equilibria of a DC network, which is verified using the methods in Section 4.3. The robustness of the GAS is shown by having the network undergo a black start and subjecting the network to various changes, including ZIP load and P source changes, topology changes, and changes in the steady-state power availability of the buses. The setup of the simulation is detailed in **Section 4.4.1**. This is followed by **Section 4.4.2**, in which the OS-EIP of the clusters are verified. Finally, the simulation results are presented in **Section 4.4.3**.

4.4.1 Simulation Setup

The DC network in Figure 4.2 comprising 21 buses is simulated in MATLAB/SIMULINK using SIMSCAPE. The network, which serves as a synthetic test network (see [MTB20, Table 3]), is operated in islanded mode, i.e., without being connected to external supply points. As depicted in Figure 4.2, the DC network is partitioned into $|\mathcal{M}| = 4$ clusters and each bus is implemented using the setup in Figure 4.5. The voltage setting controller (4.12) is active at Buses 1, 6, 13, and 17, i.e., one per cluster. At the remaining buses, the voltage following controller (4.22) is active.

The DC network parameters are listed in Table 4.2. The ZIP load parameters and the P source parameters are generated randomly within the stated value ranges (see Table 4.2), with the ZIP loads also limited to $Y_n < 3$, $n \in \mathcal{N}_{\text{el}}$.[5] Note that the ZIP load and P source values are deliberately chosen larger than typical for a low voltage DC network in order to simulate operations under strained conditions. The transmission lines all exhibit the same parameter values per unit length and the line lengths are also generated randomly within the range in Table 4.2. The ZIP load, P source, and line length values are respectively listed in Table B.1, Table B.2, and Table B.3 in Appendix B.1. Uniform DGU parameters are used for simplicity.[6] Furthermore, the base simulation results are compared to the case where the converter voltages $v_{\text{b},n}$ experience saturation above and below the limits in Table 4.2.[7]

The parameters for the voltage setting controllers are chosen based on the design guidelines in [FCS23], where a similar control structure is used. The capacitor current damping parameter δ_j is chosen to be sufficiently large compared to the ZIP load parameter Y_n. For the voltage following bus controllers, the parameters $\psi_{\text{v},k}$ and $\psi_{\text{i},k}$

---

5 A value of $Y_n = 3$ corresponds to a P load with $P_n = 212.2\,\text{kW}$ at a critical voltage of $v_{\text{crit}} = 266\,\text{V}$ (see Remark 4.2).

6 Recall from Remark 4.11 that the cluster OS-EIP properties can be verified to be robust against bus and controller parameter changes. Furthermore, such variations do not influence the enforced steady state of the controllers (see Section 4.2).

7 To prevent integrator wind-up, clamping is implemented for the integrator states of the voltage setting and voltage following controllers (see [Vis06, p. 38]).

are tuned to provide sufficient damping for the voltage and current states, respectively. The filter constant $\psi_{\varepsilon,k}$ of the low-pass filter is chosen for a settling time of approximately 5 ms. A similar δ_k is chosen for the voltage following buses.

Table 4.2: DC Network Parameters for Voltage Stabilisation

Voltages	$v_{\mathrm{Ref}} = 380\,\mathrm{V}$	$v_{\mathrm{crit}} = 266\,\mathrm{V}$
ZIP Loads (4.1a), $n \in \mathcal{N}_{\mathrm{el}}$	$-0.208\,\mathrm{S} \leq \lvert Z_n^{-1} \rvert \leq 0.416\,\mathrm{S}$	$-79\,\mathrm{A} \leq \lvert I_n \rvert \leq 158\,\mathrm{A}$
	$-30\,\mathrm{kW} \leq \lvert P_n \rvert \leq 60\,\mathrm{kW}$	$Y_n \leq 3$
P Sources (4.5), $n \in \mathcal{N}_{\mathrm{el}}$	$-10\,\mathrm{kW} \leq P_{\mathrm{ps},n} \leq 0\,\mathrm{kW}$	$-38\,\mathrm{A} \leq I_{\mathrm{ps,max},n} \leq -7.5\,\mathrm{A}$
Transmission Lines (4.7), $o \in \mathcal{E}_{\mathrm{el}}$	$R'_{\mathrm{t},o} = 0.01\,\Omega/\mathrm{km}$	$L'_{\mathrm{t},o} = 2\,\mu\mathrm{H}/\mathrm{km}$
	$C'_{\mathrm{t},o} = 22\,\mathrm{nF}/\mathrm{km}$	length $\in [0.2; 5]$ km
DGU Parameters (4.8a), $n \in \mathcal{N}_{\mathrm{el}}$	$R_n = 0.1\,\Omega$	$L_n = 1.8\,\mathrm{mH}$
	$C_n = 2.2\,\mathrm{mF}$	$v_n(0) = 0\,\mathrm{V}$
	$0\,\mathrm{V} \leq v_{\mathrm{b},n}(t) \leq 600\,\mathrm{V}$	
Voltage Setting Controllers (4.12),	$\kappa_{\varepsilon,j} = 10^6$	$\kappa_{\mathrm{v},j} = 25000$
$j \in \mathcal{S}_m, m \in \mathcal{M}$	$\kappa_{\mathrm{i},j} = 250$	$\delta_j = 100$
Voltage Following Controllers (4.22),	$\psi_{\varepsilon,k} = 100$	$\psi_{\mathrm{v},k} = 20$
$k \in \mathcal{F}_m, m \in \mathcal{M}$	$\psi^{\diamond}_{\mathrm{i},k} = 1$	$\delta_k = 100$

To test the proposed controllers and to demonstrate the robustness of the GAS, the DC network subjected to the following sequence of events.

- $t = 0\,\mathrm{s}$: The DC network in Figure 4.2 undergoes a black start with random ZIP load and P source parameters generated for each bus.
- $t = 0.2\,\mathrm{s}$: Bus 13 loses access to its steady-state power supply and switches to using a voltage following controller. For $50\,\%$ of the buses selected at random, new random ZIP load and P source parameters are generated.
- $t = 0.4\,\mathrm{s}$: Steady-state power is restored at Bus 13 and it reverts to using the voltage setting controller. The lines interconnecting the clusters are disconnected, yielding four separated networks comprising one cluster each. For $50\,\%$ of the buses selected at random, new random ZIP load and P source parameters are generated.
- $t = 0.6\,\mathrm{s}$: The lines are reconnected. The ZIP load parameters are set to $Z_n^{-1} = -0.15\,\mathrm{S}$, $I_n = -20\,\mathrm{A}$, $P_n = -30\,\mathrm{kW}$ ($\Longrightarrow$ $Y_n = 1.18$), representing large sources. For $50\,\%$ of the buses selected at random, new random P source parameters are generated.
- $t = 0.8\,\mathrm{s}$: The setpoints of the voltage following buses are changed to $v_1^\star = 384\,\mathrm{V}$, $v_6^\star = 388\,\mathrm{V}$, $v_{13}^\star = 376\,\mathrm{V}$, $v_{17}^\star = 372\,\mathrm{V}$. The ZIP loads are set to random P only loads, i.e., $Z_n^{-1} = 0\,\mathrm{S}$ and $I_n = 0\,\mathrm{A}$. For $50\,\%$ of the buses selected at random, new random P source parameters are generated.

4.4.2 Cluster Passivity Verification

Before simulating the network, the OS-EIP of each cluster is verified, with $\rho = 5 \cdot 10^{-4}$ chosen in each case. For a simplified comparison using Proposition 4.14, the OS-EIP for fictitious clusters each with a single voltage setting bus, differing numbers of voltage following buses, and connected topologies with an algebraic connectivity $\lambda_2(\mathcal{L}_m)$ are verified. By minimising $\lambda_2(\mathcal{L}_m)$ subject to the LMI in Proposition 4.10, the $\lambda_2(\mathcal{L})$ values in Table 4.3 which render clusters with complete topologies OS-EIP are found. Next, the $\lambda_2(\mathcal{L}_m)$ values for the clusters in Figure 4.2 at $t = 0\,\mathrm{s}$ are calculated, with the line resistances increased by a safety factor of $10\,\%$. The results in Table 4.3 show that Clusters 1, 3, and 4 are OS-EIP through Proposition 4.14. Only the OS-EIP of Cluster 2 is verified numerically using Proposition 4.10, and GAS then follows from Theorem 4.9. This result holds for the time in which the clusters are disconnected from each other.

Table 4.3: Comparison of λ_2 Values for Proposition 4.14

Buses	3	4	5	6	7	8	9
Complete graph minimum $\lambda_2(\mathcal{L})$	9.3	12.8	16.6	20.6	24.9	29.6	34.6
$\lambda_2(\mathcal{L}_m)$ at $t = 0\,\mathrm{s}$		30.5 (m=3)	28.4 (m=1) 67.1 (m=4)		12.0 (m=2)		
$\lambda_2(\mathcal{L}_m)$ for $0.2\,\mathrm{s} < t \leq 0.4\,\mathrm{s}$				20.8 (m=1) 32.5 (m=4)			11.9 (m=2)

For $0.2\,\mathrm{s} < t \leq 0.4\,\mathrm{s}$, in which Bus 13 has no steady-state power, the cluster partitioning is modified and their OS-EIP is again evaluated. In this period, Bus 15 is included in Cluster 1, Buses 13 and 14 in Cluster 2, and Bus 16 in Cluster 4. Note that this is done only for the sake of evaluating the cluster OS-EIP and does not entail any other changes in the network. From the $\lambda_2(\mathcal{L}_m)$ values in Table 4.3, the OS-EIP of Clusters 1 and 4 is again established via Proposition 4.14, whereas a renewed numerical OS-EIP verification of Cluster 2 is done using Proposition 4.10. GAS again follows from Theorem 4.9.

Note that the OS-EIP verification may be done offline for the possible configurations that the network can assume. Moreover, note again that OS-EIP verifications and the subsequent GAS result are independent of ZIP-load changes as long as the maximum verified Y_n limit in Table 4.2 is not exceeded online.

4.4.3 Results

The simulation results for the DC network without input saturation are shown in Figure 4.9. It can be seen in Figure 4.9a that the bus voltages asymptotically assume new steady-state values after the black start and after each change in the network. After the start-up convergence, all bus voltages in Figure 4.9a remain within an $8.8\,\%$ band of v_{Ref} (reaching extremes of 353 V and $413\,\mathrm{V}$). The steady-state voltages remain within a $6.6\,\%$ band of v_{Ref}, verifying that the DC network is stable according to Definition 2.1. Furthermore, Figure 4.9a shows that the voltage setting buses accurately regulate the bus voltage to desired setpoints when steady-state power is available. Note that none of the changes in the network result in unstable behaviour. This includes buses losing steady-state power availability, lines or clusters connecting or disconnecting, or the inclusion and modification of loads with non-monotone incremental impedances. Although the buck converter voltages in Figure 4.9b show steady-state values in the range 247 V to $482\,\mathrm{V}$, extreme transient values of $-190\,\mathrm{kV}$ and $734\,\mathrm{kV}$ are reached.

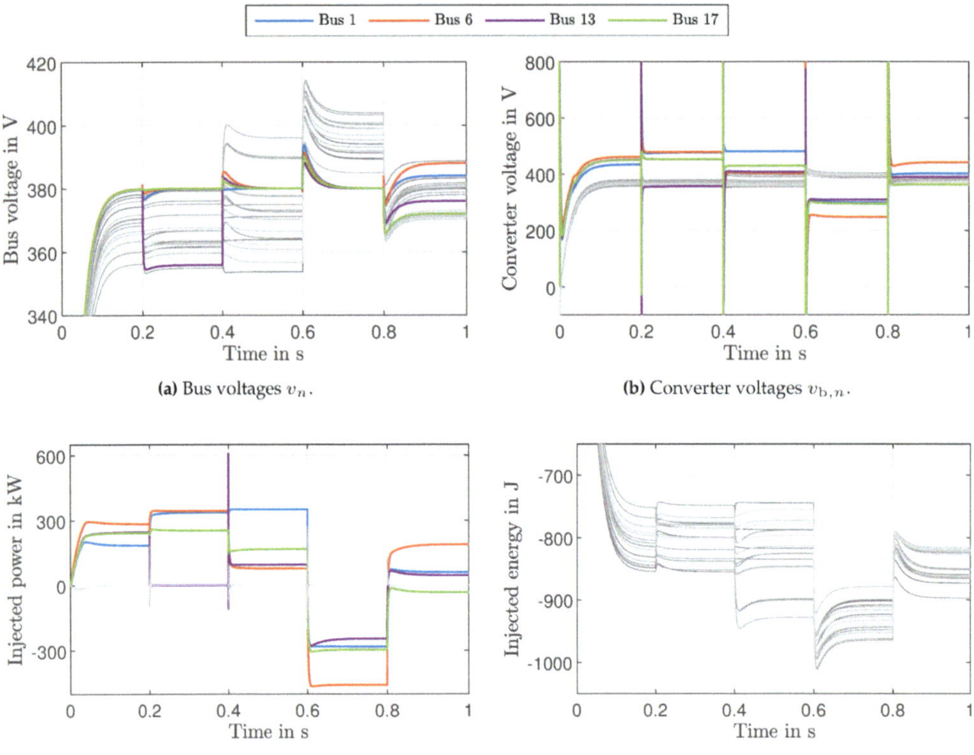

(a) Bus voltages v_n.

(b) Converter voltages $v_{\mathrm{b},n}$.

(c) Injected DGU power $v_n i_n$.

(d) Energy injected by the voltage following controllers.

Figure 4.9: Stabilised DC network results without input saturation. Voltage following buses are shown in shades of grey.

The power injected by the DGUs at each of the buses is shown in Figure 4.9c. The extreme values for the steady-state power injections of more than $351\,\mathrm{kW}$ at Bus 1 and $-459\,\mathrm{kW}$ at Bus 6 confirm the exaggerated ZIP load and P source sizes. The results in Figure 4.9c also show that only the voltage setting buses are used to power the network, with Bus 13 injecting no steady-state power in the period $0.2\,\mathrm{s} < t \leq 0.4\,\mathrm{s}$, as per the simulation setup. Although the voltage following buses do not inject steady-state power, they nevertheless require a small energy storage for transient power injection. This is demonstrated by the total energy injected in Figure 4.9d. Whereas up to 853 J is injected at the voltage following buses during the black start, the transient energy changes remain below 250 J.[8] The absolute transient powers injected by the voltage following buses remain below 200 kW, except for a single peak of 267 kW at $t = 0.6$ s.

To investigate the effect of input saturation on the simulated network, the simulation is repeated with the saturation limits specified in Table 4.2. The bus voltages in Figure 4.10a show that GAS is again achieved, although larger transient overshoots of up to 30.3 % (495 V) and undershoots of up to 25.5 % (283 V) are observed relative to v_{Ref}. Figure 4.10b shows the effects of the saturation, with all converter voltages being restricted to $0\,\text{V} \leq v_{\text{b},n} \leq 600\,\text{V}$. It is thus concluded that the controllers can achieve good transient performance in the presence of large loads and converter voltage saturation.

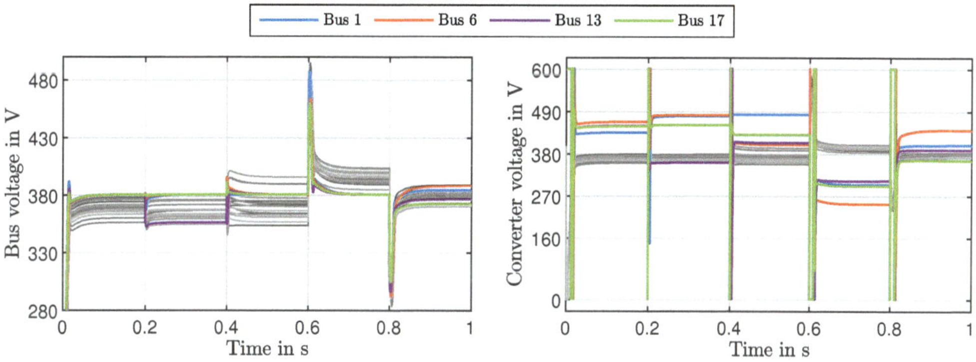

(a) Bus voltages v_n.

(b) Converter voltages $v_{\text{b},n}$.

Figure 4.10: Bus voltages of the DC network with input saturation. Voltage following buses are shown in shades of grey.

4.5 Discussion

The central benefit of the work in this chapter is the inclusion of non-monotone loads at arbitrary locations in a DC network and the subsequent stability proof corresponding to Definition 2.1. This is achieved with voltage setting controllers at nodes where steady-state power is available while voltage following controllers slow down the bus dynamics at the remaining buses using transient control action.

A key step in simplifying the OS-EIP analysis of the voltage setting buses and the clusters is the separation of the load functions into nonlinear monotone and linear parts. This allows the nonlinearities of the load to be extracted from the system, allowing for

[8] As an example, $250\,\text{J} = 0.139\,\text{W h}$ is the amount of energy released by a $10\,\text{mF}$ capacitor discharging from $400\,\text{V}$ to $331.6\,\text{V}$.

a simplified analysis of the remaining dynamics. The ability to extract a portion of the system dynamics may be performed for other systems as well, however, this process requires both the extracted function and the remaining system to interface via an EIP port. Nevertheless, this approach yields bus and cluster OS-EIP properties and a GAS result that is robust against any changes to the nonlinear load function which preserves the EIP property of the extracted part of the load. This robustness includes parameter uncertainties, piecewise-constant or time-varying parameters, entirely different voltage-current functions and even dynamical systems (see Remark 4.14).

The structure of the voltage setting controller, which is primarily responsible for the bus and cluster OS-EIP properties, resembles that of a PID controller, where the derivative term is measured instead of being computed. A single inequality with guaranteed feasibility also ensures an OS-EIP result for a closed-loop voltage setting bus. Furthermore, the PID-like structure allows conventional PID tuning rules to be applied for choosing the control parameters.

Since a bus with a non-monotone load and without steady-state power cannot be passivated around an arbitrary equilibrium, as demonstrated in Proposition 4.4, the OS-EIP of a cluster of buses is instead investigated through an LMI. However, a guaranteed feasibility similar to the one in Remark 4.6 cannot be assured for this LMI. Instead, the electrical distance between a voltage following bus and its closest voltage setting neighbour, which arises from the line topology and resistances, plays a large role in the LMI feasibility. This indicates that an OS-EIP voltage setting bus can dominate the lack of EIP at nearby voltage following buses, but not necessarily for buses that are further away. This compensation is made possible through the equality constraint in (4.32b), which only prevents cross terms from occurring in the bus voltage states. Thus, in the final stability analysis in Theorem 4.9, the storage functions are only fully separable and the stability result is therefore only fully modular at the cluster level. This compromise is necessary due to the general inability of dissipative systems to compensate for the dissipativity of other systems which are not directly linked via their inputs and outputs. In this chapter, this necessity can be traced to the indirect connection of the buses through the dynamical transmission lines.

This consideration highlights the function of the voltage following controllers. By dominating the instantaneous response of the bus, the voltage following controllers prevent voltage collapse caused by the unstable behaviour of non-monotone static loads reacting before the transmission lines can adjust to the changes in power demand. Indeed, omitting the voltage following controller at any bus with a potentially non-monotone load prevents the successful OS-EIP via Theorem 4.7 for the clusters considered in the simulation. The time duration in which power is injected is set through the control gains, while the required time depends broadly on the time constants of the transmission lines.[9] Since these controllers compensate any instantaneous

9 In this chapter, the voltage following controllers subjected to control output saturation injected power for at most 20 ms in response to load step changes.

load changes on the local bus, they must be able to supply or consume large amounts of power. The high-power requirement for a short time corresponds to a low energy requirement, making large capacitors or super-capacitors well suited to providing the necessary storage capacity.

Finally, the results in this chapter raise interesting possibilities for requirements or norms for future loads to contribute to the network stability. For example, requiring all general loads to adhere to a specified minimum incremental impedance (see Remark 4.2) or having users explicitly report the worst-case incremental impedances of sizeable loads to the network operator would fulfil the decentralised load conditions in this chapter while allowing the network operator to account for deviations to the requirement. Additionally, the function of the voltage following controllers could potentially be implemented directly into loads where the instantaneous power availability is less critical—such as electric vehicles, electronic charging devices or heat pumps—by requiring a transient response similar to the one shown in Example 4.1. This can be achieved by having the requisite energy stored locally to cover a portion of the load in the case of voltage disturbances on the bus.

4.6 Summary and Contributions

This chapter demonstrates how decentralised voltage stabilisation can be achieved for a DC network containing loads with non-monotone incremental impedances while only regulating the steady-state voltage at buses where steady-state power is available. By equipping buses with steady-state power with voltage setting controllers and those lacking steady-state power with voltage following controllers, an OS-EIP result for the entire cluster of such buses can be established by solving an LMI. This result can be made robust against DGU, ZIP load, and P source parameter changes by directly incorporating these uncertainties in the LMI. The OS-EIP clusters can then be combined into an arbitrarily large network in which any equilibrium is automatically GAS. By exploiting the timescale differences between the line and the bus dynamics, the OS-EIP cluster property is also shown to be robust against various topology changes.

The main contributions of this chapter are:

- A voltage setting controller along with analytical conditions which yield an OS-EIP property for the voltage regulated bus in the presence of non-monotone loads (Theorem 4.2);
- A voltage following controller which, together with closed-loop voltage setting buses, results in a cluster of buses that can be shown to be OS-EIP in the presence of non-monotone loads by solving an LMI (Theorem 4.7);
- A scalable and robust GAS proof for any equilibrium of a clustered DC network (Theorem 4.9);

- Several conditions which ensure the robustness of the cluster OS-EIP against topology changes (Theorem 4.11, along with Corollaries 4.12 and 4.13, and Proposition 4.14).

Together, these contributions meet the requirements of RQ 1.1 and RQ 1.2.

5 Distributed Coordination in DC Networks

In the previous chapter, it has been demonstrated how voltage stabilisation can be achieved in a DC network using decentralised controllers. While the proposed voltage setting and voltage following controllers succeed in robustly stabilising the network, even in the presence of loads with non-increasing incremental impedances, the voltage regulation requires the definition of voltage setpoints for the voltage setting controllers. Whereas the setpoints were chosen a priori for the DC network in Chapter 4, this directly raises the question of how these setpoints should be chosen, e.g., to ensure a better spread of the bus voltage around the reference voltage. Thus, the goals of robust proportional power sharing and weighted average voltage regulation, as outlined in RQ 2.1 and RQ 2.2, are considered in this chapter.

The chapter commences with **Section 5.1**, where the models of the DC network components are recalled from Chapter 4 and a power regulator is proposed to replace the voltage setting controllers at the buses where steady-state power is available. The goals of proportional power sharing and weighted average voltage regulation are then formulated as control objectives and a four-stage distributed controller is presented which is shown to impose a steady state fulfilling these objectives.

Next, in **Section 5.2**, the stability of the DC network in closed loop with the distributed power-sharing controller is investigated using EIP. In a first step, an IFOF-EIP property is investigated for the buses equipped with the power regulators. This result is then combined with the other DC network buses and lines to obtain an IFOF-EIP property for the entire network. EIP properties for the stages of the distributed power-sharing controller are then established and subsequently combined with the DC network, yielding a means of verifying the stability of closed-loop equilibrium.

The controllers proposed in this chapter are then tested in a simulation in **Section 5.3** to validate the results and demonstrate the robustness of the stability result against various topology and parameter changes. The power regulator and power-sharing controllers are also contrasted with the decentralised stabilisation controllers from Chapter 4 by comparing the steady-state results obtained when using the respective controllers for the same simulation scenario.

The chapter concludes with a discussion of the methods and results in **Section 5.4** along with a summary of the work and the contributions in **Section 5.5**.

5.1 Control Problem and Controller Design

To address the need for proportional power sharing which achieves average voltage regulation for the voltages of all buses in a DC network, a suitable model of the network along with a formulation of control goals is again required. To this end, a description of the DC network considered in this chapter is presented in **Section 5.1.1**, where a distinction is again made between the buses with and without local steady-state power availability. Thereafter, in **Section 5.1.2**, RQ 2.1 is formulated as control objectives for the distributed coordinating power-sharing controller. This power-sharing controller is then presented and its steady-state properties are verified in **Section 5.1.3**.

5.1.1 DC Network Description

As in Section 4.1, a DC network comprising buses interconnected by transmission lines is considered in this chapter, as depicted in Figure 5.1.[1] More formally, the DC network is considered by the graph $\mathcal{G}_{\text{el}} = (\mathcal{N}_{\text{el}}, \mathcal{E}_{\text{el}})$ in which the buses $\mathcal{N}_{\text{el}}$ are weakly connected by directed edges $\mathcal{E}_{\text{el}}$ representing the transmission lines. The arbitrarily assigned direction of the lines denotes the direction of positive current flow. Furthermore, to enable power sharing through a distributed controller, the physical network is augmented by a communication network. The communication is described by a connected graph $\mathcal{G}_{\text{com}} = (\mathcal{N}_{\text{el}}, \mathcal{E}_{\text{com}})$ where $\mathcal{N}_{\text{el}}$ is the set of physical buses which communicate bidirectionally via the edges in $\mathcal{E}_{\text{com}}$. Note that the sets $\mathcal{E}_{\text{el}}$ and $\mathcal{E}_{\text{com}}$ are not necessarily the same. The buses where steady-state power is available are grouped into the set $\mathcal{P} = \{1, \ldots, |\mathcal{P}|\}$ with $|\mathcal{P}| \geq 1$, whereas the buses without steady-state power are gathered into the set $\mathcal{F} = \{|\mathcal{P}|+1, \ldots, |\mathcal{P}|+|\mathcal{F}|-1\}$ with $|\mathcal{F}| \geq 0$. Note that, at any time, each bus appears in exactly one of these sets, i.e., $\mathcal{P} \cap \mathcal{F} = \emptyset$ and $\mathcal{P} \cup \mathcal{F} = \mathcal{N}_{\text{el}}$. Furthermore, a bus can be moved from one set to the other if its steady-state power availability changes. The buses in $\mathcal{F}$ with no steady-state power are considered to be uncontrolled in this chapter.

The same models as given in Section 4.1 are used for the ZIP loads (4.1), P sources (4.5), transmission lines (4.7), and the DGUs (4.8). Additionally, since no voltage following buses are used in this chapter, it is assumed that the ZIP loads in this chapter have strictly monotone incremental impedances, i.e., $Y_n < 0$ for $n \in \mathcal{N}_{\text{el}}$ (see Remark 4.2).

For the buses in $k \in \mathcal{F}$ without steady-state power, the DGUs are deactivated completely, i.e., $i_k \equiv 0$. This yields the bus dynamics

$$C_{\text{ef},k}\dot{v}_k = Y_k v_k + i_{\text{int},k} + i_{\text{ex},k}, \tag{5.1a}$$

$$i_{\text{int},k} := \sum_{o \in \mathcal{E}^+_{\text{el},k}} i_{\text{t},o} - \sum_{o \in \mathcal{E}^-_{\text{el},k}} i_{\text{t},o}, \qquad i_{\text{ex},k} := -i^{\text{m}}_{\text{l},k}(v_k) - i_{\text{ps},k}(v_k), \qquad k \in \mathcal{F}. \tag{5.1b}$$

[1] Compared to Chapter 4, a communication layer has been added, the voltage setting controllers are replaced with power regulators and the voltage following controllers are omitted.

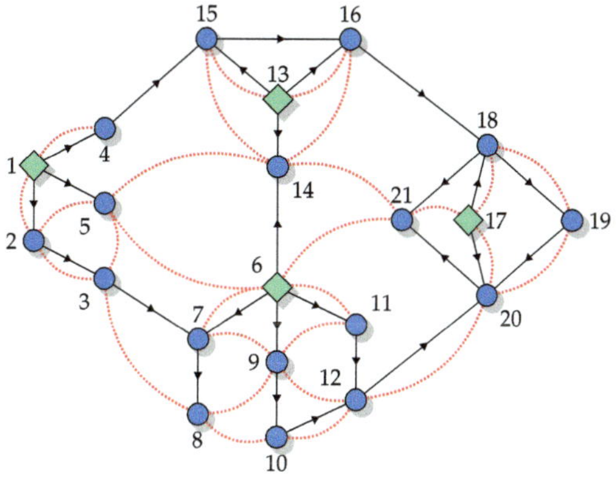

Figure 5.1: A 21-bus DC network with communication, comprising power-regulated buses ◇ in $\mathcal{P}$, buses without power injection ● in $\mathcal{F}$, lines with arbitrary directions in $\mathcal{E}_{\mathrm{el}}$, and undirected communication links in $\mathcal{E}_{\mathrm{com}}$ between the buses.

To allow a distributed controller to achieve power sharing, each bus $j \in \mathcal{P}$ where steady-state power is available is equipped with the following PI-like regulator[2] acting on the power $p_j = v_j i_j$ injected by the DGU:

$$\dot{\varepsilon}_j = (p_j - p_j^\star), \tag{5.2a}$$

$$v_{\mathrm{b},j} = -\vartheta_{\mathrm{p},j}(p_j - p_j^\star) - \vartheta_{\mathrm{i},j} i_j - \vartheta_{\varepsilon,j}\varepsilon_j - \frac{\delta_{\mathrm{p},j} L_j}{C_j} i_{\mathrm{c},j}, \quad j \in \mathcal{P}, \tag{5.2b}$$

$$p_j^\star = k_{\mathrm{p},j}\phi_j, \tag{5.2c}$$

where $\varepsilon_j \in \mathbb{R}$ is the integral error of the power regulator[3] and $\vartheta_{\mathrm{p},j}, \vartheta_{\mathrm{i},j}, \vartheta_{\varepsilon,j} \in \mathbb{R}_{>0}$ are the control gains acting on the proportional power error, the integral power error, and the filter current, respectively. The gain $\delta_{\mathrm{p},j} \in \mathbb{R}_{>0}$ injects damping onto the dynamics of the capacitor current $i_{\mathrm{c},j}$. Finally, $\phi_j \in \mathbb{R}$ is the normalised power setpoint which is weighted by the gain $k_{\mathrm{p},j} \in [\underline{k}_{\mathrm{p}}, \overline{k}_{\mathrm{p}}] \subset \mathbb{R}_{\geq 0}$ to obtain the power setpoint $p_j^\star \in \mathbb{R}$ for the power regulator at bus $j \in \mathcal{P}$. Note that the use of $i_{\mathrm{c},j}$ in (5.2) again has a derivative-like effect on the bus dynamics (see Remark 4.8).

2 The power regulator in (5.2) uses the same form as the voltage setting controller in (4.12) and simply replaces the voltage difference with the power difference.

3 A DGU in closed loop with the power regulator in (5.2) has a forced equilibrium at $\hat{p}_j = p_j^\star$ due to the addition of the integrator state ε_j.

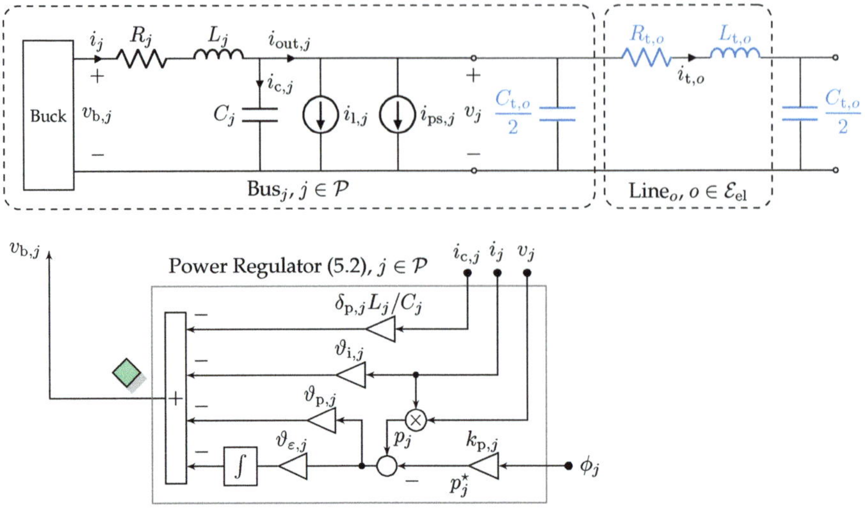

Figure 5.2: Schematic of a bus $j \in \mathcal{P}$ equipped with a power regulator.

The power regulator in (5.2) is visualised in Figure 5.2. Equipping a DGU (4.8) at a bus with the power regulator (5.2) results in the power regulated bus dynamics

$$\begin{bmatrix}\dot{\varepsilon}_j\\ L_j\dot{i}_j\\ C_{\mathrm{ef},j}\dot{v}_j\end{bmatrix} = \begin{bmatrix}0 & v_j & 0\\ -\vartheta_{\varepsilon,j} & -R_j-\vartheta_{\mathrm{i},j}-\vartheta_{\mathrm{p},j}v_j & -1\\ 0 & 1 & Y_j\end{bmatrix}\begin{bmatrix}\varepsilon_j\\ i_j\\ v_j\end{bmatrix} + \begin{bmatrix}k_{\mathrm{p},j}\phi_j\\ \vartheta_{\mathrm{p},j}k_{\mathrm{p},j}\phi_j - \frac{\delta_{\mathrm{p},j}L_j}{C_j}i_{\mathrm{c},j}\\ i_{\mathrm{int},j}+i_{\mathrm{ex},j}\end{bmatrix}, \quad (5.3a)$$

$$i_{\mathrm{int},j} := \sum_{o\in\mathcal{E}_{\mathrm{el},j}^{+}} i_{\mathrm{t},o} - \sum_{o\in\mathcal{E}_{\mathrm{el},j}^{-}} i_{\mathrm{t},o}, \qquad i_{\mathrm{ex},j} := -i_{\mathrm{l},j}^{\mathrm{m}}(v_j) - i_{\mathrm{ps},j}(v_j), \qquad j \in \mathcal{P}. \quad (5.3b)$$

Note that the bus dynamics in (5.3a) can again be transformed using (4.13) to obtain

$$\underbrace{\begin{bmatrix}\dot{\varepsilon}_j\\ L_j\dot{i}_{\mathrm{z},j}\\ C_{\mathrm{ef},j}\dot{v}_j\end{bmatrix}}_{\boldsymbol{Q}_{\mathrm{p},j}\dot{\boldsymbol{x}}_{\mathrm{p},j}} = \begin{bmatrix}v_j(i_{\mathrm{z},j}-\delta_{\mathrm{p},j}v_j)\\ -\vartheta_{\varepsilon,j}\varepsilon_j-(R_j+\vartheta_{\mathrm{i},j}+\vartheta_{\mathrm{p},j}v_j)(i_{\mathrm{z},j}-\delta_{\mathrm{p},j}v_j)-v_j\\ i_{\mathrm{z},j}+(Y_j-\delta_{\mathrm{p},j})v_j\end{bmatrix} + \begin{bmatrix}k_{\mathrm{p},j}\phi_j\\ \vartheta_{\mathrm{p},j}k_{\mathrm{p},j}\phi_j\\ i_{\mathrm{int},j}+i_{\mathrm{ex},j}\end{bmatrix}, \quad (5.4a)$$

$$\boldsymbol{x}_{\mathrm{p},j} = \begin{bmatrix}\varepsilon_j & i_{\mathrm{z},j} & v_j\end{bmatrix}^T. \quad (5.4b)$$

5.1.2 Control Problem Formulation

Using the DC network description in Section 5.1.1, which is based on the models in Section 4.1, along with the power regulator (5.2), RQ 2.1 can be formulated mathematically as objectives to be achieved by a distributed controller.

Recall from Section 2.1.2 and RQ 2.1 that an average voltage regulation is desired in which the voltages of all buses in the DC network are considered. Since voltage differences are required to achieve a flow of power over lossy lines, $v_n \to v_{\text{Ref}}$ for all $n \in \mathcal{N}_{\text{el}}$ is not practical in a network without sufficient power available at every bus. Instead, the setpoints for the controllers at buses with steady-state power should be chosen such that the average bus voltage is as close to the desired reference voltage as possible. Furthermore, recall from Definition 2.1 that it is sufficient for the steady-state bus voltages to lie within a tolerance band around the voltage reference v_{Ref}. By nonlinearly weighing the bus voltages such that voltages outside this band are penalised more strongly, the flexibility of this band can be exploited to reduce the largest difference between a bus voltage error and some global voltage setpoint $v^\star$. The first control objective thus requires finding suitable controller setpoints such that an average weighted voltage regulation is achieved.

Objective 5.1 (Average Weighted Voltage Regulation)
For the buses $j \in \mathcal{P}$ where steady-state power is available and which are equipped with the power regulator (5.2), *find ϕ_j such that*

$$\lim_{t\to\infty} \frac{1}{|\mathcal{N}_{\text{el}}|} \sum_{n\in\mathcal{N}_{\text{el}}} h(v_n(t)) = v^\star, \tag{5.5}$$

where $h : \mathbb{R} \to \mathbb{R}$ is a strictly monotonically increasing weighting function.

The second part of RQ 2.1 requires a proportional power sharing between the buses $j \in \mathcal{P}$ where steady-state power is available. Since these buses are equipped with the power regulator in (5.2) with the normalised power setpoint ϕ_j, a proportional power sharing can thus be achieved if the normalised power setpoint is equal at all buses in the steady state.

Objective 5.2 (Proportional Power Sharing)
For the buses which are equipped with the power regulator (5.2) *and where steady-state power is available, let*

$$\lim_{t\to\infty} (\phi_{j_1}(t) - \phi_{j_2}(t)) = \lim_{t\to\infty} (k_{\text{p},j_1} p^\star_{j_1}(t) - k_{\text{p},j_2} p^\star_{j_2}(t)) = 0, \quad \forall j_1, j_2 \in \mathcal{P}. \tag{5.6}$$

A distributed controller which achieves Objectives 5.1 and 5.2 thus fulfils RQ 2.1. For a controller to fulfil these objectives, however, it is assumed that the problem is well posed w.r.t. the network design.

Assumption 5.1 (Feasible DC Network)
The buses in $\mathcal{P}$ where steady-state power is available can feasibly supply the loads in the network via the electrical transmission, i.e., there exists an equilibrium for the DC network where all bus voltages lie within a tolerance band around the voltage reference v_{Ref}.

Remark 5.1. *Similar formulations for the average voltage regulation in Objective 5.1 can be found in [Nas+15; Tri+18; Tri+19; CKS07; Cuc+19; NFT20; Fan+20; SSB21; NTFT22], although some of these formulations only sum over the set $\mathcal{P}$ where steady-state control is available and often employ linear weighting functions $h(\cdot)$. The (proportional) power-sharing formulation in Objective 5.2 can similarly be found in [Tri+18; DWD18; Fan+20].*

Remark 5.2. *The gain k_{p} in* (5.2c) *may freely be interpreted as a linear, continuous cost function with the gain k_{p} linking the power setpoint $p^\star$ to a price signal ϕ. In this way, Objective 5.2 can be interpreted as finding the common price signal according to which all DGUs inject steady-state power, with a higher price signal ϕ resulting in more power being injected.*

5.1.3 Power-Sharing Controller

To achieve Objectives 5.1 and 5.2, a four-stage decentralised control structure is proposed. This control structure, visualised in Figure 5.3, uses the bus voltages of all buses $\boldsymbol{v}_{\mathrm{el}} = \mathrm{stack}_{n\in\mathcal{N}_{\mathrm{el}}}(v_n)$ to generate the proportional power setpoints $\boldsymbol{\phi} = \mathrm{stack}_{n\in\mathcal{N}_{\mathrm{el}}}(\phi_n)$ which are uniform in steady state. The setpoint $\boldsymbol{\phi}$ is generated by building the error of the bus voltage w.r.t. the global voltage setpoint $v^\star$, and nonlinearly weighing this error in Stage 1. The weighted errors are then passed through a DDA block in Stage 2, where the average weighted error is computed in a distributed fashion. Thereafter, local PI controllers are implemented for Stage 3, which generate local setpoints for regulating the average weighted error to zero. The output of Stage 3 is passed through a DDA block in Stage 4, through which the uniform proportional power setpoint $\boldsymbol{\phi}$ is determined. The Stage 4 outputs ϕ_i are then used for the power setpoints in the power regulators in (5.2). Note that each stage is implemented at each bus $n \in \mathcal{P} \cup \mathcal{F}$ in the network, even though only the outputs at the buses in $\mathcal{P}$ are used by the lower-level controllers.

In the rest of this subsection, the equations for the nonlinear weighting function, the DDA blocks, and the PI controllers are introduced. Thereafter, it is demonstrated that any equilibrium of the DC network in closed loop with the distributed power-sharing controller meets the requirements of Objectives 5.1 and 5.2.

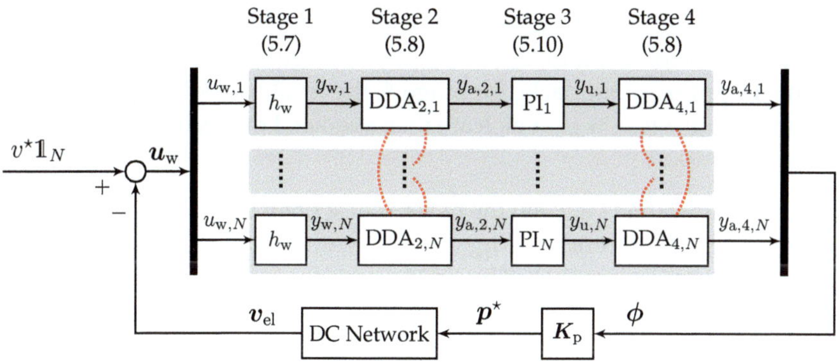

Figure 5.3: A four-stage distributed power-sharing controller for a DC network of $N = |\mathcal{N}_{el}|$ buses with communication links ···· between the buses.

Nonlinear Weighting Function

In Stage 1 of the distributed controller in Figure 5.3, the voltage errors $v^\star - v_n = u_{w,n}$ are passed through the class C^1 weighting function

$$h_w(u_{w,n}) := k_{w,1} u_{w,n} + k_{w,2} g_w(u_{w,n}) - k_{w,2} \tanh(g_w(u_{w,n})), \quad n \in \mathcal{N}_{el}, \tag{5.7a}$$

$$g_w(u_{w,n}) := \begin{cases} u_{w,n} + k_{w,3}, & u_{w,n} < -k_{w,3}, \\ 0, & |u_{w,n}| \leq k_{w,3}, \\ u_{w,n} - k_{w,3}, & u_{w,n} > k_{w,3}, \end{cases} \tag{5.7b}$$

with $k_{w,1}, k_{w,2}, k_{w,3} \in \mathbb{R}_{>0}$ and where $k_{w,1}$ is the gain for small input values $|u_{w,n}| < k_{w,3}$, $k_{w,1} + k_{w,2}$ is the gain for large input values $|u_{w,n}| \gg k_{w,3}$, as visualised in Figure 5.4.

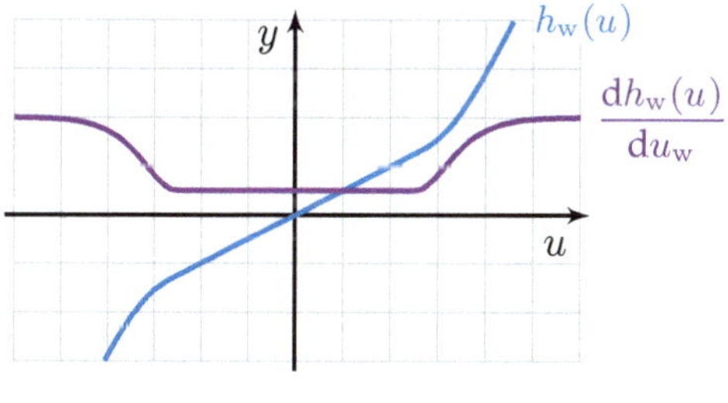

Figure 5.4: Example of the weighting function h_w (5.7) and its derivative (5.38) on a unit grid, with $k_{w,1} = 0.5$, $k_{w,2} = 1.5$ and $k_{w,3} = 2.5$.

Dynamic Distributed Averaging Subsystems

In Stages 2 and 4 of the distributed controller in Figure 5.3, the PI-DDA controller described in [FYL06] is implemented as a means of calculating the average of a specific network variable in a distributed manner. In Stage 2 of the network, which receives the weighted voltage error as input $\boldsymbol{u}_{\mathrm{a},2} = \boldsymbol{y}_{\mathrm{w}}$, the average weighted voltage error is obtained as the output $\boldsymbol{y}_{\mathrm{a},2}$. Similarly, for Stage 4, the local control outputs of the PI controllers form the input $\boldsymbol{u}_{\mathrm{a},4} = \boldsymbol{y}_{\mathrm{u}}$ which are then averaged to obtain the proportional power setpoint $\boldsymbol{y}_{\mathrm{a},4} = \boldsymbol{\phi}$.

Recall that the communication between the buses is described by the graph $\mathcal{G}_{\mathrm{com}}$ which has a Laplacian $\boldsymbol{\mathcal{L}}_{\mathrm{com}}$ arsing from the edges in $\mathcal{E}_{\mathrm{com}}$ with unit weights and arbitrarily assigned directions (see Section 3.1). The DDA in vector form is then described by [FYL06, Eq. (19)]

$$\begin{bmatrix} \dot{\boldsymbol{x}}_{\mathrm{a},s} \\ \dot{\boldsymbol{\chi}}_{\mathrm{a},s} \end{bmatrix} = \begin{bmatrix} -\delta_{\mathrm{a}} \boldsymbol{I}_{|\mathcal{N}_{\mathrm{el}}|} - k_{\mathrm{a}}^{\mathrm{P}} \boldsymbol{\mathcal{L}}_{\mathrm{com}} & k_{\mathrm{a}}^{\mathrm{I}} \boldsymbol{\mathcal{L}}_{\mathrm{com}}^{T} \\ -k_{\mathrm{a}}^{\mathrm{I}} \boldsymbol{\mathcal{L}}_{\mathrm{com}} & \boldsymbol{0} \end{bmatrix} \begin{bmatrix} \boldsymbol{x}_{\mathrm{a},s} \\ \boldsymbol{\chi}_{\mathrm{a},s} \end{bmatrix} + \begin{bmatrix} \delta_{\mathrm{a}} \boldsymbol{I}_{|\mathcal{N}_{\mathrm{el}}|} \\ \boldsymbol{0} \end{bmatrix} \boldsymbol{u}_{\mathrm{a},s}, \quad s \in \{2, 4\}, \tag{5.8a}$$

$$\boldsymbol{y}_{\mathrm{a},s} = \boldsymbol{x}_{\mathrm{a},s}, \tag{5.8b}$$

where s corresponds to the stage number in Figure 5.3, $\boldsymbol{x}_{\mathrm{a},s}, \boldsymbol{\chi}_{\mathrm{a},s} \in \mathbb{R}^{|\mathcal{N}_{\mathrm{el}}|}$ are the estimate and auxiliary integral states, respectively, $k_{\mathrm{a}}^{\mathrm{P}}, k_{\mathrm{a}}^{\mathrm{I}} \in \mathbb{R}_{>0}$ are the proportional and integral control gains, and $\delta_{\mathrm{a}} \in \mathbb{R}_{>0}$ is a global estimator parameter that determines the speed at which the estimation reacts to input changes. In [FYL06], it is shown that the output of a DDA (5.8) with a constant input $\boldsymbol{u}_{\mathrm{a},s} = \hat{\boldsymbol{u}}_{\mathrm{a},s}$ converges to the average of $\hat{\boldsymbol{u}}_{\mathrm{a},s}$, i.e.,

$$\lim_{t \to \infty} y_{\mathrm{a},s,n} = \frac{\mathbb{1}_{|\mathcal{N}_{\mathrm{el}}|}^{T} \hat{\boldsymbol{u}}_{\mathrm{a},s}}{|\mathcal{N}_{\mathrm{el}}|}, \quad \forall n \in \mathcal{N}_{\mathrm{el}}. \tag{5.9}$$

PI Controllers

In Stage 3, a control output is generated from the average weighted voltage errors at each bus using a leaky PI controller similar to the approach in [Wei+18]

$$\dot{\varepsilon}_{\mathrm{u},n} = -\delta_{\mathrm{u}} \varepsilon_{\mathrm{u},n} + u_{\mathrm{u},n}, \quad n \in \mathcal{N}_{\mathrm{el}}, \tag{5.10a}$$

$$y_{\mathrm{u},n} = k_{\mathrm{u}}^{\mathrm{I}} \varepsilon_{\mathrm{u},n} + k_{\mathrm{u}}^{\mathrm{P}} u_{\mathrm{u},n}, \tag{5.10b}$$

where $\varepsilon_{\mathrm{u},n} \in \mathbb{R}$ is the integral state, $\delta_{\mathrm{u}} \in \mathbb{R}_{\geq 0}$ is the damping factor resulting in a leaky PI controller if $\delta_{\mathrm{u}} > 0$, and $k_{\mathrm{u}}^{\mathrm{P}}, k_{\mathrm{u}}^{\mathrm{I}} \in \mathbb{R}_{>0}$ are the proportional and integral gain constants. Combining the local PI controllers (5.10) for the $|\mathcal{N}_{\mathrm{el}}|$ buses results in the vector form

$$\dot{\boldsymbol{\varepsilon}}_{\mathrm{u}} = -\delta_{\mathrm{u}} \boldsymbol{\varepsilon}_{\mathrm{u}} + \boldsymbol{u}_{\mathrm{u}}, \tag{5.11a}$$

$$\boldsymbol{y}_{\mathrm{u}} = k_{\mathrm{u}}^{\mathrm{I}} \boldsymbol{\varepsilon}_{\mathrm{u}} + k_{\mathrm{u}}^{\mathrm{P}} \boldsymbol{u}_{\mathrm{u}}. \tag{5.11b}$$

Note that for a constant input $\boldsymbol{u}_{\mathrm{u}} = \hat{\boldsymbol{\varepsilon}}_{\mathrm{u}}$, the leaky PI controller (5.11) forces an equilibrium characterised by

$$\delta_{\mathrm{u}} \hat{\boldsymbol{\varepsilon}}_{\mathrm{u}} = \hat{\boldsymbol{u}}_{\mathrm{u}}. \tag{5.12}$$

Thus, $\hat{\boldsymbol{u}}_{\mathrm{u}} = \boldsymbol{0}$ holds only if an ideal integrator with $\delta_{\mathrm{u}} = 0$ is used.

Closed-Loop Equilibrium Analysis

With the stages of the distributed power-sharing controller described by (5.7), (5.8), and (5.11), the equilibrium imposed by this power-sharing controller can be investigated. Specifically, it is verified that the proposed power-sharing controller forces an equilibrium which satisfies Objectives 5.1 and 5.2. To this end, consider the DC network comprising buses in $\mathcal{P}$ equipped with the power regulating controller (5.4a) and load buses in $\mathcal{F}$ described by (5.1). Let the buses be interconnected by the lines in $\mathcal{E}_{\text{el}}$ (4.7). The DC network is then described by

$$\begin{bmatrix} \dot{\varepsilon}_{\text{p}} \\ \boldsymbol{L}_{\text{p}}\dot{\boldsymbol{i}}_{\text{p,z}} \\ \boldsymbol{C}_{\text{p}}\dot{\boldsymbol{v}}_{\text{p}} \\ \boldsymbol{C}_{\text{f}}\dot{\boldsymbol{v}}_{\text{f}} \\ \boldsymbol{L}_{\text{t}}\dot{\boldsymbol{i}}_{\text{t}} \end{bmatrix} = \begin{bmatrix} \boldsymbol{0} & \boldsymbol{v}_{\text{p}}^T & \boldsymbol{0} & \boldsymbol{0} & \boldsymbol{0} \\ \boldsymbol{\vartheta}_{\varepsilon} & -\boldsymbol{R}_{\text{p}} - \boldsymbol{\vartheta}_{\text{i}} - \boldsymbol{v}_{\text{p}}^T\boldsymbol{\vartheta}_{\text{p}} & -\boldsymbol{I}_{|\mathcal{P}|} & \boldsymbol{0} & \boldsymbol{0} \\ \boldsymbol{0} & \boldsymbol{I}_{|\mathcal{P}|} & \boldsymbol{Y}_{\text{p}} & \boldsymbol{0} & \boldsymbol{E}_{\text{p}} \\ \boldsymbol{0} & \boldsymbol{0} & \boldsymbol{0} & \boldsymbol{Y}_{\text{f}} & \boldsymbol{E}_{\text{f}} \\ \boldsymbol{0} & \boldsymbol{0} & -\boldsymbol{E}_{\text{p}}^T & -\boldsymbol{E}_{\text{f}}^T & -\boldsymbol{R}_{\text{t}} \end{bmatrix} \begin{bmatrix} \varepsilon_{\text{p}} \\ \boldsymbol{i}_{\text{p,z}} - \boldsymbol{\delta}_{\text{p}}\boldsymbol{v}_{\text{p}} \\ \boldsymbol{v}_{\text{p}} \\ \boldsymbol{v}_{\text{f}} \\ \boldsymbol{i}_{\text{t}} \end{bmatrix} + \begin{bmatrix} \boldsymbol{K}_{\text{p}}\boldsymbol{\phi} \\ \boldsymbol{\vartheta}_{\text{p}}\boldsymbol{K}_{\text{p}}\boldsymbol{\phi} \\ \boldsymbol{i}_{\text{ex,p}} \\ \boldsymbol{i}_{\text{ex,f}} \\ \boldsymbol{0} \end{bmatrix}, \tag{5.13a}$$

$$\boldsymbol{i}_{\text{ex,p}} = \text{stack}_{j\in\mathcal{P}}(i_{\text{ex},j}), \quad \boldsymbol{i}_{\text{ex,f}} = \text{stack}_{k\in\mathcal{F}}(i_{\text{ex},k}), \tag{5.13b}$$

where ε_{p}, $\boldsymbol{i}_{\text{p,z}}$, $\boldsymbol{v}_{\text{p}}$, and $\boldsymbol{\phi}$ are the vectors obtained by stacking the states and inputs of the buses in $\mathcal{P}$, $\boldsymbol{v}_{\text{f}}$ is the stacked voltages for the buses in $\mathcal{F}$, and $\boldsymbol{i}_{\text{t}}$ is the stacked currents of the lines in $\mathcal{E}_{\text{el}}$. Similarly, $\boldsymbol{L}_{\text{p}}$, $\boldsymbol{C}_{\text{p}}$, $\boldsymbol{R}_{\text{p}}$, $\boldsymbol{Y}_{\text{p}}$, $\boldsymbol{C}_{\text{f}}$, $\boldsymbol{Y}_{\text{f}}$, $\boldsymbol{L}_{\text{t}}$, and $\boldsymbol{R}_{\text{t}}$ are diagonal matrices composed of L_j; $C_{\text{ef},j}$, R_j and Y_j for the buses $j \in \mathcal{P}$; $C_{\text{ef},k}$, R_k and Y_k for the buses $k \in \mathcal{F}$; and $L_{\text{t},o}$ and $R_{\text{t},o}$ for the lines $o \in \mathcal{E}_{\text{el}}$, respectively. The diagonal matrices $\boldsymbol{\vartheta}_{\varepsilon}$, $\boldsymbol{\vartheta}_{\text{i}}$, $\boldsymbol{\vartheta}_{\text{p}}$, $\boldsymbol{\delta}_{\text{p}}$ and $\boldsymbol{K}_{\text{p}}$ are similarly obtained by diagonally stacking the controller gains for the buses $j \in \mathcal{P}$.

Theorem 5.1 (Power-Sharing Controller Equilibrium)

Consider the DC network (5.13) *in closed loop with the four-stage distributed power-sharing controller described by* (5.7), (5.8), *and* (5.11). *Let Assumption 5.1 hold. The equilibrium of this closed loop satisfies Objective 5.2. Additionally, if ideal integrators are used for Stage 3 of the power-sharing controller, i.e.,* $\delta_{\text{u}} = 0$ *in* (5.11), *Objective 5.1 is met. If leaky PI controllers are used, the weighted average voltage error* $\boldsymbol{y}_{\text{a},2}$ *has a remaining error of*

$$\hat{\boldsymbol{y}}_{\text{a},2} = \frac{\delta_{\text{u}}}{k_{\text{u}}^{\text{I}} + \delta_{\text{u}} k_{\text{u}}^{\text{P}}} \hat{\boldsymbol{\phi}}, \tag{5.14}$$

which is dependent on the steady-state proportional power setpoint $\hat{\boldsymbol{\phi}}$.

Proof:
For the power-sharing controller in steady state, $\boldsymbol{y}_{\mathrm{u}} = \hat{\boldsymbol{y}}_{\mathrm{u}}$ is constant since $\dot{\boldsymbol{\varepsilon}}_{\mathrm{u}} = 0$. From the DDA equilibrium (5.9) for Stage 4, Objective 5.2 is achieved since the elements of $\hat{\boldsymbol{y}}_{\mathrm{a},4} = \hat{\boldsymbol{\phi}}^{\star}$ are identical. Furthermore, consider the steady state of the Stage 2 DDA

$$\hat{u}_{\mathrm{a},2,n} = h_{\mathrm{w}}(v^{\star} - \hat{v}_n), \quad n \in \mathcal{N}_{\mathrm{el}}, \tag{5.15a}$$

$$\hat{y}_{\mathrm{a},2,n} = \frac{\mathbb{1}^T_{|\mathcal{N}_{\mathrm{el}}|}\hat{\boldsymbol{u}}_{\mathrm{a},s}}{|\mathcal{N}_{\mathrm{el}}|} = \frac{1}{|\mathcal{N}_{\mathrm{el}}|}\sum_{n\in\mathcal{N}_{\mathrm{el}}} \left(v^{\star} - h(\hat{v}_n)\right), \tag{5.15b}$$

$$h(v_n) := v^{\star} - h_{\mathrm{w}}(v^{\star} - v_n), \tag{5.15c}$$

where $h(v_n)$ is obtained by shifting the Stage 1 weighting function h_{w} in (5.7). Thus, Objective 5.1 corresponds to ensuring that $\hat{\boldsymbol{y}}_{\mathrm{a},2} = \mathbf{0}$. Recalling the equilibrium of the Stage 3 leaky PI controllers in (5.12), it follows that $\hat{\boldsymbol{u}}_{\mathrm{u}} = \hat{\boldsymbol{y}}_{\mathrm{a},2} = \mathbf{0}$ for ideal integrators with $\delta_{\mathrm{u}} = 0$. For leaky integrators, substituting (5.12) into (5.11b) results in

$$\hat{\boldsymbol{y}}_{\mathrm{u}} = \frac{k^{\mathrm{I}}_{\mathrm{u}}}{\delta_{\mathrm{u}}}\hat{\boldsymbol{u}}_{\mathrm{u}} + k^{\mathrm{P}}_{\mathrm{u}}\hat{\boldsymbol{u}}_{\mathrm{u}} \implies \hat{\boldsymbol{u}}_{\mathrm{a},4} = \frac{k^{\mathrm{I}}_{\mathrm{u}} + \delta_{\mathrm{u}}k^{\mathrm{P}}_{\mathrm{u}}}{\delta_{\mathrm{u}}}\hat{\boldsymbol{u}}_{\mathrm{a},2}, \tag{5.16}$$

from which (5.14) follows. ∎

Theorem 5.1 thus confirms that the four-stage distributed power-sharing controller in Figure 5.3 meets Objectives 5.1 and 5.2 thus fulfils RQ 2.1 if ideal PI controllers are used in Stage 3 of the controller. Moreover, Theorem 5.1 also provides a formulation for the remaining average weighted voltage error in the case that leaky PI controllers are used in Stage 3.

Remark 5.3. *The proposed control structure implicitly assumes, through the PI controller in Stage 3 in particular, that a positive voltage error should be compensated by increasing the proportional power setpoint. This assumption is guaranteed in this chapter by restricting the loads to monotonically increasing functions. Notably, the exact same assumption is made in droop-based approaches, where the injected current or power is increased for a positive voltage error (see [Gao+19; Mod+23]).*

Remark 5.4. *The remaining error in* (5.14)*, which arises when using leaky PI controllers for Stage 3, is directly proportional to the power injected by the power regulated buses in $\mathcal{P}$. Moreover, $\hat{\boldsymbol{y}}_{\mathrm{a},2} > 0$ simultaneously corresponds to an injection of power as well as bus voltages which are, on average, below the desired setpoint $v^{\star}$ (see* (5.15)*). Thus, increasing the $v^{\star}$ given to the distributed power-sharing controller in Figure 5.3 also increases $\hat{\boldsymbol{y}}_{\mathrm{a},2}$ and, in turn, $\hat{\boldsymbol{\phi}}$. With more power being injected, the bus voltage will increase.[4] Thus, the error arising due to the use of leaky PI controllers can be offset by modifying the common $v^{\star}$.*

[4] Recall that strictly monotone incremental impedances are assumed for the ZIP loads in this chapter.

5.2 Passivity and Stability Analysis

The DC network (5.13) in closed loop with the four-stage distributed power-sharing controller depicted in Figure 5.3 is shown to produce a steady state which fulfils RQ 2.1 through Theorem 5.1. Next, the stability of the closed loop, i.e., the ability of the controller to steer the closed loop towards that steady state is investigated, as required by RQ 2.2.

The stability investigation in this section relies on a passivity-based approach in which a supply rate for the entire DC network is constructed and subsequently combined with the supply rates of the distributed power-sharing controller stages. To this end, an IFOF-EIP result is established in **Section 5.2.1** for the buses in $\mathcal{P}$ where steady-state power is available. This result is then used to construct an IFOF-EIP property for the DC network in **Section 5.2.2**. Thereafter, the stages of the power-sharing controller are analysed in **Section 5.2.3**. The results are then combined in **Section 5.2.4**, where the AS of the closed-loop DC network equilibria is investigated. Since several passivity indices corresponding to IFOF-EIP supply rates used in this section, an overview of these indices is provided in Table 5.1.

Table 5.1: DC Network Distributed Coordination Passivity Indices

Component	Indices	Derived via
Buses in $\mathcal{P}$	$\nu_{\mathrm{p},1}, \nu_{\mathrm{p},2}, \rho_{\mathrm{p}} \in \mathbb{R}$	Theorem 5.2, worse case in (5.26)
Buses in $\mathcal{F}$	$\rho_{\mathrm{f}} > 0$	Lemma 3.6, worse case in (5.26)
Loads and P sources in $\mathcal{N}_{\mathrm{el}}$	$\rho_{\mathrm{l}} \geq 0$	Lemma 3.5, worse case in (5.26)
Lines in $\mathcal{E}_{\mathrm{el}}$	$\rho_{\mathrm{t}} > 0$	Proposition 4.6, worse case in (5.26)
DC Network	$\nu_{\mathrm{p},1}, \rho_{\mathrm{p}} \in \mathbb{R}$	Theorem 5.5, from worst-case bus indices in $\mathcal{P}$
Weighing Function	$\nu_{\mathrm{w}}, \rho_{\mathrm{w}} \in \mathbb{R}$	Proposition 5.6, control parameter dependent
DDA	$\rho_{\mathrm{a}} = 1$	Proposition 5.7
Leaky PI	$\nu_{\mathrm{u}} > 0, \rho_{\mathrm{u}} > 0$	Proposition 5.8, control parameter dependent

5.2.1 Power Regulated Bus IFOF-EIP

Consider the nonlinear power regulated bus dynamics (5.4a) for a bus $j \in \mathcal{P}$ which seeks to regulate the power $p_j = v_j i_j$. Furthermore, notice that the regulated bus has two inputs, ϕ_j and $i_{\mathrm{int},j} + i_{\mathrm{ex},j}$. The former forms the port pair (ϕ_j, v_j) which interfaces with the distributed power-sharing controller in Figure 5.3, whereas the latter forms the electrical power $(i_{\mathrm{int},j} + i_{\mathrm{ex},j}, v_j)$ which interfaces with the lines in the DC network and the ZIP load and P source at the bus. The following theorem investigates an EIP property for a power regulated bus w.r.t. both of these ports. Note that for simplicity, the bus index j is dropped in this subsection.

Theorem 5.2 (IFOF-EIP Power Regulated Buses)
Consider the power regulated bus dynamics (5.4a) for a bus $j \in \mathcal{P}$ and let Assumption 3.1 hold. Then, the power regulated bus is simultaneously IFOF-EIP w.r.t. the input-output port $(\tilde{\phi}, \tilde{v})$ and EIP w.r.t. the input-output port $(\tilde{i}_{\text{int}} + \tilde{i}_{\text{ex}}, \tilde{v})$, i.e., it is EID w.r.t. the supply rate

$$w_{\text{p}} = (1 + \nu_{\text{p},1}\rho_{\text{p}})\tilde{\phi}\tilde{v} - \nu_{\text{p},1}\tilde{\phi}^2 - \rho_{\text{p}}\tilde{v}^2 + (\tilde{i}_{\text{int}} + \tilde{i}_{\text{ex}})\tilde{v} - \nu_{\text{p},2}(\tilde{i}_{\text{int}}^2 + \tilde{i}_{\text{ex}}^2), \tag{5.17}$$

and the storage function

$$S_{\text{p}}(\tilde{\boldsymbol{x}}_{\text{p}}) = \begin{bmatrix} \tilde{\varepsilon} \\ \tilde{i}_{\text{z}} \\ \tilde{v} \end{bmatrix}^T \boldsymbol{P}_{\text{p}} \begin{bmatrix} \tilde{\varepsilon} \\ L\tilde{i}_{\text{z}} \\ C_{\text{ef}}\tilde{v} \end{bmatrix}, \tag{5.18}$$

if a feasible solution can be found for the optimisation problem

$$\max_{\boldsymbol{P}_{\text{p}}, \nu_{\text{p},1}, \nu_{\text{p},2}, \rho_{\text{p}}} \quad \nu_{\text{p},1} + \nu_{\text{p},2} + \rho_{\text{p}} \tag{5.19a}$$

$$\text{s.t.} \quad \begin{bmatrix} \boldsymbol{D}_{\text{p}} + \rho_{\text{p}}\boldsymbol{b}_{\text{p},2}\boldsymbol{b}_{\text{p},2}^T & \boldsymbol{P}_{\text{p}}\boldsymbol{b}_{\text{p},1} - \frac{1+\nu_{\text{p},1}\rho_{\text{p}}}{2}\boldsymbol{b}_{\text{p},2} & \boldsymbol{P}_{\text{p}}\boldsymbol{b}_{\text{p},2} - \frac{1}{2}\boldsymbol{b}_{\text{p},2} \\ \boldsymbol{b}_{\text{p},1}^T\boldsymbol{P}_{\text{p}} - \frac{1+\nu_{\text{p},1}\rho_{\text{p}}}{2}\boldsymbol{b}_{\text{p},2}^T & \nu_{\text{p},1} & 0 \\ \boldsymbol{b}_{\text{p},2}^T\boldsymbol{P}_{\text{p}} - \frac{1}{2}\boldsymbol{b}_{\text{p},2}^T & 0 & \nu_{\text{p},2} \end{bmatrix} \preccurlyeq 0,$$

$$\boldsymbol{P}_{\text{p}} \succ 0, \quad \forall v \in \mathcal{V} \subseteq \mathbb{R}, \; i_{\text{z}} \in \mathcal{I} \subseteq \mathbb{R}, \; k_{\text{p}} \in [\underline{k}_{\text{p}}, \overline{k}_{\text{p}}], \tag{5.19b}$$

with $\nu_{\text{p},1}, \nu_{\text{p},2} \in \mathbb{R}$, $\rho_{\text{p}} \in \mathbb{R}_{>0}$ and where

$$\boldsymbol{D}_{\text{p}} = \boldsymbol{P}_{\text{p}}\nabla\boldsymbol{f}_{\text{p}}(v, i_{\text{z}}) + \nabla\boldsymbol{f}_{\text{p}}^T(v, i_{\text{z}})\boldsymbol{P}_{\text{p}}, \tag{5.20a}$$

$$\nabla\boldsymbol{f}_{\text{p}}(v, i_{\text{z}}) = \begin{bmatrix} 0 & v & i_{\text{z}} - 2\delta_{\text{p}}v \\ -\vartheta_{\varepsilon} & -R - \vartheta_{\text{i}} - \vartheta_{\text{p}}v & \delta_{\text{p}}(R + \vartheta_{\text{i}} + 2\vartheta_{\text{p}}v) - 1 - \vartheta_{\text{p}}i_{\text{z}} \\ 0 & 1 & Y - \delta \end{bmatrix}, \tag{5.20b}$$

$$\boldsymbol{b}_{\text{p},1}^T = \begin{bmatrix} k_{\text{p}} & \vartheta_{\text{p}}k_{\text{p}} & 0 \end{bmatrix}, \qquad \boldsymbol{b}_{\text{p},2}^T = \begin{bmatrix} 0 & 0 & 1 \end{bmatrix}. \tag{5.20c}$$

Proof:
Recall from [SP19, Remark 3.5], [AMP16, p. 26], and Lemma 3.6 that EIP can be investigated by using the Jacobian of the system dynamics for nonlinear systems. With this as a basis, use the Jacobian $\nabla\boldsymbol{f}_{\text{p}}$ (5.20b) of the bus dynamics in (5.3) to calculate the time derivative of the storage function (5.18) as

$$\dot{S}_{\text{p}} = \begin{bmatrix} \tilde{\boldsymbol{x}}_{\text{p}} \\ \tilde{\phi} \\ \tilde{i}_{\text{ex}} + \tilde{i}_{\text{int}} \end{bmatrix}^T \begin{bmatrix} \boldsymbol{D}_{\text{p}} & \boldsymbol{P}_{\text{p}}\boldsymbol{b}_{\text{p},1} & \boldsymbol{P}_{\text{p}}\boldsymbol{b}_{\text{p},2} \\ \boldsymbol{b}_{\text{p},1}^T\boldsymbol{P}_{\text{p}} & 0 & 0 \\ \boldsymbol{b}_{\text{p},2}^T\boldsymbol{P}_{\text{p}} & 0 & 0 \end{bmatrix} \begin{bmatrix} \tilde{\boldsymbol{x}}_{\text{p}} \\ \tilde{\phi} \\ \tilde{i}_{\text{ex}} + \tilde{i}_{\text{int}} \end{bmatrix}, \tag{5.21}$$

with $\boldsymbol{D}_\mathrm{p}$, $\boldsymbol{b}_{\mathrm{p},1}$ and $\boldsymbol{b}_{\mathrm{p},2}$ as defined in (5.20). Testing for EID w.r.t. w_p in (5.17) results in

$$\dot{S}_\mathrm{p} - w_\mathrm{p} = \begin{bmatrix} \tilde{\boldsymbol{x}}_\mathrm{p} \\ \tilde{\phi} \\ \tilde{i}_\mathrm{ex} + \tilde{i}_\mathrm{int} \end{bmatrix}^T \begin{bmatrix} \boldsymbol{D}_\mathrm{p} + \rho_\mathrm{p} \boldsymbol{b}_{\mathrm{p},2} \boldsymbol{b}_{\mathrm{p},2}^T & \boldsymbol{P}_\mathrm{p} \boldsymbol{b}_{\mathrm{p},1} - \frac{1+\nu_{\mathrm{p},1}\rho_\mathrm{p}}{2} \boldsymbol{b}_{\mathrm{p},2} & \boldsymbol{P}_\mathrm{p} \boldsymbol{b}_{\mathrm{p},2} - \frac{1}{2} \boldsymbol{b}_{\mathrm{p},2} \\ \boldsymbol{b}_{\mathrm{p},1}^T \boldsymbol{P}_\mathrm{p} - \frac{1+\nu_{\mathrm{p},1}\rho_\mathrm{p}}{2} \boldsymbol{b}_{\mathrm{p},2}^T & \nu_{\mathrm{p},1} & 0 \\ \boldsymbol{b}_{\mathrm{p},2}^T \boldsymbol{P}_\mathrm{p} - \frac{1}{2} \boldsymbol{b}_{\mathrm{p},2}^T & 0 & \nu_{\mathrm{p},2} \end{bmatrix} \begin{bmatrix} \tilde{\boldsymbol{x}}_\mathrm{p} \\ \tilde{\phi} \\ \tilde{i}_\mathrm{ex} + \tilde{i}_\mathrm{int} \end{bmatrix}. \tag{5.22}$$

The desired EID property for the power regulated bus thus follows if there is a matrix $\boldsymbol{P}_\mathrm{p} \succ 0$ along with passivity indices $\nu_{\mathrm{p},1}$, $\nu_{\mathrm{p},2}$ and ρ_p such that $\dot{S}_\mathrm{p} - w_\mathrm{p} \leq 0$ for all valid states. Casting this as an optimisation problem, where the largest and thus least conservative passivity indices are desired subject to the inequality in (5.22) being negative for all valid states, directly yields the problem in (5.19). ■

Theorem 5.2 allows EIP properties to be determined for a power regulated bus w.r.t. the port connecting to the power-sharing controller and the electrical port connecting to other components in the DC network. Note that since the supply rate in (5.17) does not contain all the states in $\tilde{\boldsymbol{x}}_\mathrm{p}$, an EISO property for the power regulated bus is required.

Proposition 5.3 (EISO Power Regulated Buses)
The power regulated bus dynamics (5.3) *for which Assumption 3.1 holds are EISO.*

Proof:
Consider (5.3) shifted w.r.t. the assumed equilibrium and let $\tilde{\phi} \equiv 0$, $\tilde{i}_\mathrm{int} \equiv 0$, $\tilde{i}_\mathrm{ex} \equiv 0$ as well as $\tilde{v} \equiv 0$. From the voltage dynamics, it follows that

$$C_\mathrm{ef} \dot{\tilde{v}} = \tilde{i} + Y\tilde{v} + \tilde{i}_\mathrm{int} + \tilde{i}_\mathrm{ex} \implies \tilde{i} \equiv 0. \tag{5.23}$$

The filter current dynamics then result in

$$L\dot{\tilde{i}} = -\vartheta_\varepsilon \tilde{\varepsilon} - (R - \vartheta_\mathrm{i})\tilde{i} - \tilde{v} - \vartheta_\mathrm{p}(\tilde{v}\tilde{i} + \tilde{v}\hat{i} + \hat{v}\tilde{i}) - \frac{\delta_{\mathrm{p},j} L_j}{C_j} \tilde{i}_{\mathrm{c},j} + \vartheta_\mathrm{p} k_\mathrm{p} \tilde{\phi} \implies \tilde{\varepsilon} \equiv 0, \tag{5.24}$$

since $\tilde{i}_{\mathrm{c},j} \equiv 0$ if $\dot{\tilde{v}} \equiv 0$. ■

5.2.2 DC Network IFOF-EIP

Using the IFOF-EIP result obtained by solving the optimisation problem in Theorem 5.2, a combined supply rate can now be constructed for the entire DC network w.r.t. the input $\tilde{\boldsymbol{\phi}}$ and the output $\tilde{\boldsymbol{v}}$, which describes the port used for the distributed power-sharing controller in Figure 5.3. The port variables are grouped by bus type, i.e.,

$$\tilde{\boldsymbol{\phi}}_{\mathrm{p}} = \mathrm{stack}_{j\in\mathcal{P}}(\tilde{\phi}_j), \qquad \tilde{\boldsymbol{\phi}}_{\mathrm{f}} = \mathrm{stack}_{k\in\mathcal{F}}(\tilde{\phi}_k), \qquad \tilde{\boldsymbol{\phi}} = \mathrm{stack}(\tilde{\boldsymbol{\phi}}_{\mathrm{p}}, \tilde{\boldsymbol{\phi}}_{\mathrm{f}}), \tag{5.25a}$$

$$\tilde{\boldsymbol{v}}_{\mathrm{p}} = \mathrm{stack}_{j\in\mathcal{P}}(\tilde{v}_j), \qquad \tilde{\boldsymbol{v}}_{\mathrm{f}} = \mathrm{stack}_{k\in\mathcal{F}}(\tilde{v}_k), \qquad \tilde{\boldsymbol{v}} = \mathrm{stack}(\tilde{\boldsymbol{v}}_{\mathrm{p}}, \tilde{\boldsymbol{v}}_{\mathrm{f}}). \tag{5.25b}$$

Although $\tilde{\boldsymbol{\phi}}_{\mathrm{f}}$ is not used by any bus controllers, it is nevertheless calculated by the Stage 4 DDA and therefore used in the EIP analysis in this section.

Since the components in a DC network may have differing component or parameter values, the worst-case indices for each component type are considered for the supply rate of the DC network. Recalling the summary in Table 5.1, worst-case indices for the bus and line components are

$$\rho_{\mathrm{l}} = \min_{n\in\mathcal{N}_{\mathrm{el}}} \rho_{\mathrm{l},n}, \qquad \rho_{\mathrm{f}} = -\max_{k\in\mathcal{F}} Y_k, \qquad \rho_{\mathrm{t}} = \min_{o\in\mathcal{E}_{\mathrm{el}}} R_{\mathrm{t},o}, \tag{5.26a}$$

$$\nu_{\mathrm{p},1} = \min_{j\in\mathcal{P}} \nu_{\mathrm{p},1,j}, \qquad \nu_{\mathrm{p},2} = \min_{j\in\mathcal{P}} \nu_{\mathrm{p},2,j}, \qquad \rho_{\mathrm{p}} = \min_{j\in\mathcal{P}} \rho_{\mathrm{p},j}, \tag{5.26b}$$

where $\rho_{\mathrm{l},n}$ describes the output passivity index for the sum of the strictly monotone ZIP load function $i^{\mathrm{m}}_{\mathrm{l},n}$ and the monotone P source function $i_{\mathrm{ps},n}$.[5]

Proposition 5.4 (DC Network EIP)

Consider a DC network comprising power regulated buses (5.3) *in $\mathcal{P}$, load buses* (5.1) *in $\mathcal{F}$, lines* (4.7) *in $\mathcal{E}_{\mathrm{el}}$, along with ZIP loads* (4.1) *and P sources* (4.5) *at each bus in $\mathcal{N}_{\mathrm{el}} = \mathcal{P} \cup \mathcal{F}$. Let the connected graph $\mathcal{G}_{\mathrm{el}}$ describe the interconnection of the buses via the lines and let Assumptions 3.1 and 5.1 hold for the network. Furthermore, let each bus $j \in \mathcal{P}$ be dissipative w.r.t. $w_{\mathrm{p},j}$ in* (5.17). *Then, the DC network with the storage function*

$$S_{\mathrm{el}} = \sum_{j\in\mathcal{P}} S_{\mathrm{p},j} + \sum_{k\in\mathcal{F}} S_{\mathrm{f},k} + \sum_{o\in\mathcal{E}_{\mathrm{el}}} S_{\mathrm{t},o}, \tag{5.27}$$

with $S_{\mathrm{p},j}$ in (5.18), *$S_{\mathrm{f},k} = \frac{1}{2} C_{\mathrm{ef},k} \tilde{v}_k^2$, and $S_{\mathrm{t},o} = \frac{1}{2} L_{\mathrm{t},o} \tilde{i}_{\mathrm{t},o}^2$ is dissipative w.r.t.*

$$w_{\mathrm{el,pf}} = (1 + \nu_{\mathrm{p},1}\rho_{\mathrm{p}}) \tilde{\boldsymbol{\phi}}_{\mathrm{p}}^T \tilde{\boldsymbol{v}}_{\mathrm{p}} - \nu_{\mathrm{p},1} \tilde{\boldsymbol{\phi}}_{\mathrm{p}}^T \tilde{\boldsymbol{\phi}}_{\mathrm{p}} - \rho_{\mathrm{p}} \tilde{\boldsymbol{v}}_{\mathrm{p}} \tilde{\boldsymbol{v}}_{\mathrm{p}} - \rho_{\mathrm{f}} \tilde{\boldsymbol{v}}_{\mathrm{f}}^T \tilde{\boldsymbol{v}}_{\mathrm{f}}, \tag{5.28}$$

if $\nu_{\mathrm{p},2} + \min(\rho_{\mathrm{t}}, \rho_{\mathrm{l}}) \geq 0$ holds for the worst-case indices in (5.26).

Proof:
Recall from Section 5.1.1 that the ZIP loads are strictly monotonically increasing with $Y_n < 0$ for $n \in \mathcal{N}_{\mathrm{el}}$. Thus, the load buses $k \in \mathcal{F}$ (5.1) with $S_{\mathrm{f},k}$ are OS-EIP with $\rho_{\mathrm{f},k} = -Y_k$ w.r.t. the input-output port $(\tilde{i}_{\mathrm{int},k} + \tilde{i}_{\mathrm{ex},k}, \tilde{v}_k)$ via Lemma 3.6. Additionally,

[5] Recall that $Y_n < 0$ in this chapter, thus $\rho_{\mathrm{f}} > 0$.

recall the supply rates for the power regulated buses in Theorem 5.2 and the lines in Proposition 4.6. An upper bound for $\dot{S}_{\mathrm{el}}$ can then be found by combining the EIP results of the lines and the buses, yielding

$$\begin{aligned}\dot{S}_{\mathrm{el}} \leq &(1+\nu_{\mathrm{p},1}\rho_{\mathrm{p}})\tilde{\boldsymbol{\phi}}_{\mathrm{p}}^T\tilde{\boldsymbol{v}}_{\mathrm{p}} - \nu_{\mathrm{p},1}\tilde{\boldsymbol{\phi}}_{\mathrm{p}}^T\tilde{\boldsymbol{\phi}}_{\mathrm{p}} - \rho_{\mathrm{p}}\tilde{\boldsymbol{v}}_{\mathrm{p}}^T\tilde{\boldsymbol{v}}_{\mathrm{p}} + \tilde{\boldsymbol{v}}_{\mathrm{p}}^T\boldsymbol{E}_{\mathrm{p}}\tilde{\boldsymbol{i}}_{\mathrm{t}} + \tilde{\boldsymbol{i}}_{\mathrm{ex,p}}^T\tilde{\boldsymbol{v}}_{\mathrm{p}} - \nu_{\mathrm{p},2}\tilde{\boldsymbol{i}}_{\mathrm{ex,p}}^T\tilde{\boldsymbol{i}}_{\mathrm{ex,p}}\\ &+ \tilde{\boldsymbol{v}}_{\mathrm{f}}^T\boldsymbol{E}_{\mathrm{f}}\tilde{\boldsymbol{i}}_{\mathrm{t}} - \rho_{\mathrm{f}}\tilde{\boldsymbol{v}}_{\mathrm{f}}^T\tilde{\boldsymbol{v}}_{\mathrm{f}} + \tilde{\boldsymbol{i}}_{\mathrm{ex,f}}^T\tilde{\boldsymbol{v}}_{\mathrm{f}}\\ &- \tilde{\boldsymbol{i}}_{\mathrm{t}}^T\boldsymbol{E}_{\mathrm{p}}^T\tilde{\boldsymbol{v}}_{\mathrm{p}} - \tilde{\boldsymbol{i}}_{\mathrm{t}}^T\boldsymbol{E}_{\mathrm{f}}^T\tilde{\boldsymbol{v}}_{\mathrm{f}} - (\nu_{\mathrm{p},2}+\rho_{\mathrm{t}})\tilde{\boldsymbol{i}}_{\mathrm{t}}^T\tilde{\boldsymbol{i}}_{\mathrm{t}},\end{aligned} \tag{5.29}$$

with the worst-case passivity indices from (5.26). Note that the cross terms between the bus voltages $\tilde{\boldsymbol{v}}_{\mathrm{el}}$ and the line currents $\tilde{\boldsymbol{i}}_{\mathrm{t}}$ in (5.29) sum to zero. The sum of the monotone static load and monotone source functions $i_{\mathrm{l},n}^{\mathrm{m}} + i_{\mathrm{ps},n}$ is itself monotone and thus OF-EIP with $\rho_{\mathrm{l}} \geq 0$. Additionally, since $\tilde{i}_{\mathrm{ex},n} = -\tilde{i}_{\mathrm{l},n}^{\mathrm{m}} - \tilde{i}_{\mathrm{ps},n}$ for $n \in \mathcal{N}_{\mathrm{el}}$, it holds that

$$0 \leq -\tilde{\boldsymbol{i}}_{\mathrm{ex,p}}^T\tilde{\boldsymbol{v}}_{\mathrm{p}} - \rho_{\mathrm{l}}\tilde{\boldsymbol{i}}_{\mathrm{ex,p}}^T\tilde{\boldsymbol{i}}_{\mathrm{ex,p}}, \qquad 0 \leq -\tilde{\boldsymbol{i}}_{\mathrm{ex,f}}^T\tilde{\boldsymbol{v}}_{\mathrm{f}} - \rho_{\mathrm{l}}\tilde{\boldsymbol{i}}_{\mathrm{ex,f}}^T\tilde{\boldsymbol{i}}_{\mathrm{ex,f}}, \tag{5.30}$$

which is substituted into (5.29) to obtain

$$\begin{aligned}\dot{S}_{\mathrm{el}} \leq &(1+\nu_{\mathrm{p},1}\rho_{\mathrm{p}})\tilde{\boldsymbol{\phi}}_{\mathrm{p}}^T\tilde{\boldsymbol{v}}_{\mathrm{p}} - \nu_{\mathrm{p},1}\tilde{\boldsymbol{\phi}}_{\mathrm{p}}^T\tilde{\boldsymbol{\phi}}_{\mathrm{p}} - \rho_{\mathrm{p}}\tilde{\boldsymbol{v}}_{\mathrm{p}}^T\tilde{\boldsymbol{v}}_{\mathrm{p}} - (\nu_{\mathrm{p},2}+\rho_{\mathrm{l}})\tilde{\boldsymbol{i}}_{\mathrm{ex,p}}^T\tilde{\boldsymbol{i}}_{\mathrm{ex,p}}\\ &- \rho_{\mathrm{f}}\tilde{\boldsymbol{v}}_{\mathrm{f}}^T\tilde{\boldsymbol{v}}_{\mathrm{f}} - \rho_{\mathrm{l}}\tilde{\boldsymbol{i}}_{\mathrm{ex,f}}^T\tilde{\boldsymbol{i}}_{\mathrm{ex,f}} - (\nu_{\mathrm{p},2}+\rho_{\mathrm{t}})\tilde{\boldsymbol{i}}_{\mathrm{t}}^T\tilde{\boldsymbol{i}}_{\mathrm{t}}.\end{aligned} \tag{5.31}$$

If $\nu_{\mathrm{p},2} + \rho_{\mathrm{t}} \geq 0$ and $\nu_{\mathrm{p},2} + \rho_{\mathrm{l}} \geq 0$, (5.31) verifies that $\dot{S}_{\mathrm{el}} \leq w_{\mathrm{el,pf}}$. ■

In Proposition 5.4, an EIP property for the entire DC network w.r.t. the port used by the power-sharing controller is established. The supply rate in (5.28), however, is dependent on the availability of steady-state power at the buses, with the buses in $\mathcal{P}$ and those in $\mathcal{F}$ exhibiting different EIP properties. In the next theorem, the supply rate in (5.28) is modified to obtain a supply rate that is uniform for all buses in $\mathcal{N}_{\mathrm{el}} = \mathcal{P} \cup \mathcal{F}$ by comparing the supply rates for the buses in $\mathcal{P}$ and $\mathcal{F}$ (see Remark 3.4).

Theorem 5.5 (Power-Availability Independent DC Network IFOF-EIP) *Consider the DC network in Proposition 5.4 with the storage function in (5.27) that is dissipative w.r.t. the supply rate $w_{\mathrm{el,pf}}$ in (5.28) and where $\rho_{\mathrm{t}} > 0$. If $\nu_{\mathrm{p},1} \leq 0$, then the DC network is also dissipative w.r.t. the IFOF-EIP supply rate*

$$w_{\mathrm{el}} = (1+\nu_{\mathrm{p},1}\rho_{\mathrm{p}})\tilde{\boldsymbol{\phi}}^T\tilde{\boldsymbol{v}} - \nu_{\mathrm{p},1}\tilde{\boldsymbol{\phi}}^T\tilde{\boldsymbol{\phi}} - \rho_{\mathrm{p}}\tilde{\boldsymbol{v}}^T\tilde{\boldsymbol{v}}. \tag{5.32}$$

Proof:

Split the supply rate for the DC network $w_{\mathrm{el,pf}}$ in (5.28) into the components

$$w_{\mathrm{el,pf}} = \sum_{j\in\mathcal{P}} w_{\mathrm{el,p},j} + \sum_{k\in\mathcal{F}} w_{\mathrm{el,f},k}, \quad \begin{cases} w_{\mathrm{el,p},j} = (1+\nu_{\mathrm{p},1}\rho_{\mathrm{p}})\tilde{\phi}_j\tilde{v}_j - \nu_{\mathrm{p},1}\tilde{\phi}_j^2 - \rho_{\mathrm{p}}\tilde{v}_j^2,\\ w_{\mathrm{el,f},k} = -\rho_{\mathrm{f}}\tilde{v}_k^2,\end{cases} \tag{5.33}$$

with the goal of showing that $w_{\mathrm{el,f},n} \leq w_{\mathrm{el,p},n}$ for $n \in \mathcal{N}_{\mathrm{el}}$. Recall from Remark 3.4 that supply rates can be compared via their sectors, i.e., $w_{\mathrm{el,f},n} \leq w_{\mathrm{el,p},n}$ if

$$\begin{aligned} & \left\{(\tilde{\phi}_n, \tilde{v}_n) \mid 0 \leq w_{\mathrm{el,f},n}\right\} \subseteq \left\{(\tilde{\phi}_n, \tilde{v}_n) \mid 0 \leq w_{\mathrm{el,p},n}\right\} \\ \Longrightarrow \quad & \left\{(\tilde{\phi}_n, \tilde{v}_n) \mid \tilde{v}_n = 0, \tilde{\phi}_n \in \mathbb{R}\right\} \subseteq \left\{(\tilde{\phi}_n, \tilde{v}_n) \mid 0 \leq w_{\mathrm{el,p},n}\right\}, \end{aligned} \tag{5.34}$$

where $\rho_{\mathrm{f}} > 0$. Since $w_{\mathrm{el,p},n}$ is an IFOF-EIP supply rate, the set comparison in (5.34) holds if the sector $0 \leq w_{\mathrm{el,p},n}$ includes the input $\tilde{\phi}_n$ axis. Thus, $\nu_{\mathrm{p},1} \leq 0$ ensures that $w_{\mathrm{el,f},n} \leq w_{\mathrm{el,p},n}$ holds for each bus. Substituting this into (5.33) yields

$$\dot{S}_{\mathrm{el}} \leq w_{\mathrm{el,pf}} = \sum_{j \in \mathcal{P}} w_{\mathrm{el,p},j} + \sum_{k \in \mathcal{F}} w_{\mathrm{el,f},k} \leq \sum_{n \in \mathcal{N}_{\mathrm{el}}} w_{\mathrm{el,p},n} = w_{\mathrm{el}}, \tag{5.35}$$

for the DC network with the storage function S_{el}, with w_{el} defined in (5.32). ■

Through Theorem 5.5, the previous EIP property of a DC network established in Theorem 5.2 is changed into an IFOF-EIP property that represents a more conservative supply rate. Nevertheless, the obtained supply rate w_{el} is independent of the availability of steady-state power at the buses in the DC network and thus remains valid for any bus in $\mathcal{P} \cup \mathcal{F}$. Moreover, apart from the conditions in Theorems 5.2 and 5.5, w_{el} is independent of the network size and its topology. The IFOF-EIP result with the supply rate w_{el} is also inherently robust against any parameter changes at the buses, lines, ZIP loads and P sources if these changes do not affect the worst-case indices in (5.26).

Remark 5.5. *A DC network with the input-output port* $(\tilde{\boldsymbol{\phi}}, \tilde{\boldsymbol{v}})$ *which is dissipative w.r.t. the supply rate in* (5.32) *is also EISO since: 1) the buses in* $\mathcal{P}$ *are EISO through Proposition 5.3; and 2) the outputs of the buses in* $\mathcal{F}$ *and the lines in* $\mathcal{E}_{\mathrm{el}}$ *are equal to the respective state vectors (see* (5.1) *and Proposition 4.6).*

5.2.3 Power-Sharing Controller

Having established a supply rate for the DC network using the port $(\tilde{\boldsymbol{\phi}}, \tilde{\boldsymbol{v}})$ which interconnects the DC network with the distributed power-sharing controller in Figure 5.3, the EIP properties of the stages of the power-sharing controller are investigated in this section, starting with the nonlinear weighting function and then continuing with the DDA and the leaky PI controller stages.

Weighting Function IFOF-EIP

Starting with the first stage of the power-sharing controller, an IFOF-EIP property is established for the vector form of the nonlinear weighting function

$$\tilde{\boldsymbol{y}}_{\mathrm{w}} = \tilde{\boldsymbol{h}}_{\mathrm{w}}(\tilde{\boldsymbol{u}}_{\mathrm{w}}), \tag{5.36}$$

which is obtained by stacking (5.7) for the buses $n \in \mathcal{N}_{\mathrm{el}}$ and shifting as in (3.17b).

Proposition 5.6 (Weighting Function IFOF-EIP)
The nonlinear weighting function (5.36) *with* $k_{\mathrm{w},2} > -k_{\mathrm{w},1}$ *is IFOF-EIP w.r.t.*

$$w_{\mathrm{w}} = (1 + \nu_{\mathrm{w}}\rho_{\mathrm{w}})\tilde{\boldsymbol{u}}_{\mathrm{w}}^T\tilde{\boldsymbol{y}}_{\mathrm{w}} - \nu_{\mathrm{w}}\tilde{\boldsymbol{u}}_{\mathrm{w}}^T\tilde{\boldsymbol{u}}_{\mathrm{w}} - \rho_{\mathrm{w}}\tilde{\boldsymbol{y}}_{\mathrm{w}}^T\tilde{\boldsymbol{y}}_{\mathrm{w}}, \tag{5.37a}$$

$$\nu_{\mathrm{w}} = k_{\mathrm{w},1}, \quad \rho_{\mathrm{w}} = \frac{1}{k_{\mathrm{w},1} + k_{\mathrm{w},2}}. \tag{5.37b}$$

Proof:
Since the vector form in (5.36) is found by stacking the weighting functions implemented at each bus, the IFOF-EIP of (5.36) can be determined from the scalar case (5.7), which has the derivative

$$\frac{\mathrm{d}y_{\mathrm{w}}}{\mathrm{d}u_{\mathrm{w}}} = k_{\mathrm{w},1} + k_{\mathrm{w},2}\tanh^2(k_{\mathrm{w},3}(u_{\mathrm{w}})). \tag{5.38}$$

By applying Lemma 3.5, the IFOF-EIP of (5.36) is verified with $0 \leq w_{\mathrm{w}}$ for the passivity indices in (5.37b). ■

Dynamic Distributed Averaging OS-EIP

For the DDA stages (5.8), an L_2 result is presented in [FYL06] for the input $\boldsymbol{u}_{\mathrm{a},s}$ and the output $\boldsymbol{y}_{\mathrm{a},s}$. Moreover, it is demonstrated in [FYL06] that the DDA has an unobservable and uncontrollable mode $z_{\mathrm{a},0,s} \in \mathbb{R}$ with zero dynamics, with

$$\boldsymbol{\chi}_{\mathrm{a},s} = \begin{bmatrix}\boldsymbol{\xi}_{\mathrm{a}} & \boldsymbol{\Xi}_{\mathrm{a}}\end{bmatrix}\begin{bmatrix} z_{\mathrm{a},0,s} \\ \boldsymbol{z}_{\mathrm{a},\chi,s}\end{bmatrix}, \tag{5.39}$$

and where $\boldsymbol{\xi}_{\mathrm{a}} \in \mathbb{R}^{|\mathcal{N}_{\mathrm{el}}|\times 1}$ denotes the left unit eigenvector of $\boldsymbol{\mathcal{L}}_{\mathrm{com}}$ such that $\boldsymbol{\mathcal{L}}_{\mathrm{com}}^T\boldsymbol{\xi}_{\mathrm{a}} = \boldsymbol{0}$. The remaining left unit eigenvectors $\boldsymbol{\Xi}_{\mathrm{a}} \in \mathbb{R}^{|\mathcal{N}_{\mathrm{el}}|\times(|\mathcal{N}_{\mathrm{el}}|-1)}$ thus span the image of $\boldsymbol{\mathcal{L}}_{\mathrm{com}}$. Although it is shown in [FYL06] that the reduced DDA obtained by dropping the mode $z_{\mathrm{a},0,s}$ is ISS, the following proposition establishes a stronger OS-EIP result.

Proposition 5.7 (OS-EIP and EISO DDA Stages)
Consider the DDA in (5.8) for the stages $s \in \{2,4\}$ which is shifted w.r.t. the equilibrium $\hat{\boldsymbol{u}}_{\mathrm{a},s}$, $\hat{\boldsymbol{x}}_{\mathrm{a},s}$, $\hat{\boldsymbol{z}}_{\mathrm{a},\chi,s}$, and $\hat{\boldsymbol{y}}_{\mathrm{a},s}$. The shifted DDA with the storage function

$$S_{\mathrm{a},s} = \frac{1}{2\delta_{\mathrm{a}}}\left(\tilde{\boldsymbol{x}}_{\mathrm{a},s}^T\tilde{\boldsymbol{x}}_{\mathrm{a},s} + \tilde{\boldsymbol{\chi}}_{\mathrm{a},s}^T\tilde{\boldsymbol{\chi}}_{\mathrm{a},s}\right) \tag{5.40}$$

is OS-EIP w.r.t. the input-output port $(\tilde{\boldsymbol{u}}_{\mathrm{a},s}, \tilde{\boldsymbol{y}}_{\mathrm{a},s})$ with $\rho_{\mathrm{a}} = 1$. Additionally, the reduced DDA obtained by transforming (5.8) according to (5.39) and dropping the unobservable and uncontrollable mode $\tilde{z}_{\mathrm{a},0,s}$ is EISO.

Proof:
Computing the time derivative of (5.40) and substituting in the linear dynamics (5.8) shifted to the equilibrium gives

$$\dot{S}_{\mathrm{a},s} = -\tilde{\boldsymbol{x}}_{\mathrm{a},s}^T\tilde{\boldsymbol{x}}_{\mathrm{a},s} - \frac{k_{\mathrm{a}}^{\mathrm{P}}}{\delta_{\mathrm{a}}}\tilde{\boldsymbol{x}}_{\mathrm{a},s}^T\boldsymbol{\mathcal{L}}_{\mathrm{com}}\tilde{\boldsymbol{x}}_{\mathrm{a},s} + \tilde{\boldsymbol{x}}_{\mathrm{a},s}^T\tilde{\boldsymbol{u}}_{\mathrm{a},s} \leq w_{\mathrm{a},s} := \tilde{\boldsymbol{x}}_{\mathrm{a},s}^T\tilde{\boldsymbol{u}}_{\mathrm{a},s} - \tilde{\boldsymbol{x}}_{\mathrm{a},s}^T\tilde{\boldsymbol{x}}_{\mathrm{a},s}, \tag{5.41}$$

since $\boldsymbol{\mathcal{L}}_{\mathrm{com}} \succcurlyeq 0$ and $\delta_{\mathrm{a}} > 0$. This verifies the OS-EIP property with $\rho_{\mathrm{a}} = 1$, since $\tilde{\boldsymbol{y}}_{\mathrm{a},s} = \tilde{\boldsymbol{x}}_{\mathrm{a},s}$. The EISO of the reduced DDA is investigated by setting $\tilde{\boldsymbol{u}}_{\mathrm{a},s} \equiv \boldsymbol{0}$ and $\tilde{\boldsymbol{y}}_{\mathrm{a},s} \equiv \boldsymbol{0}$, from which it follows that $\tilde{\boldsymbol{x}}_{\mathrm{a},s} \equiv \boldsymbol{0}$. From the system dynamics in (5.8), it can then be seen that $k_{\mathrm{a}}^{\mathrm{I}}\boldsymbol{\mathcal{L}}_{\mathrm{com}}^T\tilde{\boldsymbol{\chi}}_{\mathrm{a},s} = \boldsymbol{0}$ which implies that the modes $\tilde{\boldsymbol{z}}_{\mathrm{a},\chi,s}$ comprising the image of $\boldsymbol{\mathcal{L}}_{\mathrm{com}}$ are zero. Thus, the reduced DDA is EISO. ■

While the OS-EIP result applies to the full DDA, the EISO property only holds when dropping the unobservable and uncontrollable mode $z_{\mathrm{a},0,s}$. However, since this mode is constant, it has no bearing on the passivity or stability of the DDA. More formally, the full DDA with a constant input $\tilde{\boldsymbol{u}}_{\mathrm{a},s} = \boldsymbol{0}$ is thus GAS to an equilibrium in the set $\hat{\mathcal{X}}_{\mathrm{a},s} = \{\tilde{\boldsymbol{x}}_{\mathrm{a},s} = \boldsymbol{0}, \tilde{\boldsymbol{z}}_{\mathrm{a},\chi,s} = \boldsymbol{0}, z_{\mathrm{a},0,s} \in \mathbb{R}\}$.

Leaky PI IFOF-EIP

Consider the leaky PI controller in (5.11) shifted to its equilibrium, i.e., with $\tilde{\boldsymbol{u}}_{\mathrm{u}}$, $\boldsymbol{e}_{\mathrm{u}}$, and $\tilde{\boldsymbol{y}}_{\mathrm{u}}$. The leaky PI control with $\delta_{\mathrm{u}} > 0$ exhibits the following EIP property.

Proposition 5.8 (Leaky PI IFOF-EIP)
The leaky PI controller in (5.11) *with* $\delta_\mathrm{u} > 0$ *shifted to its equilibrium with the storage function*

$$S_\mathrm{u}(\tilde{\varepsilon}_\mathrm{u}) = \frac{(k_\mathrm{u}^\mathrm{I})^2}{2\delta_\mathrm{u} k_\mathrm{u}^\mathrm{P}} \tilde{\varepsilon}_\mathrm{u}^T \tilde{\varepsilon}_\mathrm{u} \tag{5.42}$$

is IFOF-EIP w.r.t. the supply rate

$$w_\mathrm{u} = (1 + \nu_\mathrm{u}\rho_\mathrm{u})\tilde{\boldsymbol{u}}_\mathrm{u}^T \tilde{\boldsymbol{y}}_\mathrm{u} - \underbrace{\frac{k_\mathrm{u}^\mathrm{I} + \delta_\mathrm{u} k_\mathrm{u}^\mathrm{P}}{\delta_\mathrm{u}}}_{\nu_\mathrm{u}} \tilde{\boldsymbol{u}}_\mathrm{u}^T \tilde{\boldsymbol{u}}_\mathrm{u} - \underbrace{\frac{1}{k_\mathrm{u}^\mathrm{P}}}_{\rho_\mathrm{u}} \tilde{\boldsymbol{y}}_\mathrm{u}^T \tilde{\boldsymbol{y}}_\mathrm{u}. \tag{5.43}$$

Proof:
Calculating the time derivative of S_u and substituting in the dynamics $\dot{\tilde{\varepsilon}}_\mathrm{u}$ in (5.11a) along with the output (5.11b) gives

$$\begin{aligned}\dot{S}_\mathrm{u} &= \frac{(k_\mathrm{u}^\mathrm{I})^2}{\delta_\mathrm{u} k_\mathrm{u}^\mathrm{P}} \tilde{\varepsilon}_\mathrm{u}^T \dot{\tilde{\varepsilon}}_\mathrm{u} = \frac{1}{\delta_\mathrm{u} k_\mathrm{u}^\mathrm{P}} (\tilde{\boldsymbol{y}}_\mathrm{u}^T - k_\mathrm{u}^\mathrm{P} \tilde{\boldsymbol{u}}_\mathrm{u}^T) \left[(k_\mathrm{u}^\mathrm{I} + \delta_\mathrm{u} k_\mathrm{u}^\mathrm{P})\tilde{\boldsymbol{u}}_\mathrm{u} - \delta_\mathrm{u}\tilde{\boldsymbol{y}}_\mathrm{u}\right] \\ &= \frac{k_\mathrm{u}^\mathrm{I} + 2\delta_\mathrm{u} k_\mathrm{u}^\mathrm{P}}{\delta_\mathrm{u} k_\mathrm{u}^\mathrm{P}} \tilde{\boldsymbol{u}}_\mathrm{u}^T \tilde{\boldsymbol{y}}_\mathrm{u} - \frac{k_\mathrm{u}^\mathrm{I} + \delta_\mathrm{u} k_\mathrm{u}^\mathrm{P}}{\delta_\mathrm{u}} \tilde{\boldsymbol{u}}_\mathrm{u}^T \tilde{\boldsymbol{u}}_\mathrm{u} - \frac{1}{k_\mathrm{u}^\mathrm{P}} \tilde{\boldsymbol{y}}_\mathrm{u}^T \tilde{\boldsymbol{y}}_\mathrm{u}.\end{aligned} \tag{5.44}$$

Thus, $\dot{S}_\mathrm{u} = w_\mathrm{u}$ with the supply rate in (5.43). ∎

Remark 5.6. *For the case that ideal integrators with* $\delta_\mathrm{u} = 0$ *are used, the PI controller can be seen to be IS-EIP with* $\nu_\mathrm{u} = k_\mathrm{u}^\mathrm{P}$ *for the storage function* $S_\mathrm{u} = k_\mathrm{u}^\mathrm{I} \tilde{\varepsilon}_\mathrm{u}^T \tilde{\varepsilon}_\mathrm{u}/2$. *Importantly, there is no excess passivity on the output in this case, which will be needed for stability in the sequel.*

5.2.4 Closed-Loop Stability

With the IFOF-EIP properties of the DC network in Theorem 5.5 along with the EIP results of the power-sharing controller stages in Propositions 5.6, 5.7 and 5.8, the stability of any equilibrium of the closed loop in Figure 5.3 can be investigated. Observe that the DC network with the port $(\tilde{\phi}, \tilde{v})$ interconnects with the stages of the power-sharing controller according to

$$\underbrace{\begin{bmatrix}\tilde{\boldsymbol{u}}_\mathrm{w} \\ \tilde{\boldsymbol{u}}_\mathrm{a,2} \\ \tilde{\boldsymbol{u}}_\mathrm{u} \\ \tilde{\boldsymbol{u}}_\mathrm{a,4} \\ \tilde{\boldsymbol{\phi}}\end{bmatrix}}_{\tilde{\boldsymbol{u}}_\Sigma} = \underbrace{\begin{bmatrix} 0 & 0 & 0 & 0 & -\boldsymbol{I}_{|\mathcal{N}_\mathrm{el}|} \\ \boldsymbol{I}_{|\mathcal{N}_\mathrm{el}|} & 0 & 0 & 0 & 0 \\ 0 & \boldsymbol{I}_{|\mathcal{N}_\mathrm{el}|} & 0 & 0 & 0 \\ 0 & 0 & \boldsymbol{I}_{|\mathcal{N}_\mathrm{el}|} & 0 & 0 \\ 0 & 0 & 0 & \boldsymbol{I}_{|\mathcal{N}_\mathrm{el}|} & 0 \end{bmatrix}}_{\boldsymbol{H}} \underbrace{\begin{bmatrix}\tilde{\boldsymbol{y}}_\mathrm{w} \\ \tilde{\boldsymbol{y}}_\mathrm{a,2} \\ \tilde{\boldsymbol{y}}_\mathrm{u} \\ \tilde{\boldsymbol{y}}_\mathrm{a,4} \\ \tilde{\boldsymbol{v}}\end{bmatrix}}_{\tilde{\boldsymbol{y}}_\Sigma}, \tag{5.45}$$

which has the form (3.7). Through this interconnection scheme, the AS of any equilibrium of the DC network in closed loop with the power-sharing controller can be verified using [AMP16, Theorem 3.1] along with Proposition 3.1 by using a weighted separable storage function as in (3.8). Since the weights p_s of the subsystems are a potential source of flexibility in the stability analysis, these weights can be exploited for the control design. Recall that the nonlinear weighting function (5.7) assigns different gains to errors inside and outside the tolerance band. It is therefore favourable to maximise the difference in gains experienced by small and large errors which corresponds to maximising the sector describing the supply rate of the nonlinear weighting function (see Proposition 5.6).

Theorem 5.9 (Designed Closed-Loop AS)

Consider the DC network described in Proposition 5.4 in closed loop with the distributed power-sharing controller comprising (5.7)*,* (5.8)*, and* (5.11) *as shown in Figure 5.3. Let the conditions of Theorem 5.5 hold for the DC network. Then, the closed-loop equilibrium described by* (5.9) *and* (5.12) *is AS with the storage function in* (3.8) *and the weighting function parameters* $k_{\mathrm{w},1} = \nu_\mathrm{w}$*,* $k_{\mathrm{w},2} = 1/\rho_\mathrm{w} - k_{\mathrm{w},1}$ *if*

$$\min_{\nu_\mathrm{w},\,\rho_\mathrm{w},\,p_i} \quad \nu_\mathrm{w} + \rho_\mathrm{w} \tag{5.46a}$$

$$\text{s.t.} \quad \boldsymbol{Q}_\Sigma \prec 0, \quad p_i > 0, \quad i = 1, \dots, 5, \tag{5.46b}$$

has a feasible solution, with

$$\boldsymbol{Q}_\Sigma = \begin{bmatrix} -\rho_\mathrm{w} p_1 & \frac{p_2}{2} & 0 & 0 & -\sigma_\mathrm{w} p_1 \\ \frac{p_2}{2} & -\rho_\mathrm{a} p_2 - \nu_\mathrm{u} p_3 & \sigma_\mathrm{u} p_3 & 0 & 0 \\ 0 & \sigma_\mathrm{u} p_3 & -\rho_\mathrm{u} p_3 & \frac{p_4}{2} & 0 \\ 0 & 0 & \frac{p_4}{2} & -\rho_\mathrm{a} p_4 - \nu_{\mathrm{p},1} p_5 & \sigma_\mathrm{p} p_5 \\ -\sigma_\mathrm{w} p_1 & 0 & 0 & \sigma_\mathrm{p} p_5 & -\rho_\mathrm{p} p_5 - \nu_\mathrm{w} p_1 \end{bmatrix}, \tag{5.47a}$$

$$\sigma_\mathrm{w} = {}^1\!/_2(1 + \nu_\mathrm{w}\rho_\mathrm{w}), \quad \sigma_\mathrm{u} = {}^1\!/_2(1 + \nu_\mathrm{u}\rho_\mathrm{u}), \quad \sigma_\mathrm{p} = {}^1\!/_2(1 + \nu_{\mathrm{p},1}\rho_\mathrm{p}). \tag{5.47b}$$

Proof:

Construct the storage function in (3.8) using (5.27), $S_\mathrm{w} = 0$, (5.40), and (5.42). Next, calculate its time derivative to obtain

$$\dot{S}_\Sigma \leq p_1 w_\mathrm{w} + p_2 w_{\mathrm{a},2} + p_3 w_\mathrm{u} + p_4 w_{\mathrm{a},4} + p_5 w_\mathrm{el}, \tag{5.48}$$

with the respective supply rates in (5.32), (5.37a), (5.41), and (5.43). Substituting in the supply rates and $\tilde{\boldsymbol{u}}_\Sigma$ in (5.45) yields $\dot{S}_\Sigma \leq \tilde{\boldsymbol{y}}_\Sigma^T(\boldsymbol{Q}_\Sigma \otimes \boldsymbol{I}_{|\mathcal{N}_\mathrm{el}|})\tilde{\boldsymbol{y}}_\Sigma$ with $\boldsymbol{Q}_\Sigma$ in (5.47a). Testing for $\dot{S}_\Sigma < 0$ thus reduces to verifying that $\boldsymbol{Q}_\Sigma \prec 0$ [HJ91, Corollary 4.2.13]. Thus, from Proposition 3.1, the equilibrium of the closed loop is AS if $\boldsymbol{Q}_\Sigma \prec 0$, since $\dot{S}_\Sigma$ is then strictly negative definite in the states of the leaky PI stage, Stage 1 is static, and the DC Network and the DDA stages are EISO (see Remark 5.5 and Proposition 5.7, respectively). This yields the constraints in (5.46b).

For the objective function in (5.46a), consider the sector $0 \leq w_{\mathrm{w}}$ for the supply rate in (5.37a), as depicted in Figure 5.5. Maximising this sector is achieved by minimising the areas corresponding to the excess passivity. Thus, the objective function in (5.46a) represents one possibility for maximising the sector size while ensuring AS.[6] ■

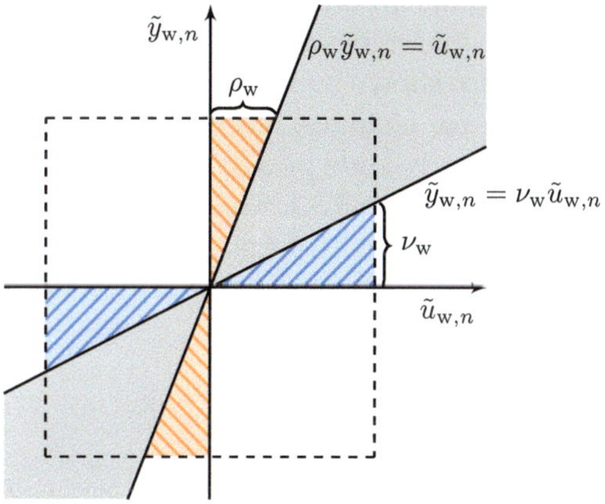

Figure 5.5: The input-output sector corresponding to $0 \leq w_{\mathrm{w}}$ □ for a single bus $n \in \mathcal{N}_{\mathrm{el}}$ along with the areas corresponding to the input passivity index ▨ and the output passivity index ▧ which are bounded by a unit square.

By applying Theorem 5.9 to the closed loop, the AS of its equilibria can be verified while simultaneously utilising the flexibility in the closed loop to find the largest sector for the nonlinear weighting function. Note that the condition in (5.46b) can also be used on its own to verify the AS result for a given set of design parameters.

Remark 5.7. *The matrix $\boldsymbol{Q}_\Sigma \in \mathbb{R}^{5\times 5}$ in* (5.46b) *has a fixed size which is independent of the number of buses or lines in the network. Furthermore, $\boldsymbol{Q}_\Sigma$ is independent of the communication and electrical topologies in the network. The AS derived by applying Theorem 5.9 is thus independent of these factors as well as being independent of the DDA, line, ZIP load, and P source changes which do not affect the worst case indices in* (5.26).

Remark 5.8. *Observe from $\boldsymbol{Q}_\Sigma$* (5.47a) *that the use of ideal integrators $\rho_{\mathrm{u}} = 0$ (see Remark 5.6) forces a zero on the diagonal of $\boldsymbol{Q}_\Sigma$. This, in turn, results in $\boldsymbol{Q}_\Sigma$ constituting an indefinite saddle-point matrix if $\delta_{\mathrm{u}} = 0$ [BGL05, Section 3], inhibiting the feasibility of* (5.46). *The use of leaky integrators is thus necessary for achieving stability via Theorem 5.9.*

Remark 5.9. *Although Theorem 5.9 yields an AS result using a quadratic storage function, a GAS result does not follow unless the optimisation problem* (5.19) *in Theorem 5.2 is solved without restricting the domains of v_j and $i_{\mathrm{z},j}$ since the resulting storage function does not necessarily constitute a Lyapunov function outside these restricted domains.*

6 The multi-objective function admits many different optimal solutions along the Pareto front [BV04, p. 181].

5.3 Simulation

The DGU power regulator and the four-stage distributed power-sharing controller, which are designed to fulfil RQ 2.1 and RQ 2.2, are tested via a simulation in this section. The simulation aims to demonstrate the achieved average voltage regulation and power sharing of a DC network which undergoes a black start. The robustness of the AS result for the controlled network is also shown by subjecting the network to various changes, including ZIP load and P source changes, electrical and communication topology changes, and changes in the steady-state power availability of the buses. The setup of the simulation is detailed in **Section 5.3.1**. This is followed by results for the leaky PI case with $\delta_{\mathrm{u}} > 0$ in **Section 5.3.2**, where the AS is verified using the methods in Section 5.2. Thereafter, in **Section 5.3.3**, results for the ideal PI controllers with $\delta_{\mathrm{u}} = 0$ are presented and compared to the leaky case.

5.3.1 Simulation Setup

The DC network in Figure 5.6 comprising 21 buses is simulated in MATLAB/SIMULINK using SIMSCAPE. The network, which serves as a synthetic test network (see [MTB20, Table 3]), is operated in an islanded mode, i.e., without being connected to external supply points. The buses are implemented as depicted in Figure 5.2 and are equipped with the distributed power-sharing controller in Figure 5.3.

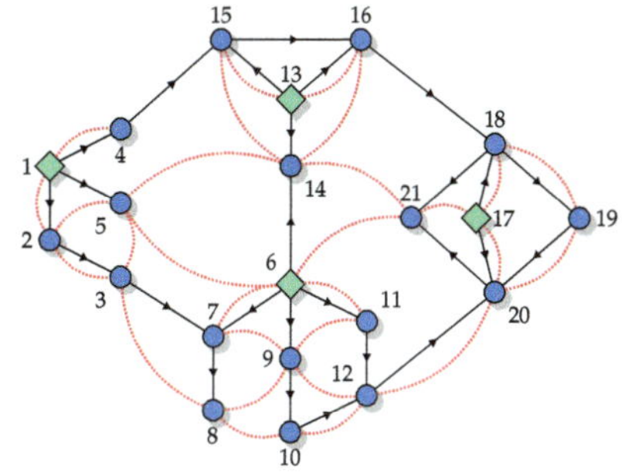

(a) State A: Base configuration.

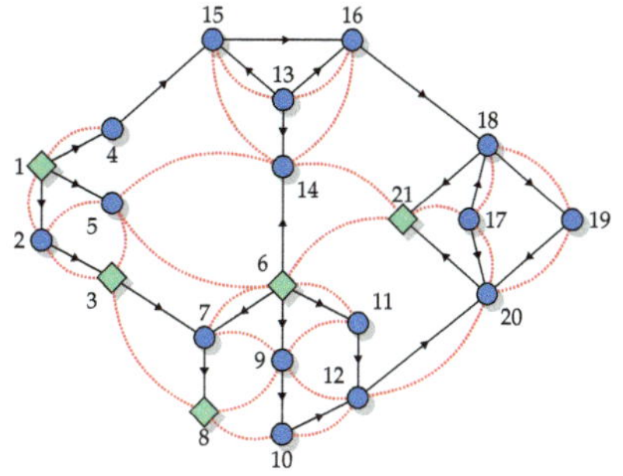

(b) State B: Alternate power availability.

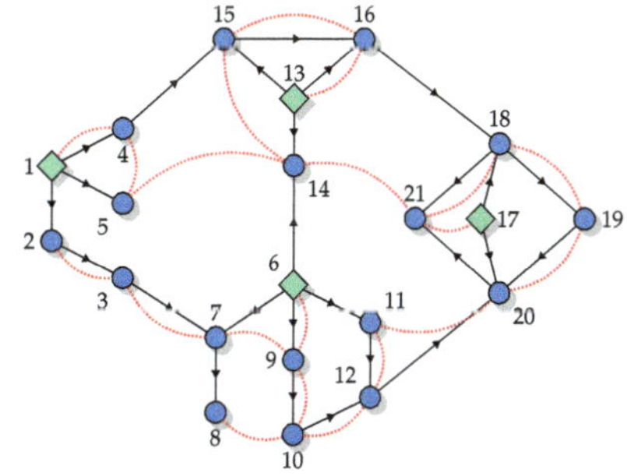

(c) State C: Sparse, radial communication.

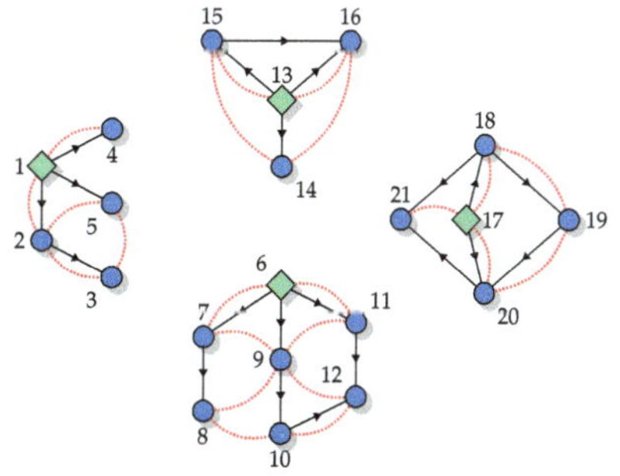

(d) State D: Separated networks.

Figure 5.6: Simulated states of a 21-Bus DC network comprising power-regulated buses ◆ in $\mathcal{P}$, buses without power injection ● in $\mathcal{F}$, lines with arbitrary directions —▸— in $\mathcal{E}_{\text{el}}$, and undirected communication links ···· in $\mathcal{E}_{\text{com}}$ between the buses.

The DC network parameters for the voltages, lines, DGUs, and P sources are taken from Table 4.2, while the ZIP load and control parameters for the simulations in this section are listed in Table 5.2. Note that saturation is again applied to the converter voltages $v_{\text{b},n}$ and that clamping is implemented for the power regulators (5.2) to prevent integrator wind-up. The power setpoint gains for the buses in $\mathcal{P}$ are listed in Table 5.3. The ZIP load parameters and the P source parameters are generated randomly within the allowed value ranges, with the ZIP loads being limited to $Y_n < 0$, $n \in \mathcal{N}_{\text{el}}$. The loads thus have strictly increasing incremental impedances (see Remark 4.2). Note that the ZIP load and P source values are deliberately chosen larger than typical for a low-voltage DC network to simulate operations under strained conditions. The same random values for transmission lines and P sources are used as in Section 4.4, which are listed in Tables B.2 and B.3 in Appendix B.1, where the time intervals for the P sources are changed to $t \in \{0\,\text{s}, 10\,\text{s}, 20\,\text{s}, 30\,\text{s}, 40\,\text{s}\}$. The generated ZIP load values are listed in Table B.4 in Appendix B.2.

Table 5.2: DC Network Parameters for Power Sharing

ZIP Loads (4.1a), $n \in \mathcal{N}_{\mathrm{el}}$	$-0.139\,\mathrm{S} \leq \lvert Z_n^{-1} \rvert \leq 0.416\,\mathrm{S}$	$-27\,\mathrm{A} \leq \lvert I_n \rvert \leq 158\,\mathrm{A}$
	$-20\,\mathrm{kW} \leq \lvert P_n \rvert \leq 20\,\mathrm{kW}$	$Y_n < 0$
PI Power Regulator (5.2), $j \in \mathcal{P}$	$\vartheta_{\mathrm{p},j} = 90$	$\vartheta_{\mathrm{i},j} = 8$
	$\vartheta_{\varepsilon,j} = 90$	$\delta_{\mathrm{p},j} = -2.41$
	$0.6 \leq k_{\mathrm{p},j} \leq 1.6$	
Weighting Function (5.7)	$k_{\mathrm{w},1} = 0.1$	$k_{\mathrm{w},2} = 1.1$
	$k_{\mathrm{w},3} = 7.5\,\mathrm{V}$	$v^\star = 380\,\mathrm{V}$
DDA Controller (5.8), $s \in \{2,4\}$	$k_{\mathrm{a}}^{\mathrm{P}} = 100$	$k_{\mathrm{a}}^{\mathrm{I}} = 200$
	$\delta_{\mathrm{a}} = 32$	
PI Controller (5.10)	$k_{\mathrm{u}}^{\mathrm{P}} = 160$	$k_{\mathrm{u}}^{\mathrm{P}} = 3000$
	$\delta_{\mathrm{u}} = 0.08$	

Table 5.3: Power Setpoint Gains

Bus $j \in \mathcal{P}$	1	3	6	8	13	17	21
$k_{\mathrm{p},j}$	0.8	1.1	1.2	1.0	0.9	1.3	0.7

The control parameters listed in Table 5.2 are successively designed. First, the parameters of the power regulator (5.2) are chosen for a fast and damped response (resulting in a 5 % settling time of 23 ms). Theorem 5.2 is then used to calculate the worst-case passivity indices $\nu_{\mathrm{p},1} = -1.0937$, $\nu_{\mathrm{p},2} = 0$ and $\rho_{\mathrm{p}} = 0.6386$ for the intervals $v \in [200\,\mathrm{V}, 500\,\mathrm{V}]$, $i_{\mathrm{z}} \in [-20\,\mathrm{A}, 100\,\mathrm{A}]$, and $k_{\mathrm{p}} \in [0.6; 1.6]$. This fulfils the conditions in Theorem 5.5 for positive line resistances and ZIP loads with $Y_n < 0$. As per Proposition 5.7, the DDA parameters do not affect its EIP property. Thus, the DDA parameters are chosen for fast settling times and good disturbance rejection (resulting in a 5 % settling time of 190 ms). Next, the PI controllers in (5.8) are tuned using the closed loop with the weighting function set to a linear gain of 1 (giving a 5 % settling of around 1.5 s).[7] Finally, the parameters of the nonlinear weighting function (5.7) are designed by using Theorem 5.9, which necessitates the use of leaky integrators ($\delta_{\mathrm{u}} > 0$).

To test the proposed controllers and to demonstrate the robustness of the resulting AS, the DC network is tested using the following sequence of events.

- $t = 0\,\mathrm{s}$: The DC network starts in State A in Figure 5.6a and undergoes a black start with random ZIP load and P source parameters generated for each bus.
- $t = 10\,\mathrm{s}$: The power availability at Buses 3, 8, 13, 17 and 21, changes and the network assumes the State B in Figure 5.6b.

[7] Note that the power regulators (5.2), the DDAs (5.7), and the PI controllers (5.8) have PI structures for which conventional tuning rules can be used.

- $t = 20\,\mathrm{s}$: The network switches to State C in Figure 5.6c, where the power availabilities of the buses are returned to their original states and the communication network is changed to a sparse, radial topology.
- $t = 30\,\mathrm{s}$: The network switches to State D in Figure 5.6d, where lines and communications links are disrupted, yielding four separated networks.
- $t = 40\,\mathrm{s}$: The network returns to its initial State A in Figure 5.6a.

At each of the time-steps, new random ZIP load and P source parameters are generated for $50\,\%$ of the buses which are selected at random. Note that the AS verification that is established using Theorem 5.9 is unaffected by the changes in the network, hence, no further verification is required.

5.3.2 Simulation Results: Leaky Integrators

The simulation results for the DC network in closed loop with the distributed power-sharing controller using leaky PI controllers are shown in Figures 5.7, 5.8 and 5.9, displaying the bus voltages and the outputs of the stages of the power-sharing controller in Figure 5.1. From the bus voltages in Figure 5.7a, it can be seen that a stability as per Definition 2.1 is achieved. Moreover, the steady-state voltages stay within $12\,\%$ ($334.4\,\mathrm{V}$ to $424.6\,\mathrm{V}$) of the reference voltage. The weighted error voltages at the output of Stage 1 of the power-sharing controller are shown in Figure 5.7b. Comparing the results in Figures 5.7a and 5.7b at Bus 17 for $30\,\mathrm{s} \leq t < 40\,\mathrm{s}$ highlights the nonlinear weighting effect, with $y_{\mathrm{w},17}$ being close to zero at the end of the interval, even though v_{17} is not equal to $380\,\mathrm{V}$.

The output of the first DDA stage shown in Figure 5.8a builds the average of the weighted errors in Figure 5.8a. It can be seen that convergence to a common value occurs slightly slower for $20\,\mathrm{s} \leq t < 30\,\mathrm{s}$ when a sparse communication topology is used. Furthermore, when the network is divided into four disconnected parts during $30\,\mathrm{s} \leq t < 40\,\mathrm{s}$, the single average is replaced by four distinct averages corresponding to the disconnected parts of the network. A single average is again formed when the communication is restored for $t \geq 40\,\mathrm{s}$. In each case, a steady-state error for $|y_{\mathrm{a},2,n}| < 11.7$ remains as a result of the use of leaky integrators. The Stage 2 DDA output is supplied to the leaky PI controllers in Stage 3, the output of which is shown in Figure 5.8b. The slight transient differences in the Stage 3 input signals result in the leaky PI output differences seen especially during $0\,\mathrm{s} \leq t < 30\,\mathrm{s}$ in Figure 5.8b, motivating the use of the DDA in Stage 4.

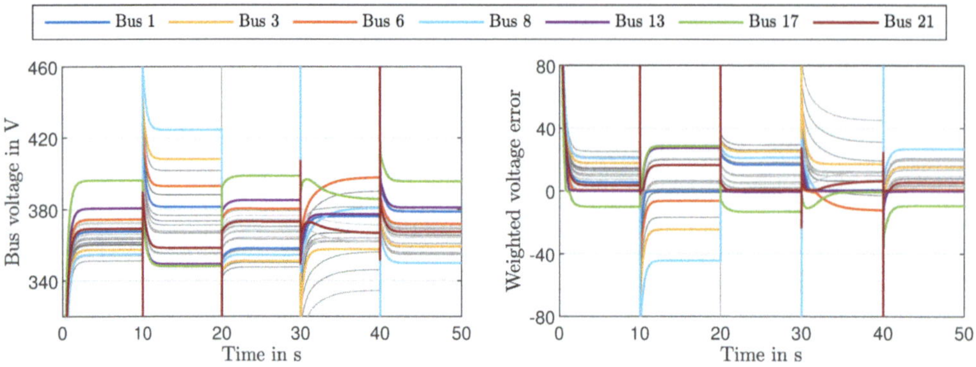

(a) Bus voltages v_n.

(b) Stage 1 output: Weighted voltage errors $y_{w,n}$.

Figure 5.7: Coordinated DC network bus voltages (a) and weighted voltage errors (b) with leaky integrators. Buses not in set $\mathcal{P}$ are shown in shades of grey.

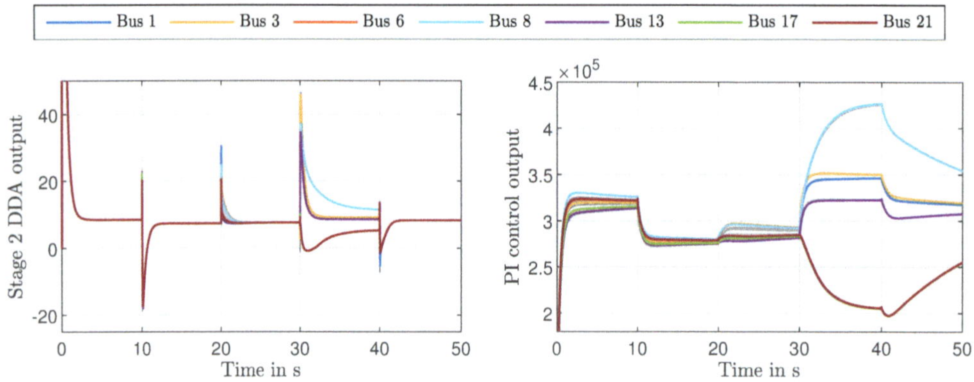

(a) Stage 2 DDA output $y_{a,2,n}$.

(b) Stage 3 PI output $y_{u,n}$.

Figure 5.8: Coordinated DC network average weighted voltages error (a) and PI output (b) with leaky integrators. Buses not in set $\mathcal{P}$ are shown in shades of grey.

The averaged leaky PI outputs are shown in Figure 5.9a, which depicts the output of the Stage 4 DDA and where the same effects arising from the communication can be seen as in Figure 5.8a. Finally, Figure 5.9b shows the power setpoints supplied to the power regulators at the buses in $\mathcal{P}$, where $p_n = 0$ for the buses $n \in \mathcal{N}_{\mathrm{el}}$ during the periods where no steady-state power is available at these buses. Notice how the setpoints are proportionally equal according to the gains in Table 5.3, except for the period $30\,\mathrm{s} \leq t < 40\,\mathrm{s}$ in which the buses in $\mathcal{P}$ are disconnected from each other.

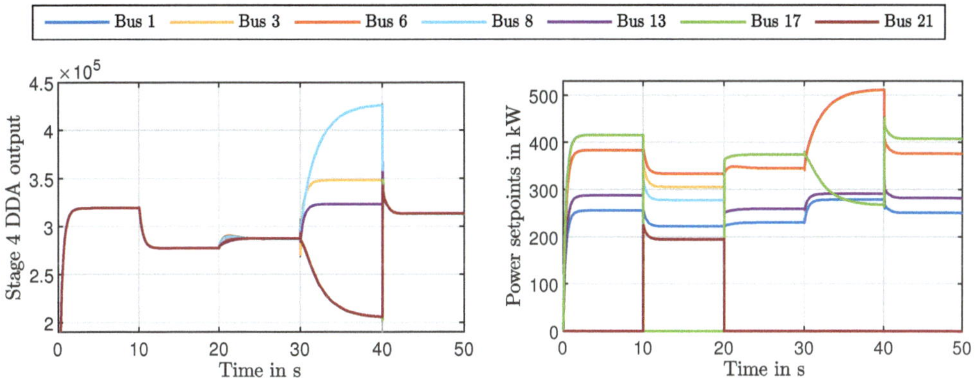

(a) Stage 4 DDA output $y_{\mathrm{a},4,n} = \psi_n$.

(b) Power setpoints $p_j^\star$, $j \in \mathcal{P}$.

Figure 5.9: Coordinated DC network average control output (a) and proportional power setpoints (b) with leaky integrators. Buses not in set $\mathcal{P}$ are shown in shades of grey.

The results in this subsection thus demonstrate the proportional power sharing along with the imperfect average voltage regulation attained by using the distributed four-stage controller along with leaky PI controllers (see Theorem 5.1). Note again that these results are obtained using DGUs with input saturation on the converter voltages. As in Chapter 4, this shows a level of robustness against such input saturation which goes beyond the stability analysis in Section 5.2.

5.3.3 Simulation Results: Ideal Integrators

The results when using leaky integrators are compared to the results obtained with ideal integrators, shown in Figures 5.10, 5.11 and 5.12. Note that since the stability analysis in Section 5.2 does not hold in this case, the stability can be investigated by linearising the entire closed-loop DC network and evaluating its eigenvalues.[8] The bus voltages in Figure 5.10a again show stability according to Definition 2.1, with the steady-state voltages staying between 345.1 V ($v_{\mathrm{Ref}} - 9.2\,\%$) and 433.6 V ($v_{\mathrm{Ref}} + 14.1\,\%$). The weighted voltage errors shown in Figure 5.10b mirror the results in Figure 5.7b.

[8] Note that such an analysis is only valid for the specific network and load configuration and only holds in a small region around the considered equilibrium.

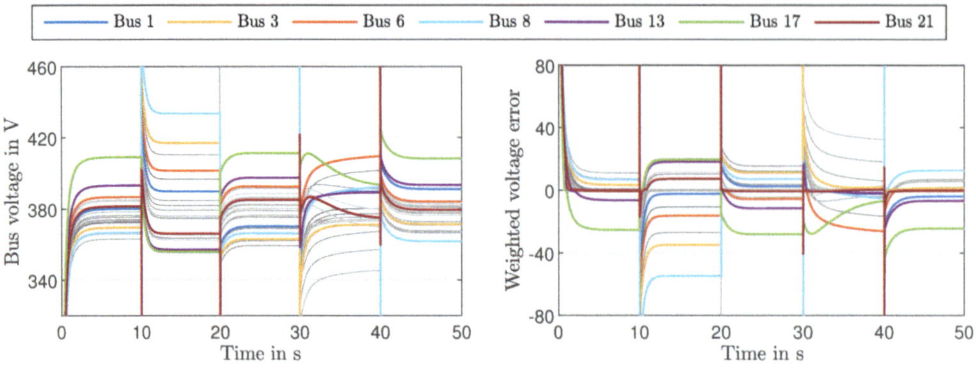

(a) Bus voltages v_n.

(b) Stage 1 output: Weighted voltage errors $y_{w,n}$.

Figure 5.10: Coordinated DC network bus voltages (a) and weighted voltage errors (b) with ideal integrators. Buses not in set $\mathcal{P}$ are shown in shades of grey.

The output of the first DDA stage with ideal integrators given in Figure 5.11a again builds the average using the weighted errors in Figure 5.10b. Compared to Figure 5.8a, it can be seen that the ideal integrators achieve $\boldsymbol{y}_{a,2} \rightarrow \boldsymbol{0}$ in each interval of the simulation, even when the network is disconnected into four isolated parts for $30\,\mathrm{s} \leq t < 40\,\mathrm{s}$. The effect of the ideal integrators is also seen in Figure 5.11b, where the ideal PI control outputs of Stage 3 are shown. Unlike in Figure 5.8b, the Stage 3 outputs in Figure 5.11b do not converge to a common value.

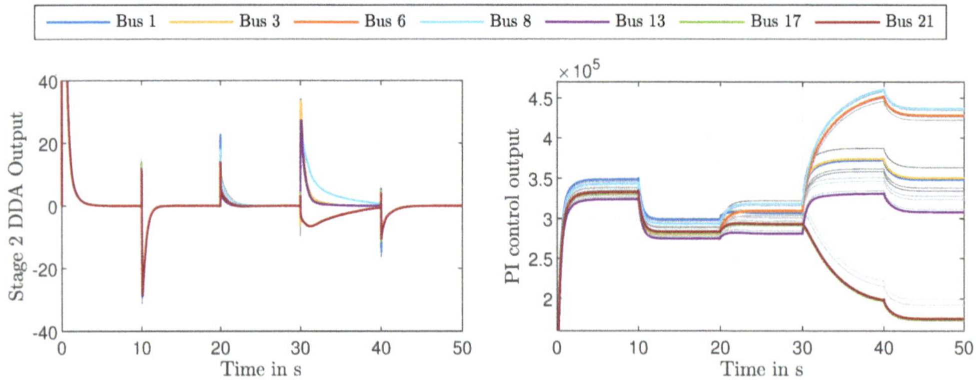

(a) Stage 2 DDA output: $y_{a,2,n}$.

(b) Stage 3 PI output $y_{u,n}$.

Figure 5.11: Coordinated DC network average weighted voltages error (a) and PI output (b) with ideal integrators. Buses not in set $\mathcal{P}$ are shown in shades of grey.

Despite the range of the Stage 3 outputs in Figure 5.11b, the output of the Stage 4 DDA in Figure 5.12a demonstrates the consensus of the distributed controller in generating the common setpoint ϕ_n which is equal at all nodes in the steady state. Indeed, convergence to a common value is achieved within 190 ms after the change at $t = 40$ s. During $20\,\text{s} \leq t < 30\,\text{s}$, when the communication topology is sparse and radial, however, the 1 % settling time is approximately 3.43 s. The power setpoints for the power regulators in Figure 5.12b again demonstrate that proportional power sharing is achieved, except for the period $30\,\text{s} \leq t < 40\,\text{s}$ in which the buses in $\mathcal{P}$ are disconnected from each other.

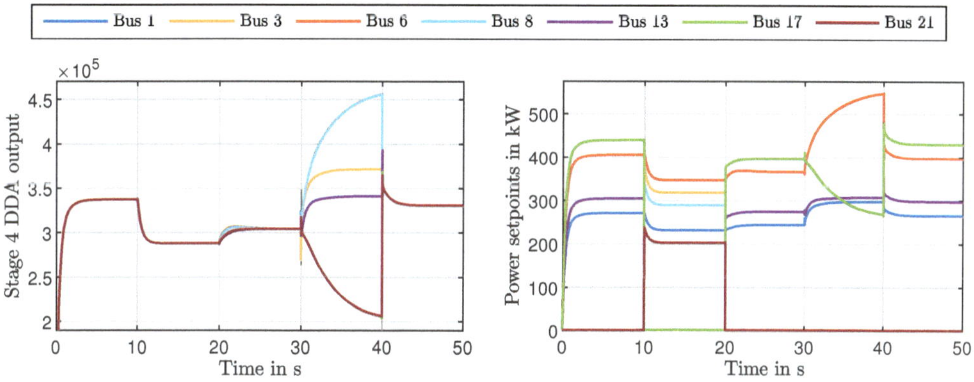

(a) Stage 2 DDA output: $y_{a,4,n} = \phi_n$.

(b) Power setpoints $p_j^\star, j \in \mathcal{P}$.

Figure 5.12: Coordinated DC network average control output (a) and proportional power setpoints (b) with ideal integrators. Buses not in set $\mathcal{P}$ are shown in shades of grey.

To better compare the effect of using leaky versus ideal integrators, the steady-state bus voltages from Figures 5.7a and 5.10a are displayed in Figure 5.13a. While the lowest steady-state values are between 1.99 and 3.11 percentage points higher for the various steady-state voltages when using ideal as opposed to leaky integrators, the highest steady-state values similarly increase by between 2.35 and 3.35 percentage points. However, comparing Figure 5.13a to Figures 5.7a and 5.10a shows in each case that the buses in $\mathcal{P}$ which inject power are responsible for the higher voltage errors. As suggested by Remark 5.4, the steady-state error arising from the use of leaky integrators, seen in Figure 5.8a, can be compensated by changing the voltage setpoint used in the power-sharing controller in Figure 5.3. Repeating the simulation with leaky controllers while setting $v^\star = 390$ V and comparing it to the ideal case with $v^\star = 380$ V yields the steady-state results in Figure 5.13b. Notably, the lowest steady-state values of the ideal case show differences of between -0.74 and 0.41 percentage points while the highest steady-state values show differences of between -0.04 and 0.87 percentage points compared to the leaky case. Setting $v^\star - v_{\text{Ref}}$ to a positive constant in the leaky case can thus yield comparable results to the ideal case.

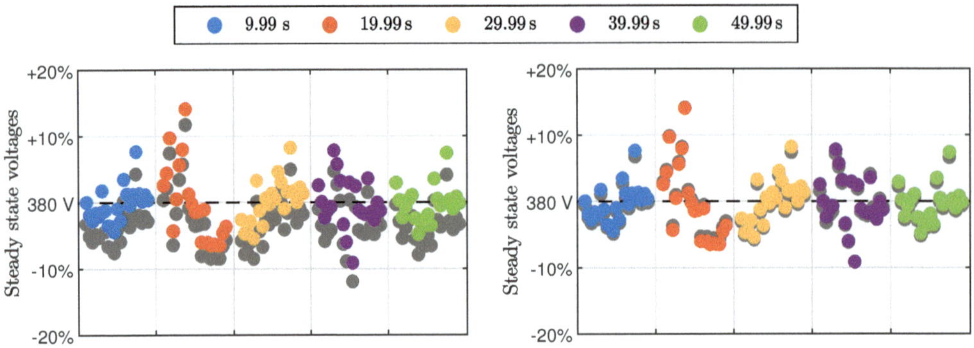

(a) Leaky $v^\star = 380\,\mathrm{V}$.

(b) Leaky $v^\star = 390\,\mathrm{V}$.

Figure 5.13: Steady-state bus voltage comparison for ideal (colour) versus leaky (grey) integrators for differing voltage setpoints in the leaky cases.

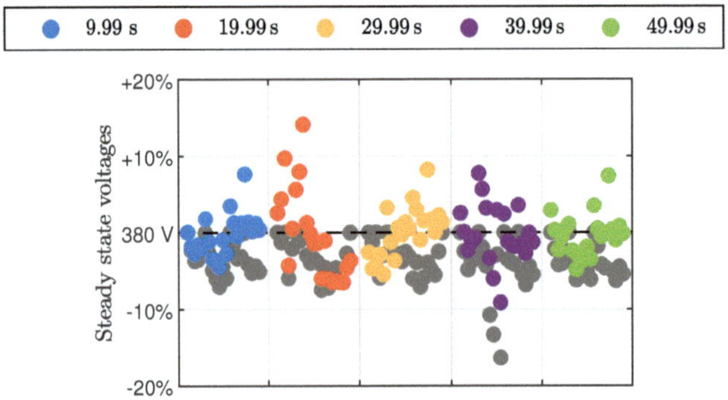

Figure 5.14: Steady-state bus voltage comparison for ideal power sharing (colour) versus voltage setting control (grey).

In a further test, the steady-state voltages achieved by the distributed power-sharing controller in this chapter are compared with the voltage setting controllers from Chapter 4 in Figure 5.14. To this end, the same simulation setup as described in Section 5.3.1 is used with the power regulators at the buses $j \in \mathcal{P}$ simply being replaced by voltage setting controllers with a uniform voltage setpoint $v_j^\star = 380\,\mathrm{V}$.[9] Again, the lowest steady-state voltages are between 0.90 and 7.10 percentage points higher when using the ideal power sharing compared to the voltage setting controllers in Figure 5.14, while the highest steady-state values exhibit increases of between 7.41 and 14.11 percentage points. Furthermore, taking the steady state at $t = 39.99\,\mathrm{s}$ as an example, the steady-state voltages lie within $[-16.32\,\%, 0.00\,\%]$ from v_{Ref} for the voltage regulation versus within $[-9.14\,\%, 7.77\,\%]$ from v_{Ref} for the distributed power sharing.

[9] See Section 4.4.1 for the control parameters of the voltage setting controllers. Note that since the ZIP loads in this chapter are assumed to have strictly increasing incremental impedances, the load buses (5.1) can be shown to be OS-EIP. The AS of the DC network equilibria then follows from Proposition 3.1 if the conditions in Theorem 4.2 hold (see also Proposition 4.6 and Remark 4.13).

Figure 5.14 thus demonstrates how the distributed power-sharing controller utilises the voltage tolerances band to improve the spread of the steady-state voltages subject to the proportional power sharing of the DDAs which inject steady-state power.

5.4 Discussion

A central feature of the proposed work in this chapter relates to the robustness of the attained stability. Indeed, though Theorems 5.5 and 5.9, the AS of the DC network equilibria, once established, is unaffected by any parameter changes which do not change the worst case indices (see Proposition 5.4 and, e.g., the load changes described in Section 5.3.1) as well as any changes to the network size or its electrical or communication topologies, as long as the network buses remains connected through the electrical edges and through the communication edges. Notably, however, this robustness pertains to piecewise-constant changes. Time-varying communication networks are also possible (see [FYL06]), although this requires suitable EIP properties to be established for the DDA stages.

The robustness of the closed-loop network stability extends to changes in the loads. Although only ZIP loads are considered in the simulations in this chapter, the EIP and stability analyses in Theorems 5.5 and 5.9 are compatible with any load model and also directly allow the inclusion of dynamic load models (see Remark 4.14). The largest restriction compared to the decentralised stabilisation results in Chapter 4, however, is the requirement that the loads have strictly increasing incremental impedances. Although the power regulators in (5.2) also include damping on the capacitor current akin to the voltage setting and voltage following controllers in the previous chapter, the nonlinear nature of the power regulators limits their ability to compensate the destabilising effects of non-monotone loads. Moreover, as discussed in Remark 5.3, the proposed control system requires a network-wide power-to-voltage relation which is monotonically increasing. This requirement is generally less restrictive than requiring loads with monotonically increasing incremental impedances. However, since the line resistance acting in series with the load function also affects the grid-wide relation, investigating such a requirement requires complex network- and load-dependent computations. This also implies that equipping buses without steady-state power with a voltage following controller as in Chapter 4 will not necessarily yield a stable coordination in the presence of large and pervasive non-monotone loads.

Beyond the need for strictly monotone loads, the EIP-based stability analysis culminating in Theorem 5.9 requires the use of leaky integrators (see Remark 5.8). This requirement represents a trade-off between the steady-state properties and the demonstrated robust stability. As argued in Remark 5.4 and demonstrated in Figure 5.13, however, a simple setpoint offset can significantly improve the quality of the achieved steady state up to the point where the differences between the leaky and ideal results become negligible. Moreover, although the stability properties of the ideal case have not been proven, the simulation results point towards the ideal case exhibiting a similar level of robustness. Both the leaky and the ideal cases may therefore be considered viable.

Although the stability proof does not extend to all practical aspects, e.g., state or input limitations, the simulations indicate that a level of robustness is achieved when the converter voltages experience saturation. The use of larger-than-typical loads and sources lends further credibility to the simulative investigation of the robustness against actuator saturation. Nevertheless, the results still rely on a suitable and feasible network design (see Assumption 5.1), and the weighted average voltage regulation does not guarantee that bus voltages will stay within a certain band in the steady state.

Despite no guarantees being made about the steady-state voltage levels, the average voltage regulation and power sharing shown in this chapter better utilise the tolerance band around the reference voltage when compared to the decentralised voltage regulation in Chapter 4 (see Figure 5.10a). Notably, the same utilisation of the tolerance band can be achieved using decentralised voltage regulation when these are supplied with appropriate setpoints. This process of finding appropriate voltage setpoints is simply automated by the power-sharing controller in this chapter.

Finally, when considering the work in this chapter in the context of grid-forming and grid-feeding DGUs, it can be seen that the proposed decentralised power regulators inject a desired amount of power in a grid-feeding fashion.[10] There is, however, no component which independently forms the grid in the work proposed in this chapter. Instead, the four-stage controller ensures stability in the DC network by coordinating the power injected by the grid-feeding DGUs. While an analysis of the effects arising from communication delays or interruptions is beyond the scope of this work, it can be noted that the network will still have AS equilibria under a full or a partial communication disruption. In this case, the weighted voltage error and setpoint averaging of DDA stages will exhibit a distinct average for each of the connected subsets of nodes in the communication graph. Thus, in the worst case, a full communication disruption will prevent a coordination of the grid-feeding DGUs, but a stable grid operation will still be ensured since the power setpoints of each DGU with steady-state power will be set such that the local voltage is regulated to the desired reference voltage.

[10] It can be argued that the combination of the power regulator and the four-stage controller operate in a grid-supporting setup.

5.5 Summary and Contributions

This chapter demonstrates how average voltage regulation and power sharing can be achieved in a distributed manner in DC networks by equipping DGUs with local power regulators and implementing a four-stage distributed controller at the buses. The EIP-based stability is derived by finding a supply rate for the power regulated buses using optimisation and then combining this result with the EIP supply rates of the stages of the power-sharing controller. This method for verifying the AS of the closed-loop network equilibria is independent of the network size and physical and communication topologies while being robust against load, bus or line parameter changes.

The main contributions of this chapter are:

- A distributed four-stage power-sharing controller which yields a steady state in which proportional power sharing and average weighted voltage regulation is achieved (Theorem 5.1);
- A decentralised power regulator along with its EIP investigation (Theorem 5.2);
- EIP properties for the DC network that are independent of the availability of steady-state power (Theorem 5.5);
- Robust AS established by combining the EIP properties of the DC network and the stages of the distributed power-sharing controller (Theorem 5.9).

Together, these contributions meet the requirements of RQ 2.1 and RQ 2.2.

6 Decentralised Stabilisation in Gas Networks

To achieve stabilisation in an entire NMES, this chapter considers the modelling and control of gas networks. For the gas network, dynamical pipelines and compressors along with static valves and loads are considered. Since the nonlinear gas pipeline dynamics greatly influence the gas network dynamics, a new high-fidelity model for gas pipelines which is conducive to control theoretical analysis methods is presented. Furthermore, to allow the extraction of excess gas from a network with local gas injection, a pressure regulating controller is designed for a compressor unit that allows bidirectional gas flows. The compressor unit, which comprises a compressor equipped with a CCV and a backflow (recycle) valve, is shown to be EIP in closed loop with its controller. By showing that the remaining gas network components are also EIP, the gas network equilibria are shown to be AS in a modular fashion using the methods established in Chapter 3. This chapter thus formulates answers to RQ 3.1 and RQ 3.2.

The chapter starts with **Section 6.1**, where a nonlinear dynamical model for a gas pipeline is derived from the governing PDEs. After invoking a few standard assumptions and drawing on parallels between gas pipeline and electrical transmission line equations, a nonlinear third-order pipeline model is presented as a response to RQ 3.1. A benchmark simulation is used to validate the model and demonstrate its fidelity. Thereafter, in **Section 6.2**, models for the remaining components of the gas network are detailed. This includes the static loads and valves as well as a surge-capable dynamical compressor model which is equipped with valves to form a compressor unit.

The compressor unit model is then used in **Section 6.3**, where a nonlinear controller is derived which regulates the inlet pressure of the compressor to a desired setpoint. This setpoint regulation allows bidirectional gas flows through the compressor unit, as needed for RQ 3.2. The section concludes by deriving conditions under which the controlled compressor unit exhibits EIP properties on its inlet and outlet ports.

In **Section 6.4**, EIP properties for the remaining components in the gas network are established, highlighting the ease of analysis of the pipeline models derived in Section 6.1. Finally, it is shown how these EIP components can be combined in arbitrary topologies to yield a gas network with AS equilibria, as required by RQ 3.2.

The proposed controllers are then tested in a simulation in **Section 6.5** and the robustness of the stability is shown by subjecting the simulated network to various changes. The chapter concludes with a discussion of the methods and results in **Section 6.6** along with a summary of the work and the contributions in **Section 6.7**.

6.1 Gas Pipeline Modelling

In this section, a third-order model with a π-structure is derived for a gas pipeline as depicted in Figure 6.1. The model is based on the isothermal Euler PDEs, which describe the flow of gas in a pipeline. These equations, along with standard assumptions made in the context of gas networks, are detailed in **Section 6.1.1**. Thereafter, in **Section 6.1.2**, the proposed pipeline model is derived by following a discretisation scheme inspired by an electrical analogy. Finally in **Section 6.1.3**, the presented model is validated using a standard benchmark setup network.

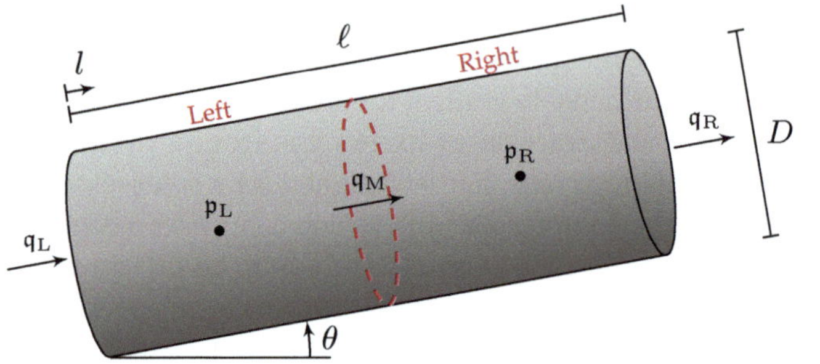

Figure 6.1: Representation of an inclined gas pipeline divided into a left-hand and a right-hand side.

6.1.1 Gas Pipeline Equations

In this subsection, the governing equations for the flow of gas in an inclined pipeline with a homogenous non-ideal gas are briefly recalled. For an extended derivation and discussion of the gas pipeline equations, the interested reader is referred to [Men05; Lur08; Koc+15; PBLD16; Pam18; Dom+21]. Unless otherwise stated, the equations in this subsection are adapted from [Koc+15; PBLD16].

Recall that the one-dimensional flow of a compressible fluid in an inclined cylindrical pipeline as in Figure 6.1 can be described by a set of PDEs known as the *Euler equations*, which comprise the conservation of mass, momentum and energy. Under isothermal conditions, where the gas temperature is considered constant, the equations governing the conservation of energy may be neglected, yielding the PDEs

$$\text{Conservation of mass:} \quad \frac{\partial \varrho}{\partial t} + \frac{\partial(\varrho \mathfrak{v})}{\partial l} = 0, \tag{6.1a}$$

$$\text{Conservation of momentum:} \quad \frac{\partial(\varrho \mathfrak{v})}{\partial t} + \frac{\partial(\varrho \mathfrak{v}^2)}{\partial l} + \frac{\partial \mathfrak{p}}{\partial l} + \frac{\zeta_{\text{ef}}\, \varrho |\mathfrak{v}| \mathfrak{v}}{2D} + \varrho g \sin(\theta) = 0, \tag{6.1b}$$

where $t \in \mathbb{R}_{\geq 0}$ is the temporal variable and $l \in [0, \ell]$ is the spatial variable for a pipe with diameter $D \in \mathbb{R}_{>0}$ and length $\ell \in \mathbb{R}_{>0}$, where $\ell \gg D$. The pipeline variables are denoted by $\mathfrak{v} = \mathfrak{v}(l,t) \in \mathbb{R}$ for the gas velocity, $\varrho = \varrho(l,t) > \mathbb{R}_{\geq 0}$ for the density, and

$\mathfrak{p} = \mathfrak{p}(l,t) > \mathbb{R}_{\geq 0}$ for the absolute pressure. Furthermore, $\zeta_{\text{ef}} = \zeta_{\text{ef}}(v, \varrho) \in \mathbb{R}_{>0}$ is the effective friction factor along the pipeline, $g \in \mathbb{R}_{>0}$ is the gravitational acceleration constant, and $\theta \in [-\pi/2, \pi/2] \subset \mathbb{R}$ is the pipe inclination angle.

Compressibility

For real gases, the pressure and density variables are related via the real gas law

$$\mathfrak{p} = ZR_{\text{spec}}T\varrho = c^2\varrho, \tag{6.2}$$

where $R_{\text{spec}} \in \mathbb{R}_{>0}$ is the specific gas constant, $T \in \mathbb{R}_{\geq 0}$ is the temperature in Kelvin, $c \in \mathbb{R}_{\geq 0}$ is the isothermal speed of sound and $Z = Z(\mathfrak{p}, T) \in \mathbb{R}_{>0}$ is the compressibility factor for real gases.[1] While the compressibility factor is a complex nonlinear function, the *Papay* approximation provides a good estimate for natural gas up to 150 bar, i.e.,

$$Z(\mathfrak{p}, T) = 1 - 3.52\frac{\mathfrak{p}}{\mathfrak{p}_{\text{crit}}}e^{-2.26\frac{T}{T_{\text{crit}}}} + 0.274\frac{\mathfrak{p}^2}{\mathfrak{p}_{\text{crit}}^2}e^{-1.878\frac{T}{T_{\text{crit}}}}, \tag{6.3}$$

with $T_{\text{crit}} \in \mathbb{R}_{>0}$ the critical temperature and $\mathfrak{p}_{\text{crit}} \in \mathbb{R}_{>0}$ the critical pressure of the gas.

Friction

The friction experienced by a fluid flowing in a pipe depends on its state of flow, characterised as being either in a laminar, transitional or turbulent state.[2] This state of flow is described by the *Reynolds* number

$$Re = \frac{\varrho|\mathfrak{v}|D}{\eta_{\text{D}}} = \frac{\varrho|\mathfrak{q}^{\text{meas}}|D}{\eta_{\text{D}}A}, \tag{6.4}$$

where η_{D} is the dynamic viscosity of the fluid, $A = \pi D^2/4$ is the cross-sectional area, and $\mathfrak{q}^{\text{meas}}$ is the measurable volumetric flow rate.[3] The laminar friction factor is characterised by the *Hagen-Poiseuille* equation

$$\zeta_{\text{Lam}} = \frac{64}{Re}, \tag{6.5}$$

whereas turbulent friction is described by the implicit *Colebrook-White* formula

$$\frac{1}{\sqrt{\zeta_{\text{Tur}}}} = -2\log_{10}\left(\frac{2.51}{Re\sqrt{\zeta_{\text{Tur}}}} + \frac{r_{\text{T}}}{3.71D}\right), \tag{6.6}$$

1 Examples of the $Z(\mathfrak{p}, T)$ map for natural gas can be found in [Men05, p. 16], [Lur08, p. 41], [Dom+21, p. 6].

2 Laminar flow generally occurs for $Re < 2000$ whereas turbulent flow occurs for $Re > 4000$. For analytical simplicity in this work, the flow will be considered as being laminar for $Re < 2300$ and turbulent for $Re \geq 2300$.

3 The Moody diagram depicts the relationship between the friction factor and the Reynolds number (see [Dom+21, p. 7]).

where r_{T} is the surface roughness of the pipe (see [Men05, p. 49] for example values of the roughness). The implicit form of the turbulent friction factor ζ_{Tur} (6.6) in gas pipelines can be approximated through the explicit *Hofer* formula

$$\zeta_{\mathrm{Tur}} = \left[2\log_{10}\left(\frac{4.518}{Re}\log_{10}\left(\frac{Re}{7}\right) + \frac{r_{\mathrm{T}}}{3.71D}\right)\right]^{-2}. \tag{6.7}$$

To account for changes in the pipe form or for curvature along the length of the pipe, an optional efficiency factor η_{ef} can be used to modify the friction factor ζ according to

$$\sqrt{\frac{1}{\zeta_{\mathrm{ef}}}} = \eta_{\mathrm{ef}}\sqrt{\frac{1}{\zeta}} \iff \zeta_{\mathrm{ef}} = \frac{\zeta}{\eta_{\mathrm{ef}}^2}, \tag{6.8}$$

where ζ is equal to ζ_{Lam} (6.5) for laminar flow and ζ_{Tur} (6.7) for turbulent flow.

Simplification and Normalisation

For gas velocities which are slow when compared to the isothermal speed of sound c, the convective term $\partial(\varrho\mathfrak{v}^2)/\partial l$ in (6.1b) is dominated by the pressure term [PBLD16, Appendix A],

$$\frac{\partial(\varrho\mathfrak{v}^2)}{\partial l} + \frac{\partial\mathfrak{p}}{\partial l} = \frac{\partial}{\partial l}\left[\mathfrak{p}\left(\frac{\mathfrak{v}^2}{c^2}+1\right)\right] \approx \frac{\partial\mathfrak{p}}{\partial l}, \tag{6.9}$$

meaning that the convective term may be neglected.[4]

Finally, recall that the steady-state gas velocity in a pipeline *increases* along the direction of flow along the pipeline in a relation that is inversely proportional to the pressure drop [SS17, p. 240]. Under the same conditions, however, the mass flow rate $\dot{m}$ is constant along the entire pipeline and is directly proportional to the *volumetric flow rate at standard conditions* $\mathfrak{q}$ via the relation

$$\dot{m} = \varrho A\mathfrak{v} = \varrho\mathfrak{q}^{\mathrm{meas}} = \varrho_{\mathrm{n}}\mathfrak{q}, \tag{6.10}$$

where ϱ_{n} is the constant density at standard conditions. Note that the unit of the standard flow rate $\mathfrak{q}$ is sm^3/s, while the unit of the measurable flow rate $\mathfrak{q}^{\mathrm{meas}}$ is m^3/s.

Applying (6.2), (6.9), and (6.10) to the isothermal Euler equations (6.1) yields the simplified pipeline PDEs [PBLD16]

$$\frac{1}{c^2}\frac{\partial\mathfrak{p}}{\partial t} = -\frac{\varrho_{\mathrm{n}}}{A}\frac{\partial\mathfrak{q}}{\partial l}, \tag{6.11a}$$

$$\frac{\varrho_{\mathrm{n}}}{A}\frac{\partial\mathfrak{q}}{\partial t} = -\frac{\partial\mathfrak{p}}{\partial l} - \frac{\zeta_{\mathrm{ef}}\varrho_{\mathrm{n}}^2c^2}{2DA^2\mathfrak{p}}|\mathfrak{q}|\mathfrak{q} - \frac{g\sin(\theta)}{c^2}\mathfrak{p}. \tag{6.11b}$$

[4] Setting $c = 300\,\mathrm{m/s}$ and letting $|\mathfrak{v}| < 15\,\mathrm{m/s}$, it follows that $\mathfrak{v}^2/c^2 < 2.5\cdot 10^{-3} \ll 1$.

For simplicity, the volumetric flow rate under standard conditions $\mathfrak{q}$ will simply be called as the flow rate $\mathfrak{q}$ from this point onwards.

Remark 6.1. *The functions for the compressibility factor* Z (6.3) *and the turbulent friction factor* ζ_{Tur} (6.7) *are two commonly used approximations for these properties. Nevertheless, many other relations have been proposed for the compressibility factor [Men05, Section 1.11],[MB+15; WYW22] and the turbulent friction factor [Men05, Chapter 2], [WC13; ZRA19] with differing accuracies and complexities.*

6.1.2 Model Derivation

To enable the derivation of a third-order model for a gas pipeline, several conditions must hold. A list of these requirements, which includes the assumptions made in Section 6.1.1, is given in the following assumption.

Assumption 6.1 (Gas Pipeline Requirements)
The gas pipeline operates under the following list of conditions.

- *The pipe is circular with a constant cross-sectional profile.*[5]
- *The pipeline dynamics are spatially one-dimensional with* $\ell \gg D$.
- *The pipe slope is constant.*[6]
- *Isothermal fluid conditions apply.*
- *The compressibility and friction factor can be approximated with* (6.3) *and* (6.7), *respectively, and the pipe has a uniform pipe roughness.*
- *The pipe experiences creeping fluid velocities with* $|\mathfrak{v}|^2/c^2 \ll 1$, *see* (6.9).
- *The fluid mixture is homogenous and does not experience phase transitions.*[7]
- *No pipe leakages occur.*[9]

---

5 Extensions to other geometries can be found in [Men05].

6 Due to the nonlinearity of the friction, consecutive pipelines with different inclinations will not necessarily produce the same result as a single pipeline with the constant average slope.

7 Non homogenous mixtures result in an additional partial differential equation (see [Dom+21, Eq. (1)]), which is similar in nature to a (lossless) thermal layer in district heating networks (see [Hau+20, Eq. (2.2)]).[8] Note that non-homogenous gas mixtures may exhibit different compressibility factors (6.3) and standard density values ϱ_{n}, which affect the pipeline dynamics through the pipeline parameters and functions in (6.14).

9 If leakage occurs, it can be modelled as an orifice at one of the endpoints of the pipelines.

To derive a model for the gas pipeline, a link to electrical transmission lines is first established. To this end, recall that the dynamics of an electrical transmission line are defined by the telegrapher's equations [UMR14, p. 57]

$$C_{\mathrm{t}}'\frac{\partial v_{\mathrm{t}}}{\partial t} = -\frac{\partial i_{\mathrm{t}}}{\partial l} - G_{\mathrm{t}}' v_{\mathrm{t}}, \quad L_{\mathrm{t}}'\frac{\partial i_{\mathrm{t}}}{\partial t} = -\frac{\partial v_{\mathrm{t}}}{\partial l} - R_{\mathrm{t}}' i_{\mathrm{t}}, \tag{6.12}$$

which describe the voltage v_{t} and current i_{t} travelling along a transmission line with the per-length capacitance $C_{\mathrm{t}}' \in \mathbb{R}_{>0}$, inductance $L_{\mathrm{t}}' \in \mathbb{R}_{>0}$, series resistance $R_{\mathrm{t}}' \in \mathbb{R}_{>0}$ and shunt conductance $G_{\mathrm{t}}' \in \mathbb{R}_{>0}$. The gas pipeline equations in (6.11) bear a clear resemblance to (6.12), with the notable differences being the nonlinearity of the friction in (6.11b) versus the resistance R_{t}' in (6.12) and the last term in (6.11b) corresponding to the effect of gravity on the gas in the inclined pipe.

Inspired by this electrical analogy and recalling the typical π-model structure used for modelling electrical transmission lines (see Figure 4.1), the PDEs in (6.11) are written as a third-order state-space model. This model is derived by discretising the PDEs in (6.11) using the method of lines on the spatial dimension [CK06, Section 6.2], giving

$$\left.\frac{\partial \mathfrak{q}}{\partial l}\right|_{l_1,l_2} \approx \frac{\mathfrak{q}_{l_2} - \mathfrak{q}_{l_1}}{\Delta l}, \qquad \left.\frac{\partial \mathfrak{p}}{\partial l}\right|_{l_1,l_2} \approx \frac{\mathfrak{p}_{l_2} - \mathfrak{p}_{l_1}}{\Delta l}, \tag{6.13a}$$

for two arbitrary points $l_1, l_2 \in [0, \ell]$ along the length of the pipeline. By dividing the pipeline into a left-hand (L) and a right-hand (R) side connected in the middle (M) as shown in Figure 6.1, a lumped-parameter π-model can be constructed. Following the electrical analogy, the π-model pipeline exhibits a *capacitive volume* $\mathfrak{C}$ and an *inductive factor* $\mathfrak{L}$. The nonlinear friction is described by a function $\mathfrak{R}(\cdot)$, while the effect of gravity in (6.11) is represented by $\mathfrak{g}(\cdot)$.

Theorem 6.1 (Third-Order π-Model Gas Pipeline)
An inclined gas pipeline (6.11) *for which Assumption 6.1 holds can be modelled by*

$$\mathfrak{C}\dot{\mathfrak{p}}_{\mathrm{L}} = \mathfrak{q}_{\mathrm{L}} - \mathfrak{q}_{\mathrm{M}}, \tag{6.14a}$$

$$\mathfrak{L}\dot{\mathfrak{q}}_{\mathrm{M}} = \mathfrak{p}_{\mathrm{L}} - \mathfrak{p}_{\mathrm{R}} - \mathfrak{R}(|\mathfrak{q}_{\mathrm{M}}|, \mathfrak{p}_{\mathrm{mean}})\mathfrak{q}_{\mathrm{M}} - \mathfrak{g}(\mathfrak{p}_{\mathrm{mean}}), \tag{6.14b}$$

$$\mathfrak{C}\dot{\mathfrak{p}}_{\mathrm{R}} = \mathfrak{q}_{\mathrm{M}} - \mathfrak{q}_{\mathrm{R}}, \tag{6.14c}$$

$$\mathfrak{C} := \frac{\ell A}{2\varrho_{\mathrm{n}} c^2}, \qquad \mathfrak{L} := \frac{\ell \varrho_{\mathrm{n}}}{A}, \tag{6.14d}$$

$$\mathfrak{R}(|\mathfrak{q}_{\mathrm{M}}|, \mathfrak{p}_{\mathrm{mean}}) := \frac{\zeta_{\mathrm{ef}}\varrho_{\mathrm{n}}^2 c^2 \ell |\mathfrak{q}_{\mathrm{M}}|}{2DA^2\mathfrak{p}_{\mathrm{mean}}}, \qquad \mathfrak{g}(\mathfrak{p}_{\mathrm{mean}}) := \frac{g\ell\sin(\theta)}{c^2}\mathfrak{p}_{\mathrm{mean}}, \tag{6.14e}$$

along with the mean pipeline pressure[10]

$$\mathfrak{p}_{\mathrm{mean}} = \frac{2}{3}\frac{\mathfrak{p}_{\mathrm{L}}^3 - \mathfrak{p}_{\mathrm{R}}^3}{\mathfrak{p}_{\mathrm{L}}^2 - \mathfrak{p}_{\mathrm{R}}^2} = \frac{2}{3}\left(\mathfrak{p}_{\mathrm{L}} + \mathfrak{p}_{\mathrm{R}} - \frac{\mathfrak{p}_{\mathrm{L}} \cdot \mathfrak{p}_{\mathrm{R}}}{\mathfrak{p}_{\mathrm{L}} + \mathfrak{p}_{\mathrm{R}}}\right). \tag{6.15}$$

Proof:
Consider the gas pipeline segmented into a left-hand and right-hand part as in Figure 6.1, where the pressures $\mathfrak{p}_{\mathrm{L}}$ and $\mathfrak{p}_{\mathrm{R}}$ of the respective segments are uniform throughout the segments and with the flow rates into the left-hand segment $\mathfrak{q}_{\mathrm{L}}$, between the segments $\mathfrak{q}_{\mathrm{M}}$, and out of the right-hand segment $\mathfrak{q}_{\mathrm{R}}$. Applying the method of lines approximation (6.13) for the flow rates in the left-hand and right-hand segments gives

$$\left.\frac{\partial \mathfrak{q}}{\partial l}\right|_{\mathrm{L,M}} \approx \frac{2}{\ell}(\mathfrak{q}_{\mathrm{M}} - \mathfrak{q}_{\mathrm{L}}), \qquad \left.\frac{\partial \mathfrak{q}}{\partial l}\right|_{\mathrm{M,R}} \approx \frac{2}{\ell}(\mathfrak{q}_{\mathrm{R}} - \mathfrak{q}_{\mathrm{M}}). \tag{6.16}$$

Substituting (6.16) into (6.11a) and multiplying through by $\ell A/(2\varrho_{\mathrm{n}})$ directly yields (6.14a) and (6.14c). The pressure approximation (6.13) for the entire length gives

$$\left.\frac{\partial \mathfrak{p}}{\partial l}\right|_{\mathrm{L,R}} \approx \frac{1}{\ell}(\mathfrak{p}_{\mathrm{R}} - \mathfrak{p}_{\mathrm{L}}). \tag{6.17}$$

Furthermore, notice that the pressure $\mathfrak{p}$ appears outside the partial differential terms in the frictional and gravitational terms in (6.11b). In these instances, the pressure $\mathfrak{p}$ can be replaced by the mean pipeline pressure $\mathfrak{p}_{\mathrm{mean}}$ in (6.15), since the entire length of the pipe is considered. Additionally, replacing $\partial\mathfrak{p}/\partial l$ in (6.11b) with (6.17) and multiplying through by ℓ yields (6.14b). ■

10 The mean pipeline pressure in (6.15) was originally derived by [Wey12, p. 203] in his derivation for the volume of gas inside a pipeline. These formulas have become standard in the gas literature [Men05, p. 37], and are equivalent to 4/3 the arithmetic mean minus 1/3 the geometric mean of the two pressures. A proof can also be found in [Koc+15, Lemma 2.3].

The π-model structure obtained in Theorem 6.1 and depicted in Figure 6.2 highlights the similarity of the underlying equations of isothermal gas pipelines and dynamical electrical transmission lines (see Figure 4.1). Differences to the electrical case comprise the nonlinear frictional resistance and the pressure (voltage) source arising due to the pipe inclination and gravity. By exploiting the electrical analogy, a pipeline can be modelled as a third-order nonlinear system. This model is well-suited for control and analysis methods (e.g., an EIP analysis), thus meeting the first part of RQ 3.1.

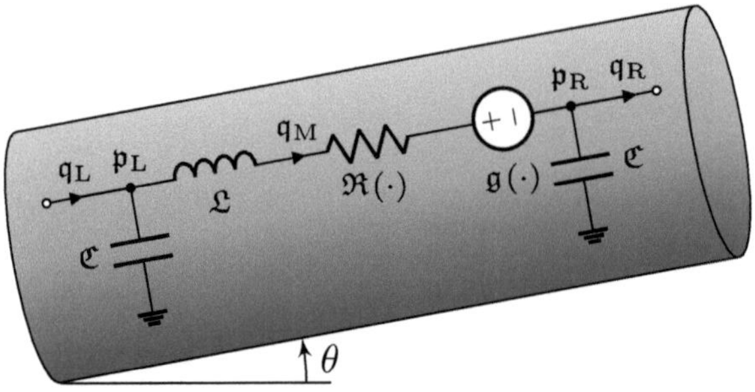

Figure 6.2: An inclined gas pipeline superimposed with its π-model electrical equivalent circuit.

Remark 6.2. *The π-structure shown in Figure 6.2 is alluded to in [PBLD16], where the capacitive volumes $\mathfrak{C}$ situated at the pipeline endpoints are similarly grouped into a nodal equation for pipelines segmented into many parts. Furthermore, [Yan+20] models a pipeline as a cascade of RC segments, neglecting the pipe inclination and the momentum dynamics in* (6.14b). *The results in Theorem 6.1 are the first to propose an RLC-like π-structure for a gas pipeline.*

Remark 6.3. *Notice from* (6.2) *and* (6.3) *that the capacitive volume $\mathfrak{C}$ exhibits a pressure dependency. This effect is negligible close to the operating point and at typical operating temperatures, with the compressibility* (6.3) *at $T = 0\,°\mathrm{C}$, e.g., decreasing with a slope no larger than* $0.3\,\%/\mathrm{bar}$.

Remark 6.4. *By making the additional assumptions that the mean pressure $\mathfrak{p}_{\mathrm{mean}}$ and the compressibility are constant (see Remark 6.3), the capacitive volume $\mathfrak{C}$ and the gravitational function $\mathfrak{g}(\cdot)$ become constant. This allows the pipeline model in* (6.14) *to be written as a third-order port-Hamiltonian model which further simplifies the model for analytical purposes (see [Mal+23] for the formal construction of this model). These assumptions are investigated in the model validation in the sequel.*

Remark 6.5. *Of the requirements listed in Assumption 6.1, only the pressure-to-density relation of the real gas law* (6.2) *along with the compressibility function pertain specifically to gas (see also the hydraulic PDEs used for district heating networks, e.g., in [Hau+20]).*

Compressible hydraulic fluids may thus be considered by replacing the real gas law with the isothermal bulk modulus for fluids [MF19, p. 6],

$$K = -V\frac{\partial \mathfrak{p}}{\partial V} = \varrho\frac{\partial \mathfrak{p}}{\partial \varrho}, \tag{6.18}$$

where $V = m/\varrho$ *is the pipe volume, from which it follows that*

$$\frac{\partial \mathfrak{p}}{\partial t} = \frac{\partial \mathfrak{p}}{\partial \varrho}\frac{\partial \varrho}{\partial t} = \frac{K}{\varrho}\frac{\partial \varrho}{\partial t}. \tag{6.19}$$

By substituting (6.19) *into* (6.1a) *and converting to standard conditions using* (6.10), *the momentum equation for compressible and incompressible* ($K \to \infty$) *fluids is obtained as*

$$\frac{\varrho}{K}\frac{\partial \mathfrak{p}}{\partial t} = -\frac{\varrho_{\mathrm{n}}}{A}\frac{\partial \mathfrak{q}}{\partial l} \quad \xrightarrow{K\to\infty} \quad 0 = -\frac{\varrho_{\mathrm{n}}}{A}\frac{\partial \mathfrak{q}}{\partial l}. \tag{6.20}$$

Theorem 6.1 can be applied using (6.20) *and* (6.11b) *to obtain a third-order model for compressible fluids or a first-order model for incompressible fluids (see the hydraulic equations used in district heating networks, e.g., [Str+24]).*

6.1.3 Model Validation

The third-order π-structure pipeline model in Theorem 6.1 is now validated with the standard three-node benchmark network depicted in Figure 6.3a (see also [Ke00; HG+09; ABNG12; PBLD16; THM17; BS19]). Both the lumped-parameter model in Theorem 6.1 and the simplified model obtained by applying the assumptions in Remark 6.4 are considered in this section.[11] The obtained simulated results are compared with two discretised models [HG+09; PBLD16], a linearised model [ABNG12], a model using a different electrical analogy [Ke00], and the results obtained using the commercial SIMONE simulation software as published in [PBLD16].

Remark 6.6. *Even for the small benchmark system, significant variations are obtained by the various approaches in the literature (see [BS19, Fig. 7]). In this work, the model in [PBLD16] is chosen as a basis for the comparison, since their model contains fewer simplifying assumptions and delivers similar results compared to the high-fidelity models implemented in the SIMONE simulation software.*

[11] For the simplified model, the mean pressure is fixed to $\mathfrak{p}_{\mathrm{mean}} = 50\,\mathrm{bar}$ whereas the compressibility is calculated from (6.3) for the standard temperature and for a pressure of $\mathfrak{p} = 50\,\mathrm{bar}$.

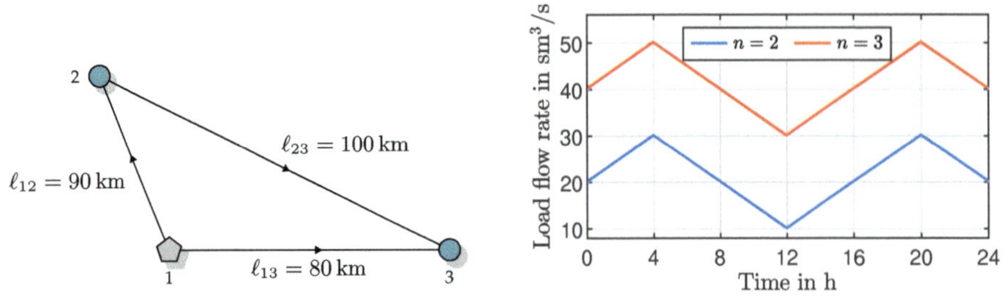

(a) Benchmark network.

(b) Load profiles $\mathfrak{q}_n$.

Figure 6.3: Benchmark gas network comprising three nodes (a), where Node 1 ⬠ has a fixed pressure $\mathfrak{p}_1 = 50\,\text{bar}$ and Nodes 2 and 3 ● have time-varying flow-rate loads (b).

Simulation Setup

The pipeline model (6.14) is implemented in SIMSCAPE, which is subsequently used in a MATLAB/SIMULINK simulation of the benchmark network in Figure 6.3a. The pressure of Node 1 is fixed to $\mathfrak{p}_1 = 50\,\text{bar}$ while Nodes 2 and 3 are subjected to the load profiles in Figure 6.3b. The parameter values of the simulation are listed in Table B.5 in Appendix B, while the pipe lengths are set as indicated in Figure 6.3a. The diameters of all pipes are set to $D_o = 0.6\,\text{m}$, $o \in \{12, 13, 23\}$. The simulation is carried out five times, once for each height $h_1 \in \{-1, -0.5, 0, 0.5, 1\}\text{km}$ of Node 1, with $h_1 = 0\,\text{km}$ constituting the nominal case.

Results

The simulation results of the pressures at Nodes 2 and 3 for the nominal case with $h_1 = 0\,\text{km}$ are shown in Figure 6.4. The absolute pressure trajectories indicate a close match to the results from [PBLD16], with absolute errors remaining within $1.28\,\text{mbar}$ for the lumped-parameter model and $4.03\,\text{mbar}$ for the simplified model. The relative pressure errors in percent, measured against [PBLD16] and shown in Figures 6.4c and 6.4d confirm the model accuracy, with the lumped-parameter model staying within $0.027\,\%$ and the simplified model within $0.083\,\%$ of the results from [PBLD16].

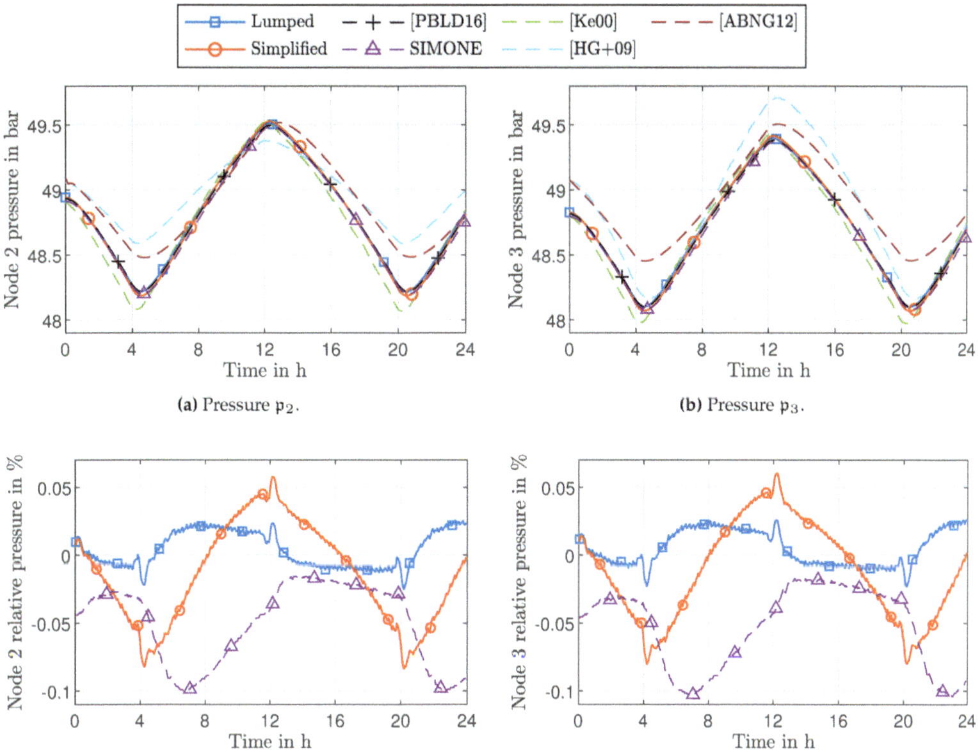

(a) Pressure $\mathfrak{p}_2$.

(b) Pressure $\mathfrak{p}_3$.

(c) Relative pressure error for Node 2.

(d) Relative pressure error for Node 3.

Figure 6.4: Pressure simulation results of the lumped-parameter and simplified models, showing the absolute pressures $\mathfrak{p}_2$ (a) and $\mathfrak{p}_3$ (b), and the relative pressure differences for $\mathfrak{p}_2$ (c) and $\mathfrak{p}_3$ (d) normalised to the data from [PBLD16]. The simulated results are compared with the results in [PBLD16; Ke00; HG+09; ABNG12] along with the SIMONE results.

The flow rates in the respective pipelines for the nominal case with $h_1 = 0\,\mathrm{km}$ are shown in Figure 6.5. Again, very similar results are obtained compared to [PBLD16], with absolute flow-rate deviations of no more than $0.199\,\mathrm{sm}^3/\mathrm{s}$ for both the lumped-parameter and the simplified models. This translates to relative flow-rate errors of no more than $1.034\,\%$, with both models performing similarly.

The pressure results of the nodes obtained when simulating the network with Node 1 at different heights $h_1 \in \{-1, -0.5, 0, 0.5, 1\}\mathrm{km}$ are shown in Figure 6.6. Compared to the nominal case with $h_1 = 0\,\mathrm{km}$, larger differences can be seen when the elevation of Node 1 changes. For both increased and decreased elevations, the proposed models yield lower pressures compared to [PBLD16], with the absolute pressure differences for the lumped-parameter model rising to $20.3\,\mathrm{mbar}$ and those of the simplified model rising to $42.7\,\mathrm{mbar}$ across all simulation runs in Figure 6.6. This equates to maximum error differences of 0.386 % for the lumped-parameter model and 0.811 % for the simplified model relative to the results from [PBLD16]. The simulation in this section demonstrates the high fidelity of the proposed model and thus answers RQ 3.1.

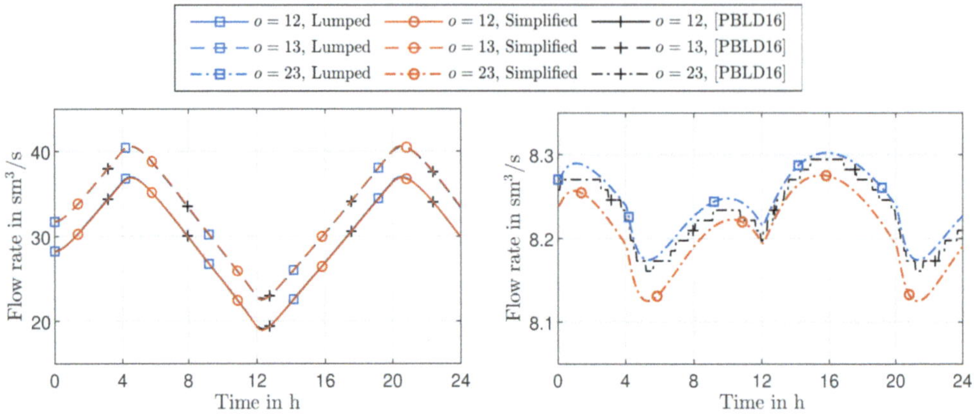

(a) Flow rates $q_{M,12}$ and $q_{M,13}$.

(b) Flow rate $q_{M,23}$.

Figure 6.5: Flow rate simulation results of the lumped-parameter and simplified models, showing $q_{M,o}$ for $o = 12$ and $o = 13$ (a) and $o = 23$ (b). The simulated results are compared with [PBLD16].

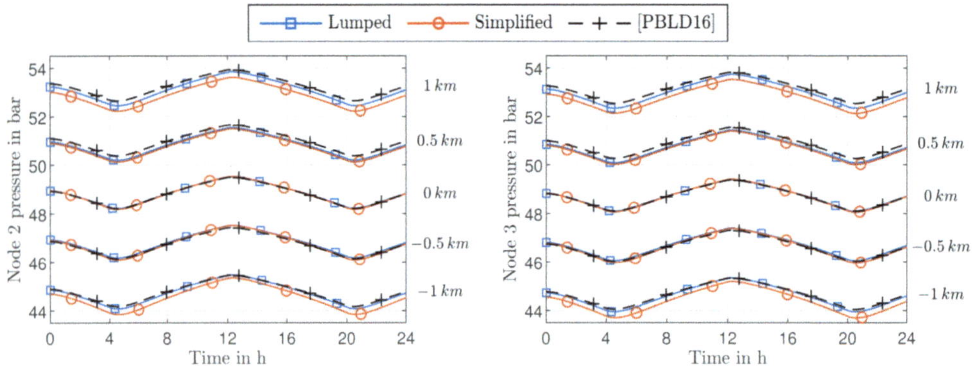

(a) Pressure p_2.

(b) Pressure p_3.

Figure 6.6: Pressure simulation results of the lumped-parameter and simplified models, showing the absolute pressures p_2 (a) and p_3 (b) when Node 1 is situated at different heights $h_1 = \{-1, -0.5, 0, 0.5, 1\}$km. The simulated results are compared with [PBLD16].

6.2 Gas Network Modelling

The results of the pipeline modelling in Section 6.1 culminating with the model validation demonstrate the high level of fidelity of the third-order model proposed in Theorem 6.1. Using this pipeline model as a component, a model for the gas network can be formulated. The section commences with **Section 6.2.1**, in which a gas network is described as a network of nodes and edges, each with different possible constitutive components which are subsequently introduced. These components include static loads and valves (**Section 6.2.2**), the edges and nodes derived from the pipeline model equations (**Section 6.2.3**), and dynamic compressors (**Section 6.2.4**).

6.2.1 Network Description

The gas network considered in this chapter is composed of nodes which represent points of equal pressure in the network. These nodes are interconnected via edges which may denote simple pipelines, compressor units, or pipelines equipped with valves, as depicted in Figure 6.7. Each node can be equipped with a load which represents the aggregate gas demand of the node (or gas supply for a negative load value). For simplicity in this work but without loss of generality, the nodes in the network are grouped into regions of equal elevation and target pressures (e.g., low- or medium-pressure regions).

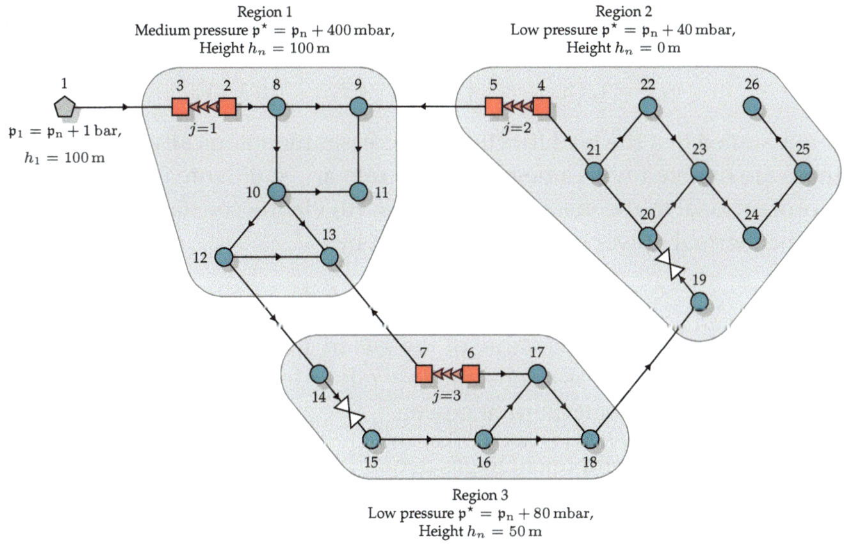

Figure 6.7: A gas network comprising nodes ● in $\mathcal{N}_\text{N}$, constant pressure nodes ⬠ in $\mathcal{N}_\text{S}$, nodes ■ in $\mathcal{N}_\text{Cs}$ adjacent to compressor units ⋙, $j \in \mathcal{E}_\text{Cs}$, and pipelines with arbitrary assigned directions of positive flow → in $\mathcal{E}_\text{T}$, and valve-equipped pipelines ⋈ in $\mathcal{E}_v$. The network comprises three regions with different elevations and target pressures relative to the atmospheric pressure $\mathfrak{p}_\text{n}$.

More formally, the gas network is represented using a directed and weakly connected graph $\mathcal{G}_\text{g} = (\mathcal{N}_\text{g}, \mathcal{E}_\text{g})$. The edges $\mathcal{E}_\text{g}$ are grouped into the set of compressor units $\mathcal{E}_\text{Cs} = \{1, \ldots, |\mathcal{E}_\text{Cs}|\}$, $|\mathcal{E}_\text{Cs}| \geq 0$; the set of pipelines $\mathcal{E}_\text{T} = \{|\mathcal{E}_\text{Cs}| + 1, \ldots, |\mathcal{E}_\text{Cs}| + |\mathcal{E}_\text{T}|\}$, $|\mathcal{E}_\text{T}| \geq 0$; and the set of valve-equipped pipelines $\mathcal{E}_v = \{|\mathcal{E}_\text{Cs}| + |\mathcal{E}_\text{T}| + 1, \ldots, |\mathcal{E}_\text{Cs}| + |\mathcal{E}_\text{T}| +$

$|\mathcal{E}_v|\}$, $|\mathcal{E}_v| \geq 0$. Furthermore, the nodes $\mathcal{N}_\mathrm{g}$ are grouped into a set of nodes supplying a constant pressure $\mathcal{N}_\mathrm{S} = \{1, \ldots, |\mathcal{N}_\mathrm{S}|\}$, $|\mathcal{N}_\mathrm{S}| \geq 1$; a set of nodes situated at the inlet or outlet of a compressor unit $\mathcal{N}_\mathrm{Cs} = \{|\mathcal{N}_\mathrm{S}| + 1, \ldots, |\mathcal{N}_\mathrm{S}| + 2|\mathcal{E}_\mathrm{Cs}|\}$; and a set comprising the remaining pipeline endpoints $\mathcal{N}_\mathrm{N} = \{|\mathcal{N}_\mathrm{S}| + 2|\mathcal{E}_\mathrm{Cs}| + 1, \ldots, |\mathcal{N}_\mathrm{S}| + 2|\mathcal{E}_\mathrm{Cs}| + |\mathcal{N}_\mathrm{N}|\}$, $|\mathcal{N}_\mathrm{N}| \geq 1$. Thus $\mathcal{E}_\mathrm{g} = \mathcal{E}_\mathrm{Cs} \cup \mathcal{E}_\mathrm{T} \cup \mathcal{E}_v$ and $\mathcal{N}_\mathrm{g} = \mathcal{N}_\mathrm{S} \cup \mathcal{N}_\mathrm{Cs} \cup \mathcal{N}_\mathrm{N}$. The arbitrarily assigned directions of the edges in $\mathcal{E}_\mathrm{T} \cup \mathcal{E}_v$ indicate the direction of positive flow for those edges, whereas the directions of the edges in $\mathcal{E}_\mathrm{Cs}$ indicate the direction in which pressure increases due to the work done by the respective compressors. Note that each node in $\mathcal{N}_\mathrm{Cs}$ is connected to exactly one edge in $\mathcal{E}_\mathrm{Cs}$ and for an edge $j \in \mathcal{E}_\mathrm{Cs}$ its source and sink nodes are denoted by the sets $\mathcal{N}^+_{\mathrm{Cs},j}$ and $\mathcal{N}^-_{\mathrm{Cs},j}$, respectively. Thus, the set $\mathcal{N}_\mathrm{Cs}$ can be divided into inlet nodes $\mathcal{N}^+_\mathrm{Cs}$ and outlet nodes $\mathcal{N}^-_\mathrm{Cs}$ according to

$$\mathcal{N}^+_\mathrm{Cs} = \bigcup_{j \in \mathcal{E}_\mathrm{Cs}} \mathcal{N}^+_{\mathrm{Cs},j}, \qquad \mathcal{N}^-_\mathrm{Cs} = \bigcup_{j \in \mathcal{E}_\mathrm{Cs}} \mathcal{N}^-_{\mathrm{Cs},j}. \tag{6.21}$$

6.2.2 Static Components

The loads in the gas network are considered to be static functions dependent on the local node pressures, i.e,

$$\mathfrak{q}_{\mathfrak{l},n} = \mathfrak{f}_n(\mathfrak{p}_n), \qquad \mathfrak{f}_n \colon \mathbb{R}_{\geq 0} \to \mathbb{R}, \qquad n \in \mathcal{N}_\mathrm{g}, \tag{6.22}$$

where it is assumed that the load function $\mathfrak{f}_n$ increases monotonically[12] in $\mathfrak{p}_n$ and where gas sources are denoted with a negative flow rate $\mathfrak{q}_{\mathfrak{l},n} < 0$. Note that the flow rate is directly proportional to the mass flow rate $\dot{m}$ (6.10) via the constant standard density ϱ_n. Thus, the thermal power of the load is given by

$$p_{\mathrm{th},n} = c_\mathrm{cal} \varrho_\mathrm{n} \mathfrak{q}_{\mathfrak{l},n}, \quad n \in \mathcal{N}_\mathrm{g}, \tag{6.23}$$

where $c_\mathrm{cal} \in \mathbb{R}_{>0}$ is the calorific value of the gas in J/kg. It is therefore common to model loads in gas networks as a constant flow rate independent of the pressure, since this corresponds to a constant thermal power.

Another important static component in gas networks is a valve, which enables flow-rate or pressure regulation through a controlled reduction in the pressure between the valve inlet and outlet. The pressure reduction is determined by the area of flow through the value A_v. This flow area is modified using the valve stem position $s_v \in [0, 1]$, with $s_v = 0$ corresponding to a fully closed valve. The flow rate through the valve $\mathfrak{q}_v \in \mathbb{R}$ can be described by the standard orifice equation [MF19, p. 49], [Qad+20, p. 16]

$$\mathfrak{q}_v = C_\mathrm{d} A_v(s_v) \sqrt{\frac{2|\Delta\mathfrak{p}_v|}{\varrho_{v,1}}} \operatorname{sign}(\Delta\mathfrak{p}_v), \quad \Delta\mathfrak{p}_v = \mathfrak{p}_{v,1} - \mathfrak{p}_{v,2}, \tag{6.24}$$

---

[12] The monotonicity of $\mathfrak{f}_n$ for practical systems is motivated by observing that the power supplied a flow of gas is directly proportional to the mass flow. A constant power demand thus requires a constant $\mathfrak{q}_{\mathfrak{l},n}$ (see (6.10)).

where $\mathfrak{p}_{v,1}, \mathfrak{p}_{v,2} \in \mathbb{R}_{\geq 0}$ are the valve inlet and outlet pressures, respectively, $\Delta\mathfrak{p}_v$ is the differential valve pressure, $C_\mathrm{d} \in \mathbb{R}_{>0}$ is the valve discharge coefficient, and $\varrho_{v,1} \in \mathbb{R}_{\geq 0}$ is the gas density at the inlet. For linear valves, which exhibit the stem-to-flow-area relation $A_v(s_v) = s_v A_{v,\max}$, and by substituting in the real gas law (6.2), the valve flow rate in (6.24) can be simplified to obtain equivalent equations for the flow rate and differential valve pressure

$$\mathfrak{q}_v = C_v s_v \sqrt{\frac{|\Delta\mathfrak{p}_v|}{\mathfrak{p}_{v,1}}}\,\mathrm{sign}(\Delta\mathfrak{p}_v), \tag{6.25a}$$

$$\Delta\mathfrak{p}_v = \frac{\mathfrak{p}_{v,1}|\mathfrak{q}_v|\mathfrak{q}_v}{C_v^2 s_v^2}, \tag{6.25b}$$

$$C_v := \sqrt{2}cA_{v,\max}C_\mathrm{d}, \tag{6.25c}$$

where $\mathrm{sign}(\Delta\mathfrak{p}_v) = \mathrm{sign}(\mathfrak{q}_v)$.

Remark 6.7. *Additional losses, such as those resulting from leakages, measuring equipment, pipe curvature, and intersections, are typically represented using linear or nonlinear resistors [Koc+15, p. 28]. Such losses, which include constant flow rates, linear resistances, frictional effects* (6.5) *or* (6.7)*, or leakages following the orifice equation [MF19, p. 49] may all be included in the monotonically increasing load function* $\mathfrak{q}_n(\mathfrak{p}_n)$.

6.2.3 Pipeline Edges and Nodes

Using the π-structure pipeline model (6.14) proposed in Theorem 6.1 in conjunction with the loads and valves formulated in Section 6.2.2, dynamical models can now be formulated for the edges which consist of gas pipelines in $\mathcal{E}_\mathrm{T}$ and $\mathcal{E}_v$ along with the nodes at the pipeline endpoints $\mathcal{N}_\mathrm{N}$. In line with the electrical analogy of the pipeline model (compare Figure 6.2 with Figure 4.1), the flow-rate dynamics of the pipelines are considered for the edges, whereas the pressure dynamics at the endpoints of the pipelines are considered for the nodes.

Starting with an edge $o \in \mathcal{E}_\mathrm{T}$, which connects a source node $n_{o,+}$ and a sink node $n_{o,-}$, the edge dynamics comprise the momentum equation of the pipeline (6.14b), i.e.,

$$\mathfrak{L}_o\dot{\mathfrak{q}}_{\mathrm{T},o} = \mathfrak{p}_{n_{o,+}} - \mathfrak{p}_{n_{o,-}} - \mathfrak{R}(|\mathfrak{q}_\mathrm{T}|, \mathfrak{p}_{\mathrm{mean},o})\mathfrak{q}_{\mathrm{T},o} - \mathfrak{g}(\mathfrak{p}_{\mathrm{mean},o}), \quad o \in \mathcal{E}_\mathrm{T},\ n_{o,+}, n_{o,-} \in \mathcal{N}_\mathrm{g}, \tag{6.26}$$

where $\mathfrak{q}_{\mathrm{T},o} \in \mathbb{R}$ is the flow rate in the pipeline and $\mathfrak{L}_o \in \mathbb{R}_{>0}$ is the inductive factor of the pipeline defined in (6.14d). The pressures $\mathfrak{p}_{n_{o,+}}, \mathfrak{p}_{n_{o,-}} \in \mathbb{R}_{\geq 0}$ correspond to the respective source and sink nodes, with the mean pressure $\mathfrak{p}_{\mathrm{mean},o}$ being derived from these pressures according to (6.15). Finally, the effects due to friction $\mathfrak{R}\colon \mathbb{R}_{\geq 0} \times \mathbb{R}_{\geq 0} \to \mathbb{R}_{\geq 0}$ and gravity $\mathfrak{g}\colon \mathbb{R}_{\geq 0} \to \mathbb{R}$ are described by the functions in (6.14e).

Consider now the valve-equipped pipelines in $\mathcal{E}_v$, where it is assumed that the pipeline is short, has no inclination, and the pressure losses due to friction are negligible compared to the pressure losses over the valves.[13] Furthermore, it is assumed that the valve regulates the pressure of the sink node. This requires $n_{o,-} \in \mathcal{N}_{\mathrm{N}}$, to ensure that the sink node pressure can be regulated. The edge dynamics then are derived from (6.26) in combination with the valve equation (6.25b), which gives

$$\mathfrak{L}_o \dot{\mathfrak{q}}_{v,o} = \mathfrak{p}_{n_{o,+}} - \mathfrak{p}_{n_{o,-}} - \frac{\mathfrak{p}_{n_{o,+}} |\mathfrak{q}_{v,o}| \mathfrak{q}_{v,o}}{C_{v,o}^2 s_{v,o}^2}, \quad o \in \mathcal{E}_v,\ n_{o,+} \in \mathcal{N}_{\mathrm{g}},\ n_{o,-} \in \mathcal{N}_{\mathrm{N}}, \tag{6.27}$$

for the valve-equipped pipeline connecting the source node $n_{o,+}$ and the sink node $n_{o,-}$ and where $s_{v,o} \in [0,1]$ is the stem position of the valve.

Remark 6.8. *Note that fully closing the valve $s_{v,o} = 0$ results in a singularity in* (6.27)*. This singularity is resolved in* (6.25a)*, from which it is clear that $s_{v,o} = 0 \implies \mathfrak{q}_{v,o} = 0$. This can also be interpreted as the damping on $\mathfrak{q}_{v,o}$ tending to infinity.*

For a node $n \in \mathcal{N}_{\mathrm{N}}$, the aggregated capacitive volume of the pipelines terminating in that node is used for the node dynamics (see (6.14a), (6.14c) and (6.14d)), i.e.,

$$\mathfrak{C}_{\mathrm{ef},n} \dot{\mathfrak{p}}_n = \mathfrak{q}_{\mathrm{ex},n}, \qquad n \in \mathcal{N}_{\mathrm{N}}, \tag{6.28a}$$

$$\mathfrak{q}_{\mathrm{ex},n} := \sum_{o \in \mathcal{E}_{\mathrm{T},n}^+ \cup \mathcal{E}_{v,n}^+} \mathfrak{q}_{\mathrm{T},o} - \sum_{o \in \mathcal{E}_{\mathrm{T},n}^- \cup \mathcal{E}_{v,n}^-} \mathfrak{q}_{\mathrm{T},o} - \mathfrak{q}_{\mathfrak{l},n}, \tag{6.28b}$$

$$\mathfrak{C}_{\mathrm{ef},n} = \sum_{o \in \mathcal{E}_{\mathrm{T},n}^+ \cup \mathcal{E}_{\mathrm{T},n}^-} \mathfrak{C}_o + \mathfrak{C}_n. \tag{6.28c}$$

where $\mathfrak{p}_n \in \mathbb{R}_{\geq 0}$ is the absolute pressure of the node, $\mathfrak{q}_{\mathrm{ex},n} \in \mathbb{R}$ is the sum of the external flow rates into and out of the node. The flow rate $\mathfrak{q}_{\mathrm{ex},n}$ consists of the pipeline flow rates $\mathfrak{q}_{\mathrm{T},o} \in \mathbb{R}$ which connect to node n, along with the aggregate load flow rate at the node $\mathfrak{q}_{\mathfrak{l},n} \in \mathbb{R}$. The effective capacitive volume of the node $\mathfrak{C}_{\mathrm{ef},n} \in \mathbb{R}_{>0}$ comprises the capacitive volumes of the pipelines $\mathfrak{C}_o \in \mathbb{R}_{\geq 0}$ (6.14d) which connect to the node, along with the capacitive volume $\mathfrak{C}_n \in \mathbb{R}_{\geq 0}$ of an optional storage tank situated at the node.

Finally, the supply nodes in $\mathcal{N}_{\mathrm{S}}$ are characterised by a constant pressure, i.e.,

$$\dot{\mathfrak{p}}_n \equiv 0, \qquad \mathfrak{p}_n = \hat{\mathfrak{p}}_n, \qquad \mathfrak{q}_{\mathrm{ex},n} \in \mathbb{R}, \qquad n \in \mathcal{N}_{\mathrm{S}}, \tag{6.29}$$

where $\hat{\mathfrak{p}}_n \in \mathbb{R}_{\geq 0}$ is the designated constant pressures of the respective supply nodes and $\mathfrak{q}_{\mathrm{ex},n}$ defined in (6.28b) is a free variable.

Remark 6.9. *While the supply nodes $n \in \mathcal{N}_{\mathrm{S}}$ are considered as having a fixed pressure, such fixed pressure nodes can also be realised practically as a normal node* (6.28) *with a large local storage tank which is connected via a valve-equipped pipeline or a compressor unit. In such a case, however, the equilibrium of the normal node will only be AS if the steady-state gas flow into that node sums to zero.*

[13] If any of these assumptions do not apply, the pipeline may simply be spilt apart to obtain a section of the pipeline which contains the valve and fulfils these requirements.

6.2.4 Compressor Units

For the final component in the gas network, consider an edge $j \in \mathcal{E}_{\mathrm{Cs}}$ which connects a source node $k \in \mathcal{N}^{+}_{\mathrm{Cs},j}$ to a sink node $l \in \mathcal{N}^{-}_{\mathrm{Cs},j}$, as depicted in Figure 6.8. The compressor unit j is made up of three capacitive volumes, one at the inlet with pressure $\mathfrak{p}_k \in \mathbb{R}_{\geq 0}$, one at the outlet with pressure $\mathfrak{p}_l \in \mathbb{R}_{\geq 0}$, and one with a constant pressure $\mathfrak{p}_{\mathrm{B},j} > \mathfrak{p}_k$.[14] The compressor unit allows pressure regulation with a bidirectional flow of gas between the inlet and outlet by splitting the operation between a forward path with flow rate $\mathfrak{q}_{\mathrm{F},j} \in \mathbb{R}$ and a backward path with flow rate $\mathfrak{q}_{\mathrm{B},j} \in \mathbb{R}$. The backward path consists of a pipe of negligible length and a recycle valve with a controllable stem position $s_{\mathrm{B},j}$. The forward path, on the other hand, comprises a centrifugal compressor along with a CCV with the stem position $s_{\mathrm{F},j}$. The backward path thus can be used to increase the rate of change of $\mathfrak{p}_k$ by increasing the flow of gas into node k, whereas the forward path allows gas to flow out of node k, decreasing the rate of change of $\mathfrak{p}_k$. This allows the pressure at $\mathfrak{p}_k$ to be regulated through a bidirectional flow of gas.

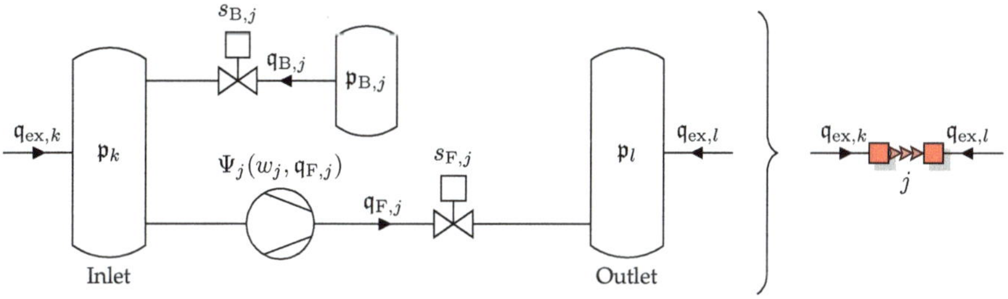

Figure 6.8: Schematic of a compressor unit comprising input and output capacitive volumes, a centrifugal compressor, two control valves and a tank with constant pressure $\mathfrak{p}_{\mathrm{B},j}$. The network representation of the compressor unit corresponding to Figure 6.7 is given on the right.

Before stating the model for the compressor unit, the operation of a centrifugal compressor is briefly recalled from [GE99b; Gra+00]. By using a machine to spin an impeller, gas at the inlet of a centrifugal compressor is imparted with radial kinetic energy. The gas moving in a radially outward direction is gathered and decelerated by a diffuser, increasing the pressure by $\Psi_j(\cdot)\mathfrak{p}_k > 0$, with the compressor characteristic

$$\Psi_j(\omega_j, \mathfrak{q}_{\mathrm{F},j}) = \left[1 + \frac{\varsigma_j r_{2,j}^2 \omega_j^2 - \frac{r_{1,j}^2}{2}(\omega_j - \gamma_{\mathrm{C},j} \varrho_{\mathrm{n}} \mathfrak{q}_{\mathrm{F},j})^2 - \zeta_{\mathrm{k},j} \varrho_{\mathrm{n}}^2 \mathfrak{q}_{\mathrm{F},j}^2}{c_{\mathrm{p}} T_k}\right]^{\frac{c_{\mathrm{p}}}{c_{\mathrm{p}} - c_{\mathrm{V}}}}, \tag{6.30}$$

where $\mathfrak{q}_{\mathrm{F},j} \in \mathbb{R}$ is the flow rate through the compressor and $\omega_j \in \mathbb{R}_{\geq 0}$ is the rotational velocity of the machine. Furthermore, $0 < \varsigma_j < 1$ is the slip factor, $\gamma_{\mathrm{C},j} \in \mathbb{R}_{>0}$ is the zero incidence loss parameter, $\zeta_{\mathrm{k},j} \in \mathbb{R}_{>0}$ is the fluid friction constant, $r_{1,j} \in \mathbb{R}_{>0}$ is inlet radius, $r_{2,j} \in \mathbb{R}_{>0}$ is the impeller radius, and $T_k \in \mathbb{R}_{>0}$ is the temperature at the

[14] The constant pressure can be maintained, e.g., through a pressure-regulating valve connected to the outlet of the compressor unit.

inlet. The constants $c_\mathrm{p}, c_\mathrm{V} \in \mathbb{R}_{>0}$ describe the specific heat at constant pressure and constant volume, respectively. As is common in the literature, it is assumed that the gas at the outlet of the centrifugal compressor is appropriately cooled. Compressor surge is characterised by the regions where $\partial\Psi_j/\partial\mathfrak{q}_{\mathrm{F},j}$ is positive, which typically occurs at low flow rates. Note that (6.30) holds for positive flow rates $\mathfrak{q}_{\mathrm{F},j} \geq 0$. While the compressor characteristic can also be defined for $\mathfrak{q}_{\mathrm{F},j} < 0$ (see [Gra+04]), this region has $\partial\Psi_j/\partial\mathfrak{q}_{\mathrm{F},j} < 0$ and is thus not generally problematic for the stability. The characteristic mapping for $\mathfrak{q}_\mathrm{F} < 0$ is omitted here for simplicity, and by assuming that the CCV in Figure 6.8 will prevent backflow through the compressor.

The dynamics of the forward path for the edge $j \in \mathcal{E}_\mathrm{Cs}$, with $k \in \mathcal{N}^+_{\mathrm{Cs},j}$ and $l \in \mathcal{N}^-_{\mathrm{Cs},j}$, comprising a machine-driven compressor and a CCV (6.25b), are (see [GE99b; Gra+00])

$$\mathfrak{L}_j\dot{\mathfrak{q}}_{\mathrm{F},j} = \Psi_j(\omega_j, \mathfrak{q}_{\mathrm{F},j})\mathfrak{p}_k - \mathfrak{p}_l - \frac{\Psi_j(\omega_j, \mathfrak{q}_{\mathrm{F},j})\mathfrak{p}_k|\mathfrak{q}_{\mathrm{F},j}|\mathfrak{q}_{\mathrm{F},j}}{C^2_{v,j}s^2_{\mathrm{F},j}}, \quad j \in \mathcal{E}_\mathrm{Cs}, \tag{6.31a}$$

$$J_j\dot{\omega}_j = -\zeta_{\omega,j}\omega_j - \varsigma_j r^2_{2,j}\varrho_\mathrm{n}\mathfrak{q}_{\mathrm{F},j}\omega_j + \tau_j, \tag{6.31b}$$

where $\mathfrak{L}_j$ is the inductive factor of the pipe calculated with (6.14d), J_j is the rotational inertia, and $\zeta_{\omega,j} \in \mathbb{R}_{>0}$ is the rotational friction factor. The term $\varsigma_j r^2_{2,j}\varrho_\mathrm{n}\mathfrak{q}_{\mathrm{F},j}\omega_j$ describes the pushback torque of the gas on the machine, and $\tau_j \in \mathbb{R}$ is the torque which represents a control input for the machine. Note that it is assumed that the pipe friction losses are dominated by the valve losses and are thus omitted for simplicity. Furthermore, Remark 6.8 also applies to the flow-rate dynamics in (6.31a).

Finally, the dynamics of the capacitive volumes are given by

$$\mathfrak{C}_{\mathrm{ef},k}\dot{\mathfrak{p}}_k = \mathfrak{q}_{\mathrm{ex},k} + \mathfrak{q}_{\mathrm{B},j} - \mathfrak{q}_{\mathrm{F},j}, \quad k \in \mathcal{N}^+_{\mathrm{Cs},j}, \tag{6.31c}$$

$$\mathfrak{C}_{\mathrm{ef},l}\dot{\mathfrak{p}}_l = \mathfrak{q}_{\mathrm{ex},l} + \mathfrak{q}_{\mathrm{F},j}, \quad l \in \mathcal{N}^-_{\mathrm{Cs},j}, \tag{6.31d}$$

with $\mathfrak{p}_k \leq \eta_\mathfrak{p}\mathfrak{p}_\mathrm{B}$, $0 < \eta_\mathfrak{p} < 1$, and where $\mathfrak{C}_{\mathrm{ef},k}, \mathfrak{C}_{\mathrm{ef},l} \in \mathbb{R}_{>0}$ are the respective effective capacitive volumes of nodes k and l, calculated using (6.28c). The external flow rates $\mathfrak{q}_{\mathrm{ex},k}, \mathfrak{q}_{\mathrm{ex},l} \in \mathbb{R}$ are derived as in (6.28b). The upper limit of $\mathfrak{p}_k$ ensures a sufficient differential pressure across the valve for regulating $\mathfrak{p}_k$ via the backflow valve with the stem position s_B.

6.3 Control Problem and Controller Design

Using the gas network modelled as a graph $\mathcal{G}_\mathrm{g}$ with nodes described by (6.29) and (6.28) along with the edges (6.26), (6.27), and (6.31), this section focuses on model-based controllers which aim to achieve a stable pressure regulation in the gas network (see Definition 2.2). To this end, the requirements of RQ 3.2 are formalised as a control objective for the valves and compressor units in **Section 6.3.1**, and an assumption about the equilibrium of the gas network is made. In **Section 6.3.2**, a passivating pressure-regulating controller is formulated for the valve-equipped pipelines, and it is verified that the controller creates a steady state characterised by the desired pressure setpoint. Then, in **Section 6.3.3**, a controller for a compressor unit is proposed that regulates the inlet pressure of the compressor unit via the CCV, the backflow valve, and the compressor torque. It is also verified that the steady state of the controlled compressor unit is characterised by the desired inlet pressure setpoint.

6.3.1 Control Problem Formulation

Using the gas network model as detailed in Section 6.2, control objectives can be formulated for the valve-equipped pipelines and the compressor units. These objectives require a decentralised pressure regulation and finally a scalable and topology-independent stability result (see RQ 3.2).

Recalling the theory from Section 3.2.2, a scalable and topology-independent stability result can be obtained by verifying EIP properties for each component in the network. For the valve-equipped pipelines, a controller is desired which can regulate the pressure at the outlet of the valve.

Objective 6.1 (Valve Pressure Regulation)
For the valve-equipped edges $o \in \mathcal{E}_v$ described by (6.27), find $s_{v,o}$ such that $\hat{\mathfrak{p}}_{n_{o,-}} = \mathfrak{p}^\star_{n_{o,-}}$, where $n_{o,-}$ is the sink of the edge o. Furthermore, the dynamics of the controlled valve-equipped pipeline should be OS-EIP w.r.t. the input-output port $(\tilde{\mathfrak{p}}_{n_{o,+}} - \tilde{\mathfrak{p}}_{n_{o,-}}, \tilde{\mathfrak{q}}_{v,o})$.

A similar objective can be formulated for the pressure regulation of the compressor units. Since these compressor units are to be used to regulate the pressure in a gas network with a bottom-up supply of gas, i.e., where excess gas in a low-pressure section should be injected into a higher-pressure region, the focus is placed on regulating the pressure on the inlet of the compressor unit. This leads to the following objective formulation.

Objective 6.2 (Compressor Unit Inlet Pressure Regulation)
For each compressor unit $j \in \mathcal{E}_{\mathrm{Cs}}$ with $k \in \mathcal{N}^+_{\mathrm{Cs},j}$ and $l \in \mathcal{N}^-_{\mathrm{Cs},j}$ described by (6.31), find $s_{\mathrm{B},j}$, $s_{\mathrm{F},j}$, and τ_j such that $\hat{\mathfrak{p}}_k = \mathfrak{p}^\star_k$. The controlled compressor unit should simultaneously be EIP w.r.t. the inlet input-output port $(\tilde{\mathfrak{q}}_{\mathrm{ex},k}, \tilde{\mathfrak{p}}_k)$ and the outlet input-output port $(\tilde{\mathfrak{q}}_{\mathrm{ex},l}, \tilde{\mathfrak{p}}_l)$, while also being EISD, and having an AS equilibrium if the inputs $\tilde{\mathfrak{q}}_{\mathrm{ex},k}$ and $\tilde{\mathfrak{q}}_{\mathrm{ex},l}$ are zero.

Remark 6.10. *Recall from the forward flow-rate dynamics (6.31a) of the compressor unit that the CCV acts on the same mode as the compressor characteristic $\Psi_j(\cdot)$, allowing for stabilisation by injecting damping onto the $\mathfrak{q}_{\mathrm{F},j}$ state. However, this damping incurs a pressure loss over the CCV which counteracts the pressure increase from the centrifugal compressor. Thus, to minimise such losses and to maximise efficiency, the stem position $s_{\mathrm{F},j}$ of the CCV should be as open as possible when the compressor is running while still injecting enough damping to prevent surge from occurring.*

By designing controllers that meet Objectives 6.1 and 6.2, EIP properties are obtained for the valve-equipped edges and the compressor units. These properties pave the way towards an AS investigation via Theorem 3.3 in the sequel. Note that it is assumed that suitable pressure setpoints are provided for the control units and that the pressure regulation of the network is feasible, i.e., all components have appropriate dimensions and parameters, as stipulated in the following assumption.

Assumption 6.2 (Feasible Gas Network)
The loads with $\mathfrak{q}_n > 0$, $n \in \mathcal{N}_{\mathrm{g}}$, in the gas network can feasibly be supplied by the sources of gas—including gas sources with $\mathfrak{q}_n < 0$, $n \in \mathcal{N}_{\mathrm{g}}$, supply nodes in $\mathcal{N}_{\mathrm{S}}$, and the gas stored in the capacitive volumes $\mathfrak{C}_{\mathrm{ef},n}$ of the nodes—over the network of pipelines and compressor units describing the edges in $\mathcal{E}_{\mathrm{g}}$.

6.3.2 Valve Control Design

Consider a valve-equipped pipeline $o \in \mathcal{E}_v$ where the outlet pressure $\mathfrak{p}_{n_{o,-}}$ is to be controlled to a desired setpoint $\mathfrak{p}^{\star}_{n_{o,-}}$ as per Objective 6.1. To this end, let the valve-equipped pipeline (6.27) be equipped with the controller

$$s_{v,o} = \begin{cases} 0, & \text{if } (\mathfrak{p}_{n_{o,-}} > \mathfrak{p}^{\star}_{n_{o,-}}) \vee (\mathfrak{p}_{n_{o,-}} \geq \mathfrak{p}_{n_{o,+}}), \\ \left[k_{v,o}(\mathfrak{p}^{\star}_{n_{o,-}} - \mathfrak{p}_{n_{o,-}}) + \dfrac{\mathfrak{q}_{v,o}}{C_{v,o}} \right] \sqrt{\dfrac{\mathfrak{p}_{n_{o,+}}}{|\mathfrak{p}_{n_{o,+}} - \mathfrak{p}_{n_{o,-}}|}}, & \text{otherwise}, \end{cases} \tag{6.32}$$

where $k_{v,o} \in \mathbb{R}_{>0}$ is the gain of the controller. Note that it is assumed that the inlet pressure is greater than the outlet pressure for the pressure-regulating valve. Thus, the valve is closed when $\mathfrak{p}_{n_{o,-}} \geq \mathfrak{p}_{n_{o,+}}$. Due to this assumption, the valve is also closed when the setpoint is lower than the outlet pressure, i.e., when $\mathfrak{p}_{n_{o,-}} > \mathfrak{p}^{\star}_{n_{o,-}}$.

The following proposition investigates the steady state of the controlled valve-equipped pipeline and shows that it meets the equilibrium requirements of Objective 6.1.

Proposition 6.2 (Pressure Regulated Valve Equilibrium)
Consider a valve-equipped pipeline $o \in \mathcal{E}_v$ with the dynamics (6.27) in closed loop with the controller (6.32). If $\mathfrak{p}_{n_{o,-}} \leq \mathfrak{p}^\star_{n_{o,-}} < \mathfrak{p}_{n_{o,+}}$, then the closed loop has an equilibrium characterised by $\hat{\mathfrak{p}}_{n_{o,-}} = \mathfrak{p}^\star_{n_{o,-}}$.

Proof:
Substitute the valve controller (6.32) into the steady-state flow rate through a valve (6.25a) with $\mathfrak{p}_{n_{o,-}} \leq \mathfrak{p}^\star_{n_{o,-}} < \mathfrak{p}_{n_{o,+}}$, to obtain

$$\hat{\mathfrak{q}}_{v,o} = C_{v,o} s_{v,o} \sqrt{\frac{|\hat{\mathfrak{p}}_{n_{o,+}} - \hat{\mathfrak{p}}_{n_{o,-}}|}{\hat{\mathfrak{p}}_{n_{o,+}}}} \tag{6.33a}$$

$$\hat{\mathfrak{q}}_{v,o} = C_{v,o} k_{v,o} (\mathfrak{p}^\star_{n_{o,-}} - \hat{\mathfrak{p}}_{n_{o,-}}) + \hat{\mathfrak{q}}_{v,o} \tag{6.33b}$$

$$0 = C_{v,o} k_{v,o} (\mathfrak{p}^\star_{n_{o,-}} - \hat{\mathfrak{p}}_{n_{o,-}}), \tag{6.33c}$$

from which it follows that $\mathfrak{p}^\star_{n_{o,-}} = \hat{\mathfrak{p}}_{n_{o,-}}$, since $C_{v,o}, k_{v,o} > 0$. ■

Remark 6.11. *Gas pressure regulation is often achieved through pressure-regulating or pressure-relief valves, in which the outlet pressure is regulated using a hydraulic-mechanical feedback loop [MF19].*

Remark 6.12. *In a practical setting, the flow-rate dependent term $\mathfrak{q}_{v,o}/C_{v,o}$ in (6.32) can be generated using an integrator $\dot{\varepsilon}_o = (\mathfrak{p}^\star_{n_{o,-}} - \mathfrak{p}_{n_{o,-}})/k^{\mathrm{I}}_{v,o}$, $k^{\mathrm{I}}_{v,o} \in \mathbb{R}_{>0}$. This circumvents the need to measure the flow rate $\mathfrak{q}_{v,o}$ while also providing robustness against uncertain $C_{v,o}$ values.*

6.3.3 Compressor Unit Control Design

Consider now a compressor unit $j \in \mathcal{E}_{\mathrm{Cs}}$, along with its inlet node $k \in \mathcal{N}^+_{\mathrm{Cs},j}$ and outlet node $l \in \mathcal{N}^-_{\mathrm{Cs},j}$. For the compressor unit, the inlet pressure $\mathfrak{p}_k$ should be regulated to a desired setpoint $\mathfrak{p}^\star_k$ while also minimising the usage of the CCV, as per Objective 6.2 and Remark 6.10. The control objective is achieved in the following design steps. First, the backflow valve with $s_{\mathrm{B},j}$ is controlled to inject damping onto the state $\mathfrak{p}_k$ relative to the desired setpoint $\mathfrak{p}^\star_k$. Next, nonlinear damping is injected onto the forward flow rate $\mathfrak{q}_{\mathrm{F},j}$ through the CCV with $s_{\mathrm{F},j}$ such that the amount of damping is high for low flow rates where surge can occur while decreasing when $\mathfrak{q}_{\mathrm{F},j}$ increases. Finally, the torque of the compressor is controlled using a PI-like structure on the error $\mathfrak{p}_k - \mathfrak{p}^\star_k$. The control output $s_{\mathrm{B},j}$ for the backflow valve is also used to decrease the torque such that the machine of the compressor slows down when gas flows along the backflow path.

The controller with the setpoint $0 < \mathfrak{p}_k^\star < \eta_{\mathfrak{p},j}\mathfrak{p}_{\mathrm{B},j}$ is given by

$$s_{\mathrm{B},j} = \left[\beta_{\mathrm{B},j}(\mathfrak{p}_k - \mathfrak{p}_k^\star) + \frac{\mathfrak{q}_{\mathrm{B},j}}{C_{v,j}}\right]\sqrt{\frac{\mathfrak{p}_{\mathrm{B},j}}{|\mathfrak{p}_{\mathrm{B},j} - \mathfrak{p}_k|}}, \qquad j \in \mathcal{E}_{\mathrm{Cs}}, k \in \mathcal{N}_{\mathrm{Cs},j}^+, \tag{6.34a}$$

$$s_{\mathrm{F},j} = \begin{cases} 0, & \mathfrak{q}_{\mathrm{F},j} < 0, \\ \sqrt{\Psi_j(\cdot)\mathfrak{p}_{k,j}}\ \beta_{\mathrm{F},j}\mathfrak{q}_{\mathrm{F},j}^{\frac{3}{4}}, & \mathfrak{q}_{\mathrm{F},j} \geq 0, \end{cases} \tag{6.34b}$$

$$\tau_j = \beta_{\mathfrak{p},j}(\mathfrak{p}_k - \mathfrak{p}_k^\star) + \varepsilon_{\mathrm{C},j} - \frac{\beta_{\mathfrak{p},j}\beta_{\mathrm{st},j}}{\beta_{\varepsilon,j}} s_{\mathrm{B},j} - \beta_{\omega,j}(\omega_j - \omega_{0,j}), \tag{6.34c}$$

$$\dot{\varepsilon}_{\mathrm{C},j} = \beta_{\varepsilon,j}(\mathfrak{p}_k - \mathfrak{p}_k^\star) - \beta_{\mathrm{st},j} s_{\mathrm{B},j}, \tag{6.34d}$$

$$\omega_{0,j} = \sqrt{\frac{c_p T_k}{\varsigma_j r_{2,j}^2 - \frac{r_{1,j}^2}{2}}\left(\Psi_{j,0}^{\frac{c_p - c_V}{c_p}} - 1\right)}, \tag{6.34e}$$

where $\beta_{\mathrm{B},j}, \beta_{\mathrm{F},j}, \beta_{\mathfrak{p},j}, \beta_{\omega,j}, \beta_{\mathrm{st},j}, \beta_{\varepsilon,j} \in \mathbb{R}_{>0}$ are the control gains, $\varepsilon_{\mathrm{C},j} \in \mathbb{R}$ is the augmented integral state, and $\omega_{0,j} \in \mathbb{R}_{\geq 0}$ is an optional term providing faster start-up for the machine. The gains $\beta_{\mathrm{B},j}$ and $\beta_{\mathrm{F},j}$ modulate the amount of damping injected onto the $\mathfrak{p}_k$ and $\mathfrak{q}_{\mathrm{F},j}$ states, respectively. Furthermore, the control law in (6.34b) compensates the inlet pressure term of the valve in (6.31a) through $\sqrt{\Psi_j(\cdot)\mathfrak{p}_{k,j}}$ and sets a flow-rate characteristic of $\mathfrak{q}_{\mathrm{F},j}^{\frac{3}{4}}$ for $s_{\mathrm{F},j}$, while also closing fully to prevent negative flow rates.[15] The torque τ_j (6.34c) comprises PI terms formed by $\beta_{\mathfrak{p},j}$, $\beta_{\varepsilon,j}$, and $\varepsilon_{\mathrm{C},j}$, while $\beta_{\omega,j}$ injects damping onto the rotational velocity state ω_j. The gain β_{st} in (6.34c) and (6.34d) leads to a reduction in the torque while the backflow value is open. Since $s_{\mathrm{B},j}$ is also included in the integral state dynamics (6.34d), the torque reduction is continual until either $\omega_j = 0$ or $s_{\mathrm{B},j} = 0$. Finally, the optional start-up term $\omega_{0,j}$ is derived from the compressor characteristic (6.30) for a chosen fixed pressure ratio $\Psi_{j,0} > 1$ with $\mathfrak{q}_{\mathrm{F},j} = 0$. Note that this term is only relevant at start-up and will be dominated by the integral state ε_{C} as time progresses. Furthermore, note that Remarks 6.11 and 6.12 also apply to (6.34a).

To analyse the controlled compressor unit in the sequel, the closed-loop dynamics are formulated as a nonlinear input-affine system. The state transformation

$$z_j := \frac{\beta_{\mathfrak{p},j}}{\beta_{\varepsilon,j}}\varepsilon_{\mathrm{C},j} - J_j\omega_j, \tag{6.35}$$

[15] Note that the term $\Psi_j(\cdot)\mathfrak{p}_{k,j}$ in (6.34b) corresponds to the pressure at the inlet of the CCV. Since this value can be obtained via a measurement, an inclusion of the complex and potentially uncertain compressor characteristic $\Psi_j(\cdot)$ in the controller can be circumvented in a practical setting.

is made to simplify the analysis in the sequel. Applying the controller (6.34) and the transformation (6.35) to the combined compressor unit dynamics (6.31) yields the controlled system in the form of a nonlinear input-affine system (3.21), i.e.,

$$\underbrace{\begin{bmatrix} \mathfrak{C}_{\mathrm{ef},k}\dot{\mathfrak{p}}_k \\ \mathfrak{C}_{\mathrm{ef},l}\dot{\mathfrak{p}}_l \\ \mathfrak{L}_j\dot{\mathfrak{q}}_{\mathrm{F},j} \\ J_j\dot{\omega}_j \\ \dot{z}_j \end{bmatrix}}_{\boldsymbol{Q}_{\mathrm{C},j}\dot{\boldsymbol{x}}_{\mathrm{C},j}} = \underbrace{\begin{bmatrix} -\frac{C_{v,j}\beta_{\mathrm{B},j}}{2}(\mathfrak{p}_k - \mathfrak{p}_k^\star) - \mathfrak{q}_{\mathrm{F},j} \\ \mathfrak{q}_{\mathrm{F},j} \\ \Psi_j(\mathfrak{q}_{\mathrm{F},j}, \mathfrak{p}_{\mathrm{mean},j})\mathfrak{p}_k - \mathfrak{p}_l - \mathfrak{R}_{\mathrm{F},j}(\mathfrak{q}_{\mathrm{F},j}) \\ f_{\omega,j}(\mathfrak{p}_k) - f_{\mathrm{z},j}(\mathfrak{q}_{\mathrm{F},j}, \omega_j, z_j) \\ f_{\mathrm{z},j}(\mathfrak{q}_{\mathrm{F},j}, \omega_j, z_j) \end{bmatrix}}_{\boldsymbol{f}_{\mathrm{C},j}(\boldsymbol{x}_{\mathrm{C},j})} + \underbrace{\begin{bmatrix} \mathfrak{q}_{\mathrm{ex},k} \\ \mathfrak{q}_{\mathrm{ex},l} \\ 0 \\ 0 \\ 0 \end{bmatrix}}_{\boldsymbol{B}_{\mathrm{C},j}\boldsymbol{u}_{\mathrm{C},j}}, \tag{6.36a}$$

$$\boldsymbol{y}_{\mathrm{C},j} = \underbrace{\begin{bmatrix} 1 & 0 & 0 & 0 & 0 \\ 0 & 1 & 0 & 0 & 0 \end{bmatrix}}_{\boldsymbol{B}_{\mathrm{C},j}^T} \boldsymbol{x}_{\mathrm{C},j} = \begin{bmatrix} \mathfrak{p}_k \\ \mathfrak{p}_l \end{bmatrix}, \tag{6.36b}$$

where $\mathfrak{q}_{\mathrm{F},j} \geq 0$, $\boldsymbol{x}_{\mathrm{C},j}^T = [\mathfrak{p}_k, \mathfrak{p}_l, \mathfrak{q}_{\mathrm{F},j}, \omega_j, z_j]$, $\boldsymbol{u}_{\mathrm{C},j}^T = [\mathfrak{q}_{\mathrm{ex},k}, \mathfrak{q}_{\mathrm{ex},l}]$, $\Psi_j(\mathfrak{q}_{\mathrm{F},j}, \mathfrak{p}_{\mathrm{mean},j})$ is given in (6.30), and

$$\mathfrak{R}_{\mathrm{F},j}(\mathfrak{q}_{\mathrm{F},j}) := \frac{\sqrt{\mathfrak{q}_{\mathrm{F},j}}}{C_{v,j}^2\beta_{\mathrm{F},j}^2}, \tag{6.37a}$$

$$f_{\omega,j}(\mathfrak{p}_k) := \beta_{\mathfrak{p},j}\left(1 - \frac{\beta_{\mathrm{B},j}\beta_{\mathrm{st},j}}{2\beta_{\varepsilon,j}}\sqrt{\frac{\mathfrak{p}_{\mathrm{B},j}}{|\mathfrak{p}_{\mathrm{B},j} - \mathfrak{p}_k|}}\right)(\mathfrak{p}_k - \mathfrak{p}_k^\star), \tag{6.37b}$$

$$f_{\mathrm{z},j}(\mathfrak{q}_{\mathrm{F},j}, \omega_j, z_j) := \left(\beta_{\omega,j} + \zeta_{\omega,j} + \varsigma_j r_{2,j}^2 \varrho_{\mathrm{n}} \mathfrak{q}_{\mathrm{F},j} - \frac{\beta_{\varepsilon,j}J_j}{\beta_{\mathfrak{p},j}}\right)\omega_j + \beta_{\omega,j}\omega_{0,j} - \frac{\beta_{\varepsilon,j}}{\beta_{\mathfrak{p},j}}z_j. \tag{6.37c}$$

A full proof of the derivation of (6.36) and (6.37) can be found in Appendix A.2. Note that the output in (6.36b) is chosen such that the input-output ports $(\mathfrak{q}_{\mathrm{ex},k}, \mathfrak{p}_k)$ and $(\mathfrak{q}_{\mathrm{ex},l}, \mathfrak{p}_l)$ correspond to the ports of the pipelines and loads which may connect to the respective inlet and outlet sides of the compressor unit. Next, it is verified that the closed-loop equilibrium is set by the pressure regulation as required by Objective 6.2.

Proposition 6.3 (Steady-State Compressor Unit Pressure Regulation) *Consider the controlled compressor unit in* (6.36) *with* $j \in \mathcal{E}_{\mathrm{Cs}}$ *and* $k \in \mathcal{N}^{+}_{\mathrm{Cs},j}$. *If*

$$2\beta_{\varepsilon,j} < \beta_{\mathrm{st},j}\beta_{\mathrm{B},j} \quad \vee \quad 2\beta_{\varepsilon,j} > \beta_{\mathrm{st},j}\beta_{\mathrm{B},j}\sqrt{\frac{1}{1-\eta_{\mathfrak{p},j}}}, \tag{6.38}$$

then the steady state of the controlled compressor unit has $\hat{\mathfrak{p}}_k = \mathfrak{p}_k^\star$.

Proof:
To determine the steady state of (6.36), set $\dot{z}_j = f_{z,j}(\cdot) = 0$. Next, set $\mathfrak{L}_j\dot{\omega}_j = 0$ which implies that $f_{\omega,j}(\cdot) = 0$. Substituting this into (6.37b) leads to

$$f_{\omega,j}(\hat{\mathfrak{p}}_k) = 0 \implies \frac{2\beta_{\varepsilon,j}}{\beta_{\mathrm{B},j}\beta_{\mathrm{st},j}} = \sqrt{\frac{\mathfrak{p}_{\mathrm{B},j}}{|\mathfrak{p}_{\mathrm{B},j} - \hat{\mathfrak{p}}_k|}} \quad \vee \quad \hat{\mathfrak{p}}_k = \mathfrak{p}_k^\star. \tag{6.39}$$

Now, since $0 < \mathfrak{p}_k \le \eta_{\mathfrak{p},j}\mathfrak{p}_{\mathrm{B},j}$, a bound for the square root can be found, which in turn places a bound on the first solution of (6.39), i.e.,

$$1 \le \sqrt{\frac{\mathfrak{p}_{\mathrm{B},j}}{|\mathfrak{p}_{\mathrm{B},j} - \hat{\mathfrak{p}}_k|}} \le \sqrt{\frac{1}{1-\eta_{\mathfrak{p},j}}} \implies 1 \le \frac{2\beta_{\varepsilon,j}}{\beta_{\mathrm{B},j}\beta_{\mathrm{st},j}} \le \sqrt{\frac{1}{1-\eta_{\mathfrak{p},j}}}. \tag{6.40}$$

By choosing the control parameters according to (6.38), (6.40) and therefore the first solution of (6.39) becomes impossible. This forces the steady-state solution $\hat{\mathfrak{p}}_k = \mathfrak{p}_k^\star$. ■

Through Proposition 6.3, it is shown how the pressure regulated steady state for the controlled compressor unit, as required by Objective 6.2, is ensured by choosing the controller gains such that (6.38) holds.

Remark 6.13. *Careful observation of* (6.36) *and* (6.37) *reveals that the control gains* $\beta_{\mathrm{B},j}$, $\beta_{\mathrm{F},j}$, $\beta_{\omega,j}$, *and* $\beta_{\varepsilon,j}/\beta_{\mathfrak{p},j}$ *modulate the damping injected onto the states* $\mathfrak{p}_k$, $\mathfrak{q}_{\mathrm{F},j}$, ω_j, *and* z_j, *respectively, while the state* $\mathfrak{p}_l$ *is undamped. This allows some manner of intuition when tuning the control gains, with higher damping factors generally yielding faster convergence through more aggressive control outputs. Note that* $\mathfrak{R}_{\mathrm{F},j}$ *in* (6.37a) *is inversely proportional to* $\beta_{\mathrm{F},j}$. *Thus, decreasing* $\beta_{\mathrm{F},j}$ *increases the damping injected onto* $\mathfrak{q}_{\mathrm{F},j}$.

6.4 Passivity and Stability Analysis

By applying the controller proposed in Section 6.3 to the valve-equipped pipelines and the compressor unit, the desired steady state is produced for the respective pressures at the pipeline outlets and compressor unit inlets. In this section, it is investigated under which conditions a gas network as modelled in Section 6.2 will converge to an equilibrium characterised by the desired pressure regulation. To this end, EIP properties for gas pipeline nodes and edges (**Section 6.4.1**), for controlled valve-equipped pipelines (**Section 6.4.2**), and for controlled compressor units (**Section 6.4.3**) are investigated. Finally, these results are combined in **Section 6.4.4**, where the EIP properties of the gas network components are used to derive the AS of the gas network equilibria.

6.4.1 Pipeline Nodes and Edges EIP

Starting with the nodes representing the endpoints of pipelines in $\mathcal{N}_{\mathrm{N}}$ described by (6.28), the EIP of its hydraulic port is investigated.

Proposition 6.4 (Pipeline Endpoint EIP)
A node $n \in \mathcal{N}_{\mathrm{N}}$ *with the dynamics* (6.28) *and the storage function*

$$S_{\mathrm{N},n}(\tilde{\mathfrak{p}}_n) = \frac{\mathfrak{C}_{\mathrm{ef},n}}{2}\tilde{\mathfrak{p}}_n^2, \quad n \in \mathcal{N}_{\mathrm{N}}, \tag{6.41}$$

is EIP w.r.t. the input-output port $(\tilde{\mathfrak{q}}_{\mathrm{ex},n}, \tilde{\mathfrak{p}}_n)$.

Proof:
The result follows as a direct consequence of applying Lemma 3.6 with $\rho = 0$ for the linear, scalar system (6.28), where $f_{\mathrm{N},n}(\mathfrak{p}_n) = 0$ and thus $\nabla f_{\mathrm{N},n} = 0$. ∎

Although Proposition 6.4 demonstrates that the nodes in $\mathcal{N}_{\mathrm{N}}$ are inherently EIP, it is noted that the equilibria of these undamped integrating nodes are GAS if $\mathfrak{q}_{\mathrm{ex},n} \equiv 0$, but not necessarily if $\tilde{\mathfrak{q}}_{\mathrm{ex},n} \equiv 0$. Continuing on to the pipeline edges in $\mathcal{E}_{\mathrm{T}}$, a stronger OS-EIP property can be shown for its hydraulic ports

Theorem 6.5 (Pipeline Edge OS-EIP and EIID)
Consider a pipeline edge $o \in \mathcal{E}_\mathrm{T}$ *with the dynamics (6.26) and where the mean pressure* $\mathfrak{p}_{\mathrm{mean},o}$ *is assumed to be constant.*[16] *Such a pipeline with the storage function*

$$S_{\mathrm{T},o}(\tilde{\mathfrak{q}}_{\mathrm{T},o}) = \frac{\mathfrak{L}_o}{2}\tilde{\mathfrak{q}}_{\mathrm{T},o}^2, \quad o \in \mathcal{E}_\mathrm{T}, \tag{6.42}$$

is OS-EIP w.r.t. the input-output port $(\tilde{\mathfrak{p}}_{n_{o,+}} - \tilde{\mathfrak{p}}_{n_{o,-}}, \tilde{\mathfrak{q}}_{\mathrm{T},o})$. *Moreover, the pipeline is EIID for the input* $\tilde{u}_{\mathrm{T},o} = \tilde{\mathfrak{p}}_{n_{o,+}} - \tilde{\mathfrak{p}}_{n_{o,-}}$.

Proof:
Applying Lemma 3.6 to the nonlinear, scalar system (6.26) results in the nonlinear system dynamics $f_{\mathrm{T},o}(\mathfrak{q}_{\mathrm{T},o}) = -\mathfrak{R}_o(|\mathfrak{q}_{\mathrm{T},o}|, \mathfrak{p}_{\mathrm{mean},o})\mathfrak{q}_{\mathrm{T},o}$, where $f_{\mathrm{T},o}(\mathfrak{q}_{\mathrm{T},o}) = -f_{\mathrm{T},o}(-\mathfrak{q}_{\mathrm{T},o})$. Dropping the positive constant multiplicative terms of $\mathfrak{R}_o(\cdot)$ in (6.14e), the condition (3.22) for OS-EIP, where $\mathfrak{p}_{\mathrm{mean},o}$ is a positive constant, reduces to

$$\exists \rho > 0\colon \frac{\partial(\zeta_{\mathrm{ef},o}|\mathfrak{q}_{\mathrm{T},o}|\mathfrak{q}_{\mathrm{T},o})}{\partial \mathfrak{q}_{\mathrm{T},o}} - \rho \geq 0, \quad \forall \mathfrak{q}_{\mathrm{T},o} \in \mathbb{R}_{>0}, \tag{6.43}$$

which entails finding the derivative for the friction factor under laminar and turbulent conditions (see (6.5), (6.7) and (6.14e)). Although analysing the derivative of the friction factor in (6.43) is complex analytically, it can easily be done numerically. To this end, consider the following function found by substituting the $\mathfrak{q}_{\mathrm{T},o}$ for its associated Reynolds number Re_o in (6.4) and dropping the positive constant multiplicative terms to obtain

$$g_{\mathrm{T},o}(Re_o, \tfrac{r_{\mathrm{T},o}}{D_o}) = \zeta_{\mathrm{ef},o}(Re_o, \tfrac{r_{\mathrm{T},o}}{D_o}) \cdot Re_o^2 \tag{6.44a}$$

$$\Longrightarrow \text{(6.43) holds} \iff \exists \rho > 0\colon \frac{\partial g_{\mathrm{T},o}(Re_o)}{\partial Re_o} \geq \rho, \quad \forall Re_o \in \mathbb{R}_{>0}. \tag{6.44b}$$

Note that $g_{\mathrm{T},o}(\cdot)$ in (6.44a) is only a function of the Reynolds number and the surface-roughness-to-pipe-diameter ratio $r_{\mathrm{T},o}/D_o$ (see (6.7)). Since $g_{\mathrm{T},o}(\cdot)$ is dependent on the friction factor $\zeta_{\mathrm{ef},o}(\cdot)$, the condition in (6.44b) is verified if $g_{\mathrm{T},o}(\cdot)$ increases for both laminar, transitional, and turbulent friction. A numerical computation for $g_{\mathrm{T},o}(\cdot)$ along with its derivative is shown in Figure 6.9, where it can be seen that the laminar and turbulent cases are strictly monotone in Re_o.

[16] Recall that the simplified pipeline model in Section 6.1.3 assumes a constant $\mathfrak{p}_{\mathrm{mean},o}$ (see Remark 6.4) while also showing a high fidelity in the benchmark system.

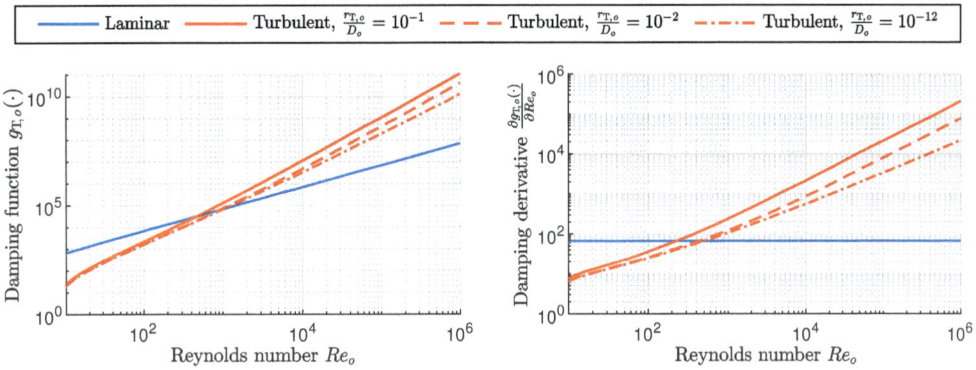

(a) Damping factor $g_{T,o}(\cdot)$. (b) Derivative $\partial g_{T,o}(\cdot)/\partial Re_o$.

Figure 6.9: Scaled pipeline damping factor $g_T(\cdot)$ (a) caused by friction along with its derivative (b) for laminar and turbulent cases.

In the transitional region between laminar to turbulent flow, which typically occurs in $2000 \leq Re_o \leq 4000$, the friction factor transitions between the laminar and the turbulent curves.[17] Monotonicity of the transitional case can thus be assured if it can be ensured that $g_{T,\mathrm{Lam},o}(\cdot)$ for the laminar case is less than $g_{T,\mathrm{Tur},o}(\cdot)$ for the turbulent case when the transition occurs, i.e., that $g_{T,o}(\cdot)$ will increase when changing from the laminar to the turbulent case. In Figure 6.10, the friction type which yields a higher $g_{T,o}(\cdot)$ is plotted to show the Re_o number at which the turbulent factor starts to dominate the laminar factor for a range of surface-roughness-to-pipe-diameter ratios $r_{T,o}/D_o$.

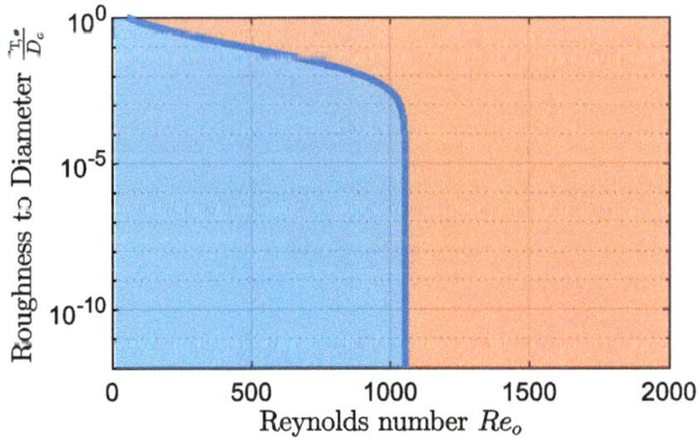

Figure 6.10: Plot of the transition from a dominating laminar damping function $g_{T,\mathrm{Lam},o}(\cdot)$ to a dominating turbulent damping factor $g_{T,\mathrm{Tur},o}(\cdot)$.

17 Although no exact description for transitional friction has been found to date, experimental data suggests that an interpolation between the laminar and turbulent factors takes place in the transitional region [Cer+18, Fig. 1].

As indicated in Figure 6.10, $g_{\mathrm{T,Tur},o}(\cdot)$ dominates $g_{\mathrm{T,Lam},o}(\cdot)$ when $Re_o \geq 1055.61$, with the threshold for the Reynolds number decreasing as $r_{\mathrm{T},o}/D_o$ increases. Thus $g_{\mathrm{T},o}(\cdot)$ is strictly monotone in Re_o, which implies that $f_{\mathrm{T},o}(\mathfrak{q}_{\mathrm{T},o})$ is strictly monotone in $\mathfrak{q}_{\mathrm{T},o}$, completing the OS-EIP requirements for Lemma 3.6. Finally, EIID follows directly from shifting the pipeline dynamics (6.26) w.r.t. an assumed equilibrium and setting $\tilde{\mathfrak{q}}_{\mathrm{T},o} = \tilde{y}_{\mathrm{T},o} \equiv 0$ to obtain $0 = \tilde{\mathfrak{p}}_{n_{o,+}} - \tilde{\mathfrak{p}}_{n_{o,-}}$ since $\mathfrak{p}_{\mathrm{mean}}$ is constant (see Remark 6.4). ■

Through Theorem 6.5, the pipeline edges are shown to be OS-EIP by numerically demonstrating the monotonicity of the pipeline friction function. Despite the numerical nature of the result, the conditions for monotonicity are verified for all practical circumstances (with a surface-roughness-to-pipe-diameter ratio $10^{-12} \leq r_{\mathrm{T},o}/D_o \leq 1$). Indeed, this result extends to any fluid where the laminar and turbulent friction factors can be represented by (6.5) and (6.7), respectively.

Remark 6.14. *The monotonicity of friction has previously been investigated in the context of gas networks in [Dvi+15; VMC12; Zlo+06; MVZ20], where the monotonicity property proves useful for ensuring the tractability of optimisation problems. Nevertheless, the previous approaches assume a constant friction factor $\zeta_{\mathrm{ef}} = const$. To the best of the author's knowledge, the monotonicity of the fluid friction has not previously been demonstrated across all flow rates with laminar, transitional and turbulent friction.*

Remark 6.15. *From the proof of Theorem 6.5 and from Figure 6.9b in particular, it can be seen that the derivative of $g_{\mathrm{T}}(\cdot)$ is lower-bounded by the laminar friction case when the laminar and turbulent cases are combined in their respective domains. Thus, the OS-EIP index can be calculated explicitly by substituting ζ_{Lam} (6.5), Re_o (6.4) for the normalised flow rate $\mathfrak{q}_{\mathrm{T},o}$ (6.10) into $\mathfrak{R}(\cdot)$ (6.14e). This yields the output passivity index*

$$\rho_{\mathrm{T},o} = \frac{8\eta_{\mathrm{ef},o}\eta_{\mathrm{D}}\varrho_{\mathrm{n}}c^2\ell_o}{\pi D_o^4 \mathfrak{p}_{\mathrm{mean},o}}. \tag{6.45}$$

6.4.2 Valve-Equipped Pipeline OS-EIP

After establishing the inherent OS-EIP of the pipelines, a similar OS-EIP property is investigated for the valve-equipped pipelines in closed loop with the controller in Section 6.3.2. Compared to the pipeline analysis in Theorem 6.5, however, the damping injected via the valve dominates the frictional losses in the dynamics of (6.27). The following proposition investigates an OS-EIP property for this closed-loop system w.r.t. its inlet and outlet hydraulic ports.

Proposition 6.6 (Valve-equipped Pipeline OS-EIP and EIID)
Consider a valve-equipped pipeline $o \in \mathcal{E}_v$ with the dynamics (6.27) in closed loop with the controller (6.32). This closed-loop system with the storage function

$$S_{v,o}(\tilde{\mathfrak{q}}_{v,o}) = \frac{\mathfrak{L}_o}{2}\tilde{\mathfrak{q}}_{v,o}^2, \quad o \in \mathcal{E}_v, \tag{6.46}$$

is OS-EIP w.r.t. the input-output port $(\tilde{\mathfrak{p}}_{n_{o,+}} - \tilde{\mathfrak{p}}_{n_{o,-}}, \tilde{\mathfrak{q}}_{v,o})$. Moreover, the valve-equipped pipeline is EIID for the input $\tilde{u}_{v,o} = \tilde{\mathfrak{p}}_{n_{o,+}} - \tilde{\mathfrak{p}}_{n_{o,-}}$.

Proof Sketch:
The proof entails applying Lemma 3.6 for the controlled valve-equipped pipeline and demonstrating that the control law in (6.32) ensures that the conditions in Lemma 3.6 are met. The EIID property is obtained directly from analysing (6.27). The full proof can be found in Appendix A.2. ■

Thus, through the combination of Propositions 6.2 and 6.6, the criteria of Objective 6.1 for the valve-equipped pipelines are met.

6.4.3 Compressor Unit EIP

The controlled compressor unit in (6.36) is now analysed for suitable EIP and AS properties as required in Objective 6.2. To this end, the dynamics of the controlled system are analysed and design criteria for the control parameters are derived under which the desired properties are guaranteed. Note that the index j is omitted where clear from context for notational clarity in this subsection.

Theorem 6.7 (Controlled Compressor Unit EIP, EISD, and AS)
Consider the controlled compressor unit $j \in \mathcal{E}_{\mathrm{Cs}}$ with its inlet $k \in \mathcal{N}^+_{\mathrm{Cs},j}$ and outlet $l \in \mathcal{N}^-_{\mathrm{Cs},j}$ nodes, which together have the dynamics (6.36) and (6.37). The controlled compressor unit with the storage function

$$S_{\mathrm{C},j}(\tilde{\boldsymbol{x}}_{\mathrm{C},j}) = \tfrac{1}{2}\tilde{\boldsymbol{x}}_{\mathrm{C},j}^T \boldsymbol{Q}_{\mathrm{C},j}\tilde{\boldsymbol{x}}_{\mathrm{C},j}, \quad j \in \mathcal{E}_{\mathrm{Cs}}, k \in \mathcal{N}^+_{\mathrm{Cs},j}, l \in \mathcal{N}^-_{\mathrm{Cs},j}, \tag{6.47}$$

is simultaneously EIP w.r.t. the input-output ports $(\tilde{\mathfrak{q}}_{\mathrm{ex},k}, \tilde{\mathfrak{p}}_k)$ and $(\tilde{\mathfrak{q}}_{\mathrm{ex},l}, \tilde{\mathfrak{p}}_l)$, EISD, and has equilibria that are AS for $\tilde{\boldsymbol{u}}_{\mathrm{C},j} = [\tilde{\mathfrak{q}}_{\mathrm{ex},k}\tilde{\mathfrak{q}}_{\mathrm{ex},l}]^T = \mathbf{0}$ if the following inequalities

$$0 < C_v\beta_{\mathrm{B}} - |1 - \Psi| - \beta_{\mathfrak{p}}\varkappa_{\mathrm{C}}, \tag{6.48a}$$
$$0 < 2(\nabla_{\mathfrak{q}_{\mathrm{F}}}\mathfrak{R}_{\mathrm{F}} - \mathfrak{p}_k\nabla_{\mathfrak{q}_{\mathrm{F}}}\Psi) - |1 - \Psi| - |\mu_{\mathrm{C}}\omega - \mathfrak{p}_k\nabla_{\omega}\Psi| - \mu_{\mathrm{C}}\omega, \tag{6.48b}$$

$$0 < \beta_\omega + \zeta_\omega + \mu_\mathrm{C}\mathfrak{q}_\mathrm{F} - \frac{\beta_\varepsilon(1+J)}{\beta_\mathfrak{p}} - \beta_\mathfrak{p}\varkappa_\mathrm{C} - |\mu_\mathrm{C}\omega - \mathfrak{p}_k\nabla_\omega\Psi|\,, \tag{6.48c}$$

$$0 < \frac{2\beta_\varepsilon(1+J)}{\beta_\mathfrak{p}} - \mu_\mathrm{C}(\omega + \mathfrak{q}_\mathrm{F}) - \beta_\omega - \zeta_\omega, \tag{6.48d}$$

hold for all $\boldsymbol{x}_\mathrm{C} \in \mathcal{X}_\mathrm{C}$*, with* $0 < \mathfrak{p}_k^\star \leq \eta_\mathfrak{p}\mathfrak{p}_\mathrm{B}$ *and where*

$$\mathcal{X}_\mathrm{C} = \{\boldsymbol{x}_\mathrm{C} \mid 0 < \mathfrak{p}_k \leq \eta_\mathfrak{p}\mathfrak{p}_\mathrm{B}, 0 < \mathfrak{p}_l, 0 \leq \mathfrak{q}_\mathrm{F} \leq \mathfrak{q}_{\mathrm{F}\max}, 0 \leq \omega \leq \omega_{\max}\}, \tag{6.49a}$$

$$\nabla_{\mathfrak{q}_\mathrm{F}}\Psi = \frac{1}{(c_\mathrm{p}-c_\mathrm{V})T_k}\left[\gamma_\mathrm{C} r_1^2 \varrho_\mathrm{n}\omega - (r_1^2\gamma_\mathrm{C}^2 + 2\zeta_\mathrm{k})\varrho_\mathrm{n}\mathfrak{q}_\mathrm{F}\right]\Psi^{\frac{c_\mathrm{V}}{c_\mathrm{p}}}, \tag{6.49b}$$

$$\nabla_\omega\Psi = \frac{1}{(c_\mathrm{p}-c_\mathrm{V})T_k}\left[(2\varsigma r_2^2 - r_1^2)\omega + r_1^2\gamma_\mathrm{C}\varrho_\mathrm{n}\mathfrak{q}_\mathrm{F}\right]\Psi^{\frac{c_\mathrm{V}}{c_\mathrm{p}}}, \tag{6.49c}$$

$$\nabla_{\mathfrak{q}_\mathrm{F}}\mathfrak{R}_\mathrm{F} = \frac{1}{2C_v^2\beta_\mathrm{F}^2\sqrt{\mathfrak{q}_\mathrm{F}}}, \tag{6.49d}$$

$$\varkappa_\mathrm{C} = \max\left\{\left|1 - \frac{\beta_\mathrm{st}\beta_\mathrm{B}(2-\eta_\mathfrak{p})}{4\beta_\varepsilon}\right|, \left|1 - \frac{\beta_\mathrm{st}\beta_\mathrm{B}(2-\eta_\mathfrak{p})}{4\beta_\varepsilon(1-\eta_\mathfrak{p})^{\frac{3}{2}}}\right|\right\}, \tag{6.49e}$$

$$\mu_\mathrm{C} = \varsigma r_2^2 \varrho_\mathrm{n}. \tag{6.49f}$$

Proof Sketch:
The proof for EIP entails applying Lemma 3.6 to the controlled compressor unit dynamics and verifying the resulting LMI by deriving four inequalities using the diagonal dominance of the matrix in question. Next, it is shown how a strict requirement for the four inequalities ensures the AS of the equilibria of the autonomous shifted system. The strict inequalities are simplified, resulting in (6.48), and EISD follows as a consequence of the AS result. The full proof can be found in Appendix A.2. ■

The control requirements in Objective 6.2 are met through Proposition 6.3 and Theorem 6.7. While the inequalities in (6.48) are nonlinear, they remain scalar. Moreover, as discussed in Remark 6.13, each inequality has a gain which modulates the amount of damping injected and which can therefore be increased to ensure that the respective inequalities hold. The rationale for the nonlinear control law for the CCV in (6.34b) also becomes apparent from (6.48b) and (6.49d), where it can be seen how more damping is injected at low flow rates $\mathfrak{q}_{\mathrm{F},j}$ which corresponds to the areas where surge is likely to occur. Finally, note that the set $\mathcal{X}_\mathrm{C}$ in (6.49a) corresponds to the physically feasible region of operation for the compressor unit, i.e., $\boldsymbol{x}_\mathrm{C} \in \mathcal{X}_\mathrm{C}$ for all $t \geq 0$. This is assured by the following assumption.[18]

[18] Assumption 6.3 can be achieved by controlling an additional inlet exhaust valve and the compressor torque to ensure that $\mathfrak{p}_k \leq \eta_{\mathfrak{p},j}\mathfrak{p}_{\mathrm{B},j}$ and $0 \leq \omega_j \leq \omega_{\max,j}$, respectively. Furthermore, $0 < \mathfrak{p}_k$ and $0 < \mathfrak{p}_l$ hold, since negative absolute pressures are non-physical in gases, while $\mathfrak{q}_{\mathrm{F},j} \leq \mathfrak{q}_{\mathrm{F}\max,j}$ is ensured practically due to the effect of choking (see [GE99b, p. 163]).

> **Assumption 6.3 (Constrained Compressor State Space)**
> *The states $\boldsymbol{x}_{\mathrm{C},j}$ of the controlled compressor unit with the dynamics (6.36) are constrained to $\mathcal{X}_{\mathrm{C},j}$ in (6.49a) for all time.*

Remark 6.16. *The following observations help to simplify the verification of the inequalities in Theorem 6.7. Although $\Psi_j(\cdot)$ is a complex function, the verification for* (6.48a) *can be simplified by considering only the extremes of $\Psi_j(\cdot)$. Furthermore,* (6.48d) *simply needs to be verified for $\omega_{\max,j}$ and $\mathfrak{q}_{\mathrm{Fmax},j}$. While analytical verifications of* (6.48b) *and* (6.48c) *are cumbersome, these scalar inequalities can efficiently be verified using constrained optimisation. Finally, the verification of the conditions of Theorem 6.7 can be made robust by considering uncertainties for the parameters and functions in* (6.48). *This can include, e.g., considering upper and lower ranges for parameters or by using different functions to describe the compressor characteristic.*

6.4.4 Gas Network Stability

With EIP properties for each of the gas network components established in Propositions 6.4 and 6.6, and Theorems 6.5 and 6.7, the AS of any equilibrium of the entire gas network can now be established in a modular fashion. To this end, the inputs and outputs of the respective components are stacked as follows

$$
\begin{aligned}
\tilde{\boldsymbol{u}}_{\mathrm{S}} &= \mathrm{stack}_{n\in\mathcal{N}_{\mathrm{S}}}(\tilde{\mathfrak{q}}_{\mathrm{ex},n}), & \tilde{\boldsymbol{y}}_{\mathrm{S}} &= \mathrm{stack}_{n\in\mathcal{N}_{\mathrm{S}}}(\tilde{\mathfrak{p}}_{n}), && \text{(6.50a)}\\
\tilde{\boldsymbol{u}}_{\mathrm{N}} &= \mathrm{stack}_{n\in\mathcal{N}_{\mathrm{N}}}(\tilde{\mathfrak{q}}_{\mathrm{ex},n}), & \tilde{\boldsymbol{y}}_{\mathrm{N}} &= \mathrm{stack}_{n\in\mathcal{N}_{\mathrm{N}}}(\tilde{\mathfrak{p}}_{n}), && \text{(6.50b)}\\
\tilde{\boldsymbol{u}}_{\mathrm{Cs+}} &= \mathrm{stack}_{n\in\mathcal{N}_{\mathrm{Cs}}^{+}}(\tilde{\mathfrak{q}}_{\mathrm{ex},n}), & \tilde{\boldsymbol{y}}_{\mathrm{Cs+}} &= \mathrm{stack}_{n\in\mathcal{N}_{\mathrm{Cs}}^{+}}(\tilde{\mathfrak{p}}_{n}), && \text{(6.50c)}\\
\tilde{\boldsymbol{u}}_{\mathrm{Cs-}} &= \mathrm{stack}_{n\in\mathcal{N}_{\mathrm{Cs}}^{-}}(\tilde{\mathfrak{q}}_{\mathrm{ex},n}), & \tilde{\boldsymbol{y}}_{\mathrm{Cs-}} &= \mathrm{stack}_{n\in\mathcal{N}_{\mathrm{Cs}}^{-}}(\tilde{\mathfrak{p}}_{n}), && \text{(6.50d)}\\
\tilde{\boldsymbol{u}}_{\mathrm{Cs}} &= \mathrm{stack}(\tilde{\boldsymbol{u}}_{\mathrm{Cs+}}, \tilde{\boldsymbol{u}}_{\mathrm{Cs-}}), & \tilde{\boldsymbol{y}}_{\mathrm{Cs}} &= \mathrm{stack}(\tilde{\boldsymbol{y}}_{\mathrm{Cs+}}, \tilde{\boldsymbol{y}}_{\mathrm{Cs-}}), && \text{(6.50e)}\\
\tilde{\mathfrak{q}}_{\mathrm{g}} &= \mathrm{stack}(\tilde{\boldsymbol{u}}_{\mathrm{S}}, \tilde{\boldsymbol{u}}_{\mathrm{N}}, \tilde{\boldsymbol{u}}_{\mathrm{Cs}}), & \tilde{\mathfrak{p}}_{\mathrm{g}} &= \mathrm{stack}(\tilde{\boldsymbol{y}}_{\mathrm{S}}, \tilde{\boldsymbol{y}}_{\mathrm{N}}, \tilde{\boldsymbol{y}}_{\mathrm{Cs}}), && \text{(6.50f)}\\
\tilde{\boldsymbol{u}}_{\mathfrak{l}} &= \tilde{\mathfrak{q}}_{\mathrm{g}}, & \tilde{\boldsymbol{y}}_{\mathfrak{l}} &= \mathrm{stack}_{n\in\mathcal{N}_{\mathrm{S}}\cup\mathcal{N}_{\mathrm{N}}\cup\mathcal{N}_{\mathrm{Cs}}}(\tilde{\mathfrak{q}}_{\mathfrak{l},n}), && \text{(6.50g)}\\
\tilde{\boldsymbol{u}}_{\mathrm{T}} &= \mathrm{stack}_{o\in\mathcal{E}_{\mathrm{T}}}(\tilde{\mathfrak{p}}_{n_{o,+}} - \tilde{\mathfrak{p}}_{n_{o,-}}), & \tilde{\boldsymbol{y}}_{\mathrm{T}} &= \mathrm{stack}_{o\in\mathcal{E}_{\mathrm{T}}}(\tilde{\mathfrak{q}}_{\mathrm{T},o}), && \text{(6.50h)}\\
\tilde{\boldsymbol{u}}_{v} &= \mathrm{stack}_{o\in\mathcal{E}_{v}}(\tilde{\mathfrak{p}}_{n_{o,+}} - \tilde{\mathfrak{p}}_{n_{o,-}}), & \tilde{\boldsymbol{y}}_{v} &= \mathrm{stack}_{o\in\mathcal{E}_{v}}(\tilde{\mathfrak{q}}_{v,o}). && \text{(6.50i)}
\end{aligned}
$$

Recalling Assumption 6.2, AS is investigated w.r.t. the gas network equilibria

$$
\left\{
\begin{aligned}
&\tilde{\mathfrak{p}}_{\mathrm{g}} = \mathbf{0},\quad \tilde{\mathfrak{q}}_{\mathrm{g}} = \mathbf{0}, && \\
&\hat{\mathfrak{p}}_{n_{o,-}} = \mathfrak{p}^{\star}_{n_{o,-}}, && \forall o \in \mathcal{E}_v \text{ where } \mathfrak{p}_{n_{o,-}} \le \mathfrak{p}^{\star}_{n_{o,-}} < \mathfrak{p}_{n_{o,+}}\\
&\tilde{\mathfrak{q}}_{\mathrm{F},j} = 0,\quad \tilde{\omega}_j = 0,\quad \tilde{z}_j = 0,\quad \hat{\mathfrak{p}}_k = \mathfrak{p}^{\star}_k, && \forall j \in \mathcal{E}_{\mathrm{Cs}}, k \in \mathcal{N}^{+}_{\mathrm{Cs},j}
\end{aligned}
\right\}
\tag{6.51}
$$

Moreover, the gas network components are interconnected skew-symmetrically as per Figure 6.11. The incidence matrices $\boldsymbol{E}_{\mathrm{T,S}}$, $\boldsymbol{E}_{\mathrm{T,N}}$, $\boldsymbol{E}_{\mathrm{T,Cs+}}$, and $\boldsymbol{E}_{\mathrm{T,Cs-}}$ describe the connections between the edges in $\mathcal{E}_{\mathrm{T}}$ to the nodes in $\mathcal{N}_{\mathrm{S}}$, $\mathcal{N}_{\mathrm{N}}$, $\mathcal{N}_{\mathrm{Cs}}^{+}$, and $\mathcal{N}_{\mathrm{Cs}}^{-}$, and similarly $\boldsymbol{E}_{v,\mathrm{S}}$, $\boldsymbol{E}_{v,\mathrm{N}}$, $\boldsymbol{E}_{v,\mathrm{Cs+}}$, and $\boldsymbol{E}_{v,\mathrm{Cs-}}$ for the edges in $\mathcal{E}_v$ and the same node sets, such that

$$\boldsymbol{E}_\mathrm{T} = \begin{bmatrix} \boldsymbol{E}_{\mathrm{T,S}} \\ \boldsymbol{E}_{\mathrm{T,N}} \\ \boldsymbol{E}_{\mathrm{T,Cs}} \end{bmatrix}, \quad \boldsymbol{E}_v = \begin{bmatrix} \boldsymbol{E}_{v,\mathrm{S}} \\ \boldsymbol{E}_{v,\mathrm{N}} \\ \boldsymbol{E}_{v,\mathrm{Cs}} \end{bmatrix}, \quad \boldsymbol{E}_{\mathrm{T,Cs}} = \begin{bmatrix} \boldsymbol{E}_{\mathrm{T,Cs+}} \\ \boldsymbol{E}_{\mathrm{T,Cs-}} \end{bmatrix}, \quad \boldsymbol{E}_{v,\mathrm{Cs}} = \begin{bmatrix} \boldsymbol{E}_{v,\mathrm{Cs+}} \\ \boldsymbol{E}_{v,\mathrm{Cs-}} \end{bmatrix}. \tag{6.52}$$

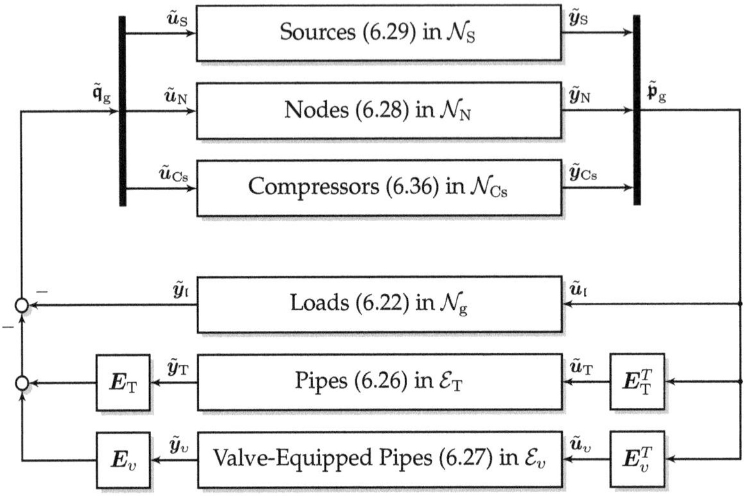

Figure 6.11: Skew-symmetric interconnection of the gas sources, nodes and compressors with the loads and pipes using the input and output variables defined in (6.50).

Theorem 6.8 (Gas Network AS Equilibria)
Consider a gas network $\mathcal{G}_\mathrm{g} = (\mathcal{N}_\mathrm{g}, \mathcal{E}_\mathrm{g})$, $\mathcal{N}_\mathrm{g} = \mathcal{N}_\mathrm{S} \cup \mathcal{N}_\mathrm{N} \cup \mathcal{N}_\mathrm{Cs}$, $\mathcal{E}_\mathrm{g} = \mathcal{E}_\mathrm{Cs} \cup \mathcal{E}_\mathrm{T} \cup \mathcal{E}_v$ which is interconnected according to Figure 6.11 and which consists of the supply nodes (6.29), the pipeline endpoint nodes (6.28), the controlled compressor dynamics (6.36), the gas load functions (6.22), the pipeline edge dynamics (6.26), and the valve-equipped pipelines (6.27) in closed loop with the valve controller (6.32). Then, the gas network equilibrium (6.51) is AS with the storage function

$$S_\mathrm{g} = \sum_{n\in\mathcal{N}_\mathrm{N}} S_{\mathrm{N},n}(\tilde{\mathfrak{p}}_n) + \sum_{j\in\mathcal{E}_\mathrm{Cs}} S_{\mathrm{C},j}(\tilde{\boldsymbol{x}}_{\mathrm{C},j}) + \sum_{o\in\mathcal{E}_\mathrm{T}} S_{\mathrm{T},o}(\tilde{\mathfrak{q}}_{\mathrm{T},o}) + \sum_{o\in\mathcal{E}_v} S_{v,o}(\tilde{\mathfrak{q}}_{\mathrm{T},o}), \tag{6.53}$$

if the conditions in Proposition 6.3, Theorem 6.7, and Assumption 6.3 hold for each controlled compressor unit $j \in \mathcal{E}_\mathrm{Cs}$ and if every node in $\mathcal{N}_\mathrm{N} \cup \mathcal{N}_\mathrm{Cs}^-$ is weakly connected to a node in $\mathcal{N}_\mathrm{S} \cup \mathcal{N}_\mathrm{Cs}^+$ via edges in $\mathcal{E}_\mathrm{T} \cup \mathcal{E}_v$.

Proof:
The interconnection matrix of the inputs and outputs corresponding to Figure 6.11 is described by

$$\begin{bmatrix} \tilde{\boldsymbol{u}}_\mathrm{T} \\ \tilde{\boldsymbol{u}}_v \\ \tilde{\boldsymbol{u}}_\mathrm{N} \\ \tilde{\boldsymbol{u}}_\mathrm{S} \\ \tilde{\boldsymbol{u}}_\mathrm{Cs} \\ \tilde{\boldsymbol{u}}_\mathfrak{l} \end{bmatrix} = \begin{bmatrix} \boldsymbol{0} & \boldsymbol{0} & \begin{bmatrix} \boldsymbol{E}_\mathrm{T}^T \end{bmatrix} & \boldsymbol{0} \\ \boldsymbol{0} & \boldsymbol{0} & \begin{bmatrix} \boldsymbol{E}_v^T \end{bmatrix} & \boldsymbol{0} \\ \begin{bmatrix} -\boldsymbol{E}_\mathrm{T} \end{bmatrix} & \begin{bmatrix} -\boldsymbol{E}_v \end{bmatrix} & \begin{matrix} \boldsymbol{0} & \boldsymbol{0} & \boldsymbol{0} \\ \boldsymbol{0} & \boldsymbol{0} & \boldsymbol{0} \\ \boldsymbol{0} & \boldsymbol{0} & \boldsymbol{0} \end{matrix} & \begin{bmatrix} -\boldsymbol{I}_{|\mathcal{N}_\mathrm{g}|} \end{bmatrix} \\ \boldsymbol{0} & \boldsymbol{0} & \begin{bmatrix} \boldsymbol{I}_{|\mathcal{N}_\mathrm{g}|} \end{bmatrix} & \boldsymbol{0} \end{bmatrix} \begin{bmatrix} \tilde{\boldsymbol{y}}_\mathrm{T} \\ \tilde{\boldsymbol{y}}_v \\ \tilde{\boldsymbol{y}}_\mathrm{N} \\ \tilde{\boldsymbol{y}}_\mathrm{S} \\ \tilde{\boldsymbol{y}}_\mathrm{Cs} \\ \tilde{\boldsymbol{y}}_\mathfrak{l} \end{bmatrix}, \tag{6.54}$$

where the sources with $(\tilde{\boldsymbol{u}}_\mathrm{S}, \tilde{\boldsymbol{y}}_\mathrm{S}) = (\tilde{\boldsymbol{u}}_\mathrm{S}, \boldsymbol{0})$ are static, the nodes with $(\tilde{\boldsymbol{u}}_\mathrm{N}, \tilde{\boldsymbol{y}}_\mathrm{N})$ are EIP (see Proposition 6.4), and the compressor units with $(\tilde{\boldsymbol{u}}_\mathrm{Cs}, \tilde{\boldsymbol{y}}_\mathrm{Cs})$ are EIP and have AS equilibria (see Theorem 6.7) with trajectories that are constrained to the region of attraction (see Assumption 6.3 and Remark 3.9). Furthermore, the static gas loads with $(\tilde{\boldsymbol{u}}_\mathfrak{l}, \tilde{\boldsymbol{y}}_\mathfrak{l})$ are monotone and therefore EIP (see Lemma 3.5), the pipelines with $(\tilde{\boldsymbol{u}}_\mathrm{T}, \tilde{\boldsymbol{y}}_\mathrm{T})$ are OS-EIP and EIID (see Theorem 6.5), and the controlled valve-equipped pipelines with $(\tilde{\boldsymbol{u}}_v, \tilde{\boldsymbol{y}}_v)$ are OS-EIP and EIID (see Proposition 6.6). Since each subsystem of the gas network is at least EIP and since the interconnection matrix in (6.54) is skew-symmetric, stability of the equilibria in (6.51) follows from Proposition 3.1.

Next, AS is demonstrated by applying Theorem 3.3 along with Algorithm 1, where

$$\tilde{\boldsymbol{u}}_{-,\mathrm{EIID}} = \mathrm{stack}(\tilde{\boldsymbol{u}}_\mathrm{T}, \tilde{\boldsymbol{u}}_v), \qquad \tilde{\boldsymbol{y}}_{-,\mathrm{EIID}} = \mathrm{stack}(\tilde{\boldsymbol{y}}_\mathrm{T}, \tilde{\boldsymbol{y}}_v), \tag{6.55a}$$

$$\tilde{\boldsymbol{u}}_\mathfrak{n} = \tilde{\boldsymbol{u}}_\mathrm{N}, \qquad \tilde{\boldsymbol{y}}_\mathfrak{n} = \tilde{\boldsymbol{y}}_\mathrm{N}, \tag{6.55b}$$

$$\tilde{\boldsymbol{u}}_{0,\mathrm{GAS}} = \mathrm{stack}(\tilde{\boldsymbol{u}}_\mathrm{S}, \tilde{\boldsymbol{u}}_{\mathrm{Cs}+}, \tilde{\boldsymbol{u}}_{\mathrm{Cs}-}, \tilde{\boldsymbol{u}}_\mathfrak{l}), \qquad \tilde{\boldsymbol{y}}_{0,\mathrm{GAS}} = \mathrm{stack}(\tilde{\boldsymbol{y}}_\mathrm{S}, \tilde{\boldsymbol{y}}_{\mathrm{Cs}+}, \tilde{\boldsymbol{y}}_{\mathrm{Cs}-}, \tilde{\boldsymbol{y}}_\mathfrak{l}). \tag{6.55c}$$

The static sources and loads are grouped in the set $\mathcal{J}_{0,\mathrm{GAS}}$ (see Remark 3.12). Recall from (6.29) that $\tilde{\boldsymbol{y}}_\mathrm{S} = \boldsymbol{0}$. Moreover, recall from Theorem 6.7, and (A.23) in particular, that $\dot{S}_{\mathrm{C}}$ and thus $\dot{S}_{\mathrm{g}}$ are negative definite in $\tilde{\boldsymbol{y}}_{\mathrm{Cs+}}$, i.e., $\lim_{t\to\infty}\tilde{\boldsymbol{y}}_{\mathrm{Cs+}} = \boldsymbol{0}$. Following the notation in (3.9), the interconnection in (6.56) reduces to

$$\begin{bmatrix} \boldsymbol{0} \\ \boldsymbol{0} \\ \tilde{\boldsymbol{u}}_{\mathrm{N}} \\ \tilde{\boldsymbol{u}}_{\mathrm{S}} \\ \tilde{\boldsymbol{u}}_{\mathrm{Cs+}} \\ \tilde{\boldsymbol{u}}_{\mathrm{Cs-}} \\ \tilde{\boldsymbol{u}}_{\mathfrak{l}} \end{bmatrix} = \begin{bmatrix} \tilde{\boldsymbol{u}}_{\mathrm{T}} & \boldsymbol{0} & \begin{bmatrix} & \boldsymbol{E}_{\mathrm{T}}^{T} & \end{bmatrix} & \boldsymbol{0} \\ \tilde{\boldsymbol{u}}_{\upsilon} & \boldsymbol{0} & \begin{bmatrix} & \boldsymbol{E}_{\upsilon}^{T} & \end{bmatrix} & \boldsymbol{0} \\ \begin{bmatrix} \\ -\boldsymbol{E}_{\mathrm{T}} \\ \\ \end{bmatrix} & \begin{bmatrix} \\ -\boldsymbol{E}_{\upsilon} \\ \\ \end{bmatrix} & \begin{matrix} \boldsymbol{0} & \boldsymbol{0} & \boldsymbol{0} & \boldsymbol{0} \\ \boldsymbol{0} & \boldsymbol{0} & \boldsymbol{0} & \boldsymbol{0} \\ \boldsymbol{0} & \boldsymbol{0} & \boldsymbol{0} & \boldsymbol{0} \\ \boldsymbol{0} & \boldsymbol{0} & \boldsymbol{0} & \boldsymbol{0} \end{matrix} & \begin{bmatrix} \\ -\boldsymbol{I}_{|\mathcal{N}_{\mathrm{g}}|} \\ \\ \end{bmatrix} \\ \boldsymbol{0} & \boldsymbol{0} & \begin{bmatrix} & \boldsymbol{I}_{|\mathcal{N}_{\mathrm{g}}|} & \end{bmatrix} & \boldsymbol{0} \end{bmatrix} \begin{bmatrix} \boldsymbol{0} \\ \boldsymbol{0} \\ \tilde{\mathfrak{y}}_{\mathrm{N}} \\ \boldsymbol{0} \\ \boldsymbol{0} \\ \tilde{\mathfrak{y}}_{\mathrm{Cs-}} \\ \tilde{\mathfrak{y}}_{\mathfrak{l}} \end{bmatrix}. \tag{6.56}$$

The first two rows of (6.56) yield

$$\tilde{\boldsymbol{u}}_{-,\mathrm{EIID}} = \begin{bmatrix} \boldsymbol{E}_{\mathrm{T,N}}^{T} \\ \boldsymbol{E}_{\upsilon,\mathrm{N}}^{T} \end{bmatrix} \tilde{\mathfrak{y}}_{\mathrm{N}} + \begin{bmatrix} \boldsymbol{E}_{\mathrm{T,Cs-}}^{T} \\ \boldsymbol{E}_{\upsilon,\mathrm{Cs-}}^{T} \end{bmatrix} \tilde{\mathfrak{y}}_{\mathrm{Cs-}}. \tag{6.57}$$

Notice that the rows of $\boldsymbol{E}_{\mathrm{T}}^{T}$ and $\boldsymbol{E}_{\upsilon}^{T}$ describe the connections between the nodes in $\mathcal{N}_{\mathrm{g}}$. Thus, if there is a direct connection between an element of $\mathrm{stack}(\tilde{\mathfrak{y}}_{\mathrm{N}}, \tilde{\mathfrak{y}}_{\mathrm{Cs-}})$ to an element of $\mathrm{stack}(\tilde{\mathfrak{y}}_{\mathrm{S}}, \tilde{\mathfrak{y}}_{\mathrm{Cs+}}) = \boldsymbol{0}$, the element of the former can be set to zero through one of the rows in (6.57). Furthermore, a connection via a row in (6.57) between a zero and a non-zero element of $\mathrm{stack}(\tilde{\mathfrak{y}}_{\mathrm{N}}, \tilde{\mathfrak{y}}_{\mathrm{Cs-}})$ likewise allows the non-zero element to be set to zero (see Lines 12–16 of Algorithm 1). By continuing this process, it is seen that an element of $\mathrm{stack}(\tilde{\mathfrak{y}}_{\mathrm{N}}, \tilde{\mathfrak{y}}_{\mathrm{Cs-}})$ may be set to zero if there is a path described by the edges in $\mathcal{E}_{\mathrm{T}} \cup \mathcal{E}_{\upsilon}$ to any element in $\mathrm{stack}(\tilde{\mathfrak{y}}_{\mathrm{S}}, \tilde{\mathfrak{y}}_{\mathrm{Cs+}}) = \boldsymbol{0}$. Finally, $\tilde{\boldsymbol{u}}_{\mathfrak{l}}$ can be set to zero when $\mathrm{stack}(\tilde{\mathfrak{y}}_{\mathrm{N}}, \tilde{\mathfrak{y}}_{\mathrm{S}}, \tilde{\mathfrak{y}}_{\mathrm{Cs+}}, \tilde{\mathfrak{y}}_{\mathrm{Cs-}}) = \boldsymbol{0}$ and $\tilde{\mathfrak{y}}_{\mathfrak{l}} = \boldsymbol{0}$ then follows since the loads are static. ■

Through Theorem 6.8, the AS of the equilibria (6.51) of an entire gas network containing the interconnected components as in Figure 6.11 can be established, thus providing an answer to RQ 3.2. This process requires only a connectivity condition for the nodes along with the decentralised requirements for the controlled compressor units in Proposition 6.3 and Theorem 6.7. This allows the AS of the gas network equilibria to be verified in a scalable manner for large networks.

Notice furthermore that the connectivity requirement in Theorem 6.8—and indeed the result in Theorem 3.2—can be interpreted as the AS property propagating through the network, moving directly and indirectly from the node pressures in which $\dot{S}_{\mathrm{g}}$ is negative definite via the OS-EIP and EIID pipelines to the node pressures in which $\dot{S}_{\mathrm{g}}$ is only negative semi-definite. This process is visualised in Figure 6.12, where Theorem 6.8 is applied to the gas network in Figure 6.7 under the assumption that the requirements for the controlled compressor units are met.

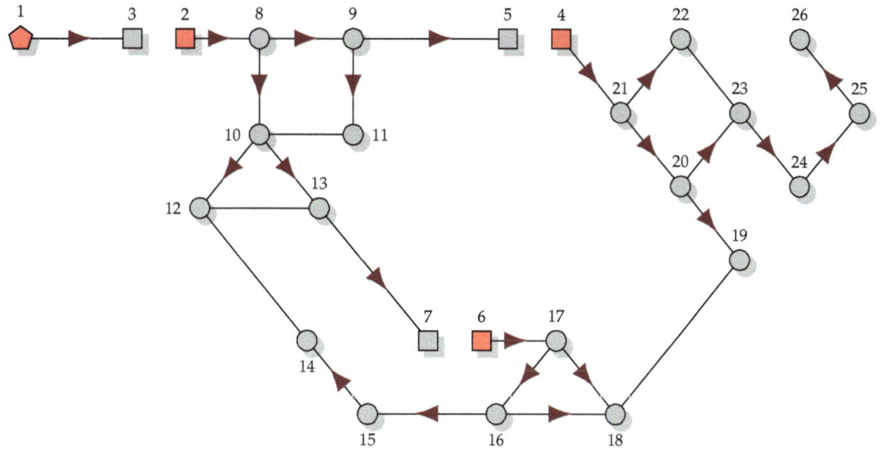

Figure 6.12: Propagation of asymptotic convergence according to Theorem 6.8 for the gas network in Figure 6.7, where the convergence propagates from the nodes ⬟ in $\mathcal{N}_\mathrm{S}$ and ■ in $\mathcal{N}_{\mathrm{Cs}}^{+}$ to the nodes ● in $\mathcal{N}_\mathrm{N}$ and □ in $\mathcal{N}_{\mathrm{Cs}}^{-}$ via the edges in $\mathcal{E}_\mathrm{T} \cup \mathcal{E}_v$. The propagation is indicated by the arrows ►.

Finally, for the case that no compressors are present in the gas network, which represents traditional gas networks with top-down supply directions, the following corollary can be used to establish AS as a direct consequence of Theorem 6.8. Note that the AS result is inherent for compressor-less gas networks and, unlike Theorem 6.8, does not require a connectivity test for the nodes.

Corollary 6.9 (Compressor-less Gas Network Inherent AS)
The equilibria of a gas network as in Theorem 6.8 with a weakly connected topology described by $\mathcal{G}_\mathrm{g} = (\mathcal{N}_\mathrm{g}, \mathcal{E}_\mathrm{g})$ are AS if $\mathcal{E}_{\mathrm{Cs}} = \emptyset$.

Proof:
From $\mathcal{E}_{\mathrm{Cs}} = \emptyset$, it follows that $\mathcal{N}_{\mathrm{Cs}}^{-} = \mathcal{N}_{\mathrm{Cs}}^{+} = \emptyset$. Thus if $\mathcal{E}_{\mathrm{Cs}} = \emptyset$, the conditions of Theorem 6.8 are met if every node in $\mathcal{N}_\mathrm{N}$ is weakly connected to a node in $\mathcal{N}_\mathrm{S}$ via the edges in $\mathcal{E}_\mathrm{T} \cup \mathcal{E}_v$ which is equivalent to the weakly connected requirement for $\mathcal{G}_\mathrm{g}$. ■

Remark 6.17. *Theorem 6.8 only demonstrates an AS result for the gas network equilibria since the compressor units are restricted to $\boldsymbol{x}_{\mathrm{C},j} \in \mathcal{X}_{\mathrm{C},j} \subset \mathbb{R}^5$, $j \in \mathcal{E}_{\mathrm{Cs}}$ (see Assumption 6.3) and since the absolute pressures are non-negative. Since these state restrictions hold for all time, the AS result may be considered global w.r.t. the restricted state space of the gas network since the storage functions are radially unbounded.*

Remark 6.18. *The AS results in Theorem 6.8 and Corollary 6.9 use the same underlying theory as in Chapters 4 and 5, thus yielding similar robustness properties. Specifically, the*

AS results in this section are robust against any load, parameter, or topology changes, as long as the loads remain monotone in the node pressures $\mathfrak{p}_n$*, the compressor unit requirements remain valid after the parameter changes, and the required network connectivity is maintained after the topology change. Naturally, the network should remain feasible after the change (see Assumption 6.2).*

Remark 6.19. *Consider the case where the nodes in* $\mathcal{N}_{\mathrm{S}} \cup \mathcal{N}_{\mathrm{Cs}}^{+}$ *are treated as a single root vertex. The connectivity condition in Theorem 6.8 is then equivalent to establishing whether the graph described by the root vertex and the vertices* $\mathcal{N}_{\mathrm{Cs}}^{-} \cup \mathcal{N}_{\mathrm{N}}$ *along with the edges* $\mathcal{E}_{\mathrm{T}} \cup \mathcal{E}_{v}$ *has a* spanning rooted tree *using the designated root vertex.*

6.5 Simulation

Using the proposed π-structure gas pipeline model along with the valve and compressor unit controllers, a gas network is simulated in this section to demonstrate the AS of the gas network equilibria. Specifically, it is shown how the valve-equipped pipelines and compressor units achieve AS pressure regulation according to Definition 2.2. The simulated gas network has a meshed topology with different pressure regions and elevations. Moreover, the robustness of the AS against various load and setpoint changes is demonstrated.

This section commences with a description of the simulation setup in **Section 6.5.1**, where the choice of the controller gains is also discussed. Thereafter, the simulation results are presented in **Section 6.5.2**.

6.5.1 Simulation Setup

The gas network in Figure 6.7 comprising 26 nodes and 31 edges is simulated in MATLAB/SIMULINK using SIMSCAPE. The network is inspired by the gas networks in [Ste+21; Sch+17] and has been chosen to investigate the effects that a local surplus of gas has on the compressor-equipped network as a whole. The network, which serves as a synthetic test network (see [MTB20, Table 3]), is connected to a single supply node and is further divided into three regions with different pressure setpoints and elevations. Two valve-equipped pipelines and three compressor units interconnect the three regions and the supply node. These components are equipped with the controllers in (6.32) and (6.34), respectively.

The gas and the pipeline parameters are listed in Appendix B.3 in Table B.5 while the network component and controller parameters are given in Table 6.1. The gas loads are considered to be piecewise constant with randomly selected values in a symmetrical range around 0 (see Table B.7 in Appendix B.3 for the specific load values used).[19] The pipelines all have the same diameter. The pipeline lengths are randomly selected in the specified range, while the length of the pipeline connecting the supply node ($n = 1$) to Node 3 is manually set to 5 km. The lengths for the simulated pipelines can be found in Table B.6 in Appendix B.3. The inductive factors for the valve-equipped pipelines are all fixed to the same value, and no nodes are explicitly equipped with extra storage tanks. The parameters of the compressor units are chosen based on the parameter sets in [GE99b; Gra+00; Gra+04; BG08]. The compressor map $\Psi(\cdot)$ corresponding to the compressor values in Table 6.1 is shown along with its surge line in Figure 6.13.[20] For each compressor unit, the pressure $\mathfrak{p}_{\mathrm{B},j}$ is kept constant through a pressure-regulating valve connected to the respective outlet. Furthermore, the valve and torque control outputs are saturated beyond their respective upper and lower ranges, and clamping is implemented on the integrator states to prevent wind-up during saturation. The torque is also restricted as required to prevent the maximum rotational velocity of the machine from being exceeded.

Table 6.1: Gas Network Parameters for Pressure Regulation

Nodes in each region	$\mathcal{N}_{\mathrm{g},1} = \{2, 3, 8, 9, 10, 11, 12, 13\}$	
	$\mathcal{N}_{\mathrm{g},2} = \{4, 5, 19, 20, 21, 22, 23, 24, 25, 26\}$	
	$\mathcal{N}_{\mathrm{g},3} = \{6, 7, 14, 15, 16, 17, 18\}$	
Gas Loads (6.22), $n \in \mathcal{N}_\mathrm{g}$	$-0.01\,\mathrm{sm}^3/\mathrm{s} \leq \mathfrak{q}_{\mathfrak{l},n} \leq 0.01\,\mathrm{sm}^3/\mathrm{s}$	
Gas Pipelines (6.26), $o \in \mathcal{E}_\mathrm{T}$	$D_o = 0.1\,\mathrm{m}$	$\ell_o \in [0.5; 3]\,\mathrm{km}$
Valve Parameters (6.27), $o \in \mathcal{E}_\upsilon$	$C_{\upsilon,o} = 0.34\,\mathrm{sm}^3/\mathrm{s}$	$\mathfrak{L}_o = 0.01\,\mathrm{kg}/\mathrm{m}^4$
Node Parameters (6.28), $n \in \mathcal{N}_\mathrm{N}$	$\mathfrak{C}_n = 0\,\mathrm{m}^4\mathrm{s}^2/\mathrm{kg}$	
Compressor Unit Parameters (6.31), $j \in \mathcal{E}_\mathrm{Cs}$	$r_{1,j} = 47.7\,\mathrm{mm}$	$r_{2,j} = 135\,\mathrm{mm}$
	$\gamma_{\mathrm{C},j} = 1969.4$	$\zeta_{\mathrm{k},j} = 10^5$
	$\varsigma_j = 0.95$	$C_{\upsilon,j} = 0.34\,\mathrm{sm}^3/\mathrm{s}$
	$\ell_j = 4\,\mathrm{m}$	$D_j = 0.025\,\mathrm{m}$
	$J_j = 0.9\,\mathrm{kgm}^2$	$\zeta_{\omega,j} = 0.01\,\mathrm{N\,m\,s/rad}$
	$\mathfrak{q}_{\mathrm{Fmax},j} = 12\,\mathrm{sm}^3/\mathrm{s}$	$\omega_{\mathrm{max},j} = 3 \cdot 10^4\,\mathrm{rpm}$
	$-40\,\mathrm{N\,m} \leq \tau \leq 40\,\mathrm{N\,m}$	$\eta_{\mathfrak{p},j} = 0.82$
	$\mathfrak{C}_k = 0\,\mathrm{m}^4\mathrm{s}^2/\mathrm{kg}$	$\mathfrak{C}_l = 0\,\mathrm{m}^4\mathrm{s}^2/\mathrm{kg}$
Valve Control Parameters (6.32), $o \in \mathcal{E}_\upsilon$	$k_{\upsilon,j} = 5 \cdot 10^{-6}$	
Compressor Unit Control Parameters (6.34), $j \in \mathcal{E}_\mathrm{Cs}$	$\beta_{\mathrm{B},j} = 5.5$	$\beta_{\mathrm{F},j} = 0.016$
	$\beta_{\mathfrak{p},j} = 10^{-5}$	$\beta_{\mathfrak{e},j} = 80$
	$\beta_{\mathrm{st},j} = 5 \cdot 10^4$	$\beta_{\omega,j} = 640$
	$\Psi_{j,0} = 1.2$	

[19] For reference, a load of $0.01\,\mathrm{sm}^3/\mathrm{s}$ equates to $p_\mathrm{th} = 338\,\mathrm{kW}$ for a calorific value of $c_\mathrm{cal} = 47.1\,\mathrm{MJ/kg}$ and a nominal density of $\varrho_\mathrm{n} = 0.7179\,\mathrm{kg/m}^3$ (see (6.23)).

[20] Note that the stonewall phenomenon (see [GE99b, p. 163]), in which the pressure ratio Ψ drops rapidly for flow rates above the choke line, is omitted for simplicity.

The control parameters of the compressor units are successively designed such that the requirements of Theorem 6.7 hold over the state ranges as listed in Table 6.1. Firstly, $\beta_{\mathrm{p},j}$ is set for a consistent SI unit usage.[21] Next, $\beta_{\omega,j}$ and $\beta_{\varepsilon,j}$ are selected such that (6.48c) and (6.48d) hold. Next, the smallest possible $\beta_{\mathrm{q},j}$ is selected such that (6.48b) holds (see Remark 6.10). Finally, $\beta_{\mathrm{B},j}$ and $\beta_{\mathrm{st},j}$ are selected such that (6.48a) holds. In each case, the inequalities are verified by finding the minima of the respective inequality functions.

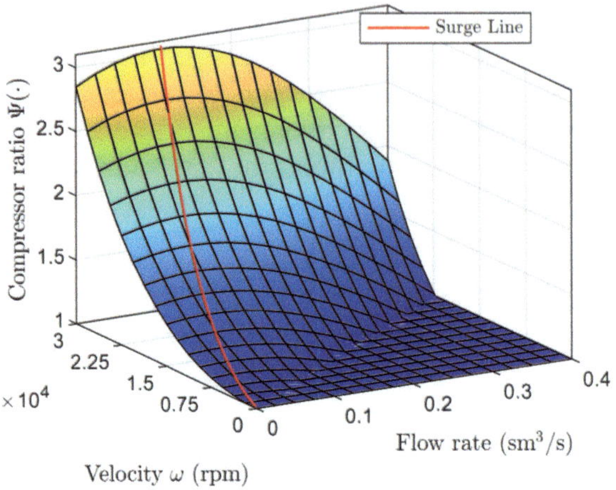

Figure 6.13: Compressor pressure ratio map $\Psi(\cdot)$ along with its surge line, to the left of which unstable surge behaviour can occur.

To test the proposed controllers and to demonstrate the robustness of the AS result, the gas network is tested using the following sequence of events.

- $t = 0\,\mathrm{s}$: The gas network in Figure 6.7 starts with the node pressures initialised to the pressure setpoints of the respective regions and with random load parameters generated for each node. The initial term $\omega_{0,j}$ is set according to (6.34e) with $\Psi_{j,0} = 1.2$ for the compressors $j = \{2, 3\}$, while $\omega_{0,1} = 0$.
- $t = 500\,\mathrm{s}$: The pressure setpoints for Region 3 are stepped up to $\mathfrak{p}_n^\star = \mathfrak{p}_n + 120\,\mathrm{mbar}$, $n \in \mathcal{N}_{\mathrm{g},3}$.
- $t = 1000\,\mathrm{s}$: The pressure setpoints for Region 2 are stepped up to $\mathfrak{p}_n^\star = \mathfrak{p}_n + 80\,\mathrm{mbar}$, $n \in \mathcal{N}_{\mathrm{g},2}$.
- $t = 1500\,\mathrm{s}$: The pressure setpoints for Region 3 are stepped down to $\mathfrak{p}_n^\star = \mathfrak{p}_n + 100\,\mathrm{mbar}$, $n \in \mathcal{N}_{\mathrm{g},3}$.

21 Recall that $1\,\mathrm{Pa} = 0.98692 \cdot 10^{-5}\,\mathrm{bar}$.

At each time step, 50 % of the nodes selected at random receive new random gas load values. Furthermore, the random seed is selected such that the various regions and the gas network as a whole switch between being net consumers and net producers of gas, as indicated in Table 6.2. From these regional aggregate load values, it can be seen that Regions 2 and 3 have an oversupply of gas ($\sum \mathfrak{q}_{\mathfrak{l},n} < 0$) for one and three simulation periods, respectively, while the network as a whole is a net supplier of gas for two simulation periods.

Table 6.2: Rounded Total Gas Load Per Region

Region m	$\sum_{n \in \mathcal{N}_{g,m}} \mathfrak{q}_{\mathfrak{l},n}$ (sm^3/s)			
	$t = 0$ s	$t = 500$ s	$t = 1000$ s	$t = 1500$ s
1	0.0288	0.0196	0.0158	0.0075
2	−0.0141	0.0089	0.0051	0.0000
3	−0.0193	0.0022	−0.0103	−0.0167
Total	−0.0045	0.0307	0.0106	−0.0093

6.5.2 Simulation Results: Gas Network

The simulated pressures and flow rates are shown in Figure 6.14, where the pressures are subdivided into figures for the high-, medium- and low-pressure ranges. In each time interval, stability according to Definition 2.2 is observed, where the pressures converge to within 35 mbar (high pressure), 26 mbar (medium pressure), and 10 mbar (low pressure) of the desired pressure for the respective regions. In each of the time periods, the compressor unit inlet pressures (Nodes 2, 4 and 6) are regulated to the desired setpoints, whereas the outlet pressures (Nodes 3, 5 and 7) can be seen to increase or decrease depending on the direction of flow through the compressor units. Note that the pressure is not regulated during the first 165 s (Node 2) and 50 s (Nodes 4 and 5) of the simulation, respectively, since the machines of the compressor units require time to reach a rotational velocity where the compressor ratio $\Psi(\cdot)$ is large enough to overcome the pressure difference between the inlet and outlet nodes. During this time when $\mathfrak{q}_{F,j}$ remains zero, the oversupply of gas causes a steady increase in the pressure until the compressor reaches a speed at which gas starts to flow, allowing the inlet pressure to be decreased. Similar increases can be seen at Node 2 at $t = 1594$ s and Node 6 at $t = 1177$ s. The flow rates in Figure 6.14d corroborate this behaviour, with the flow rates in the pipelines connected to the compressor outlets rapidly changing when the compressor units reach their operating speeds (e.g., see the flow rate for the pipeline connecting Nodes 7 and 13 at $t = 50$ s and $t = 1247$ s). Furthermore, large peaks of $\mathfrak{q}_{v,1} = 0.0589\,\text{sm}^3/\text{s}$ and $\mathfrak{q}_{v,2} = 0.0757\,\text{sm}^3/\text{s}$ can be seen at $t = 506$ s and $t = 1004$ s, respectively. These flow-rate peaks, which can be seen between Nodes 14–15 and 19–20 in (6.14d), coincide with the setpoint increases of the respective regions and arise due to the valves opening to their outlet pressures.

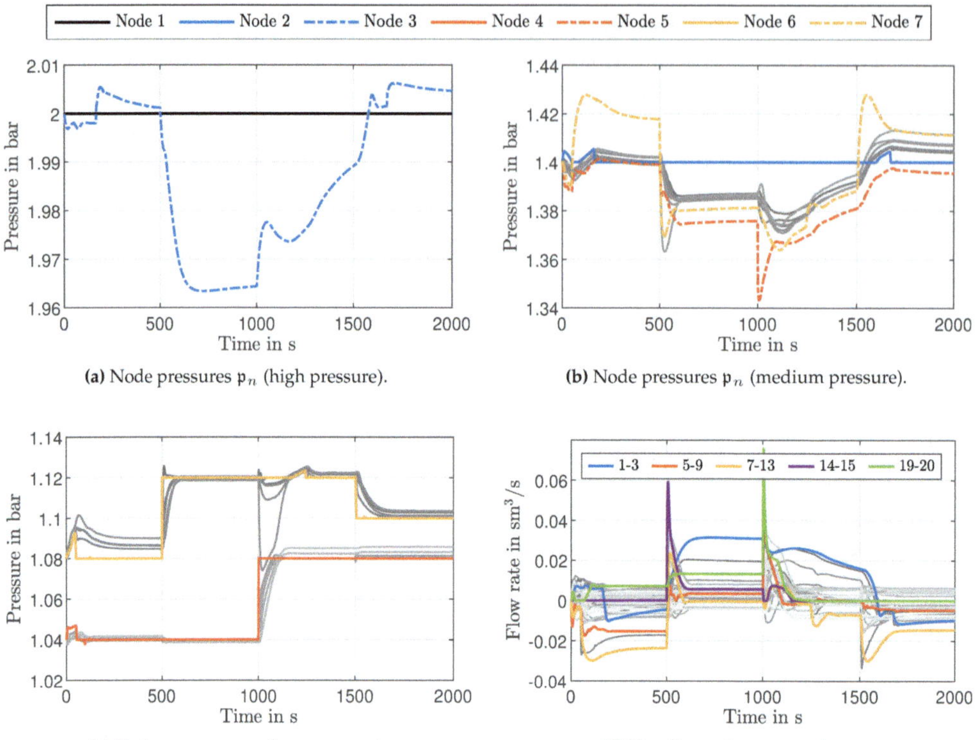

(a) Node pressures $\mathfrak{p}_n$ (high pressure).

(b) Node pressures $\mathfrak{p}_n$ (medium pressure).

(c) Node pressures $\mathfrak{p}_n$ (low pressure).

(d) Pipe flow rates $\mathfrak{q}_{\mathrm{T},o}$ and $\mathfrak{q}_{v,o}$.

Figure 6.14: Simulated pressures and flow rates for the gas network split into high pressure (a), medium pressure (b) and low pressure (c) figures and the flow rates in the pipelines (d). Nodes and edges not denoted in the respective legends are shown in shades of grey.

The behaviour of the compressor units discussed above can also be seen in Figure 6.15, where the compressor states, valve stem positions, and the resulting compressor ratios are shown. Notably, by comparing the angular velocities ω_j with the forward flow rates $\mathfrak{q}_{\mathrm{F},j}$, it can be seen that the increase in $\mathfrak{q}_{\mathrm{F},j}$ is preceded by the spin-up of the compressor machine (e.g., see the period $1177\,\mathrm{sec} \leq t \leq 1247\,\mathrm{s}$ from compressor $j = 3$). Furthermore, from the flow rates, it can be seen that $\mathfrak{q}_{\mathrm{F},1} > 0$ in two steady states ($t = \{499, 1999\}\,\mathrm{s}$), $\mathfrak{q}_{\mathrm{F},2} > 0$ in one steady state ($t = 499\,\mathrm{s}$), and $\mathfrak{q}_{\mathrm{F},3} > 0$ in three steady states ($t = \{499, 1499, 1999\}\,\mathrm{s}$). From the stem positions of the valves, it can be seen how both the CCV and backflow valve are simultaneously open during transients to ensure a stabilisation and regulation of the pressure (e.g., see $j = 1$ for $158\,\mathrm{sec} \leq t \leq 198\,\mathrm{s}$ and for $1668\,\mathrm{sec} \leq t \leq 1704\,\mathrm{s}$, where the backflow valve reaches peaks of $s_{\mathrm{B},1} = 0.656$ while the CCV is fully open). Nevertheless, upon approaching and reaching the steady state, only one valve remains open for each of the compressor units. Despite $s_{\mathrm{F},j}$ not opening fully in the steady state for any of the compressor units, the maximum steady-state pressure loss over the CCV for all compressor units is less than $50\,\mathrm{mbar}$ or $12.6\,\%$ of the pressure gained by the compressors. Recalling Remark 6.10, this can be deemed acceptable, since the low flow rates indicate a steady state within the surge region.

When considering the results in Figures 6.14 and 6.15 together with the regional gas demand in Table 6.2, the effects of interconnecting the various regions via compressor units also become apparent. Although Region 3 has a gas surplus for the period $1000\,\text{sec} \leq t < 1500\,\text{s}$ that more than covers the demand in Region 3 during the same period, the compressor only starts spinning at $t = 1177\,\text{s}$. This is due to the setpoint increase for Region 2 at $t = 1000\,\text{s}$, which requires a greater accumulation of gas and thus a greater initial inflow of gas into that region. A similar cascaded effect can be seen for $1500\,\text{sec} \leq t < 2000\,\text{s}$, where decreasing the setpoint of Region 3 results in immediate action at compressor unit $j = 3$. However, it is only at $t = 1594\,\text{s}$ that the excess gas from Region 3 reaches Node 2 and compressor unit $j = 1$ starts to react to the pressure increase. This demonstrates the typical timescales of the flow of gas in the example network that comes as a result of the spin-up times required by the compressors along with the slower flow rates of gas when compared to, e.g., the electrical case.

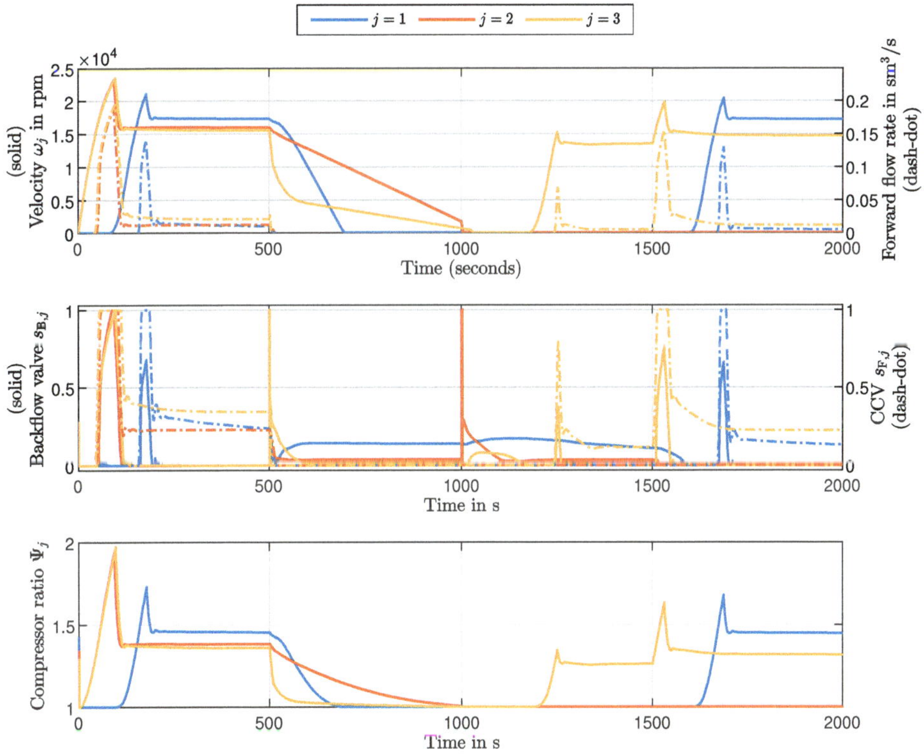

Figure 6.15: Compressor unit angular velocity ω_j and forward flow rate $q_{\mathrm{F},j}$ (top), backflow valve $s_{\mathrm{B},j}$ and CCV $s_{\mathrm{F},j}$ stem positions (middle), and the compressor ratio Ψ_j (bottom) for the three compressor units.

The trajectories and steady states of the compressor are shown in Figure 6.16, where the compressor ratios Ψ_j versus the forward flow rates $q_{F,j}$ are superimposed over the compressor characteristic from Figure 6.13. This shows the stable operation and stable equilibria of all compressor units in the surge region.

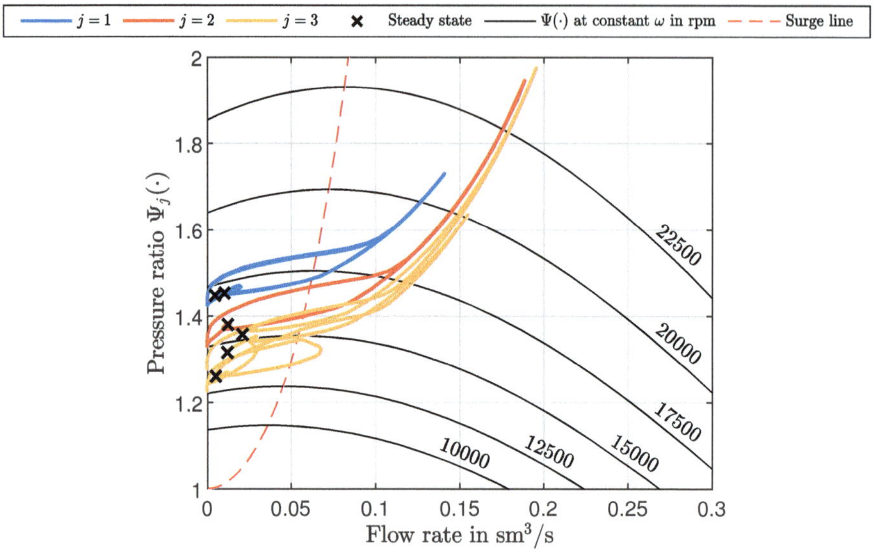

Figure 6.16: Compressor pressure ratio $\Psi_j(\cdot)$ versus forward flow rate $q_{F,j}$ along with the compressor characteristic for various constant angular velocities ω. The locations of the steady states are marked with a cross.

Finally, the torque control output for the compressor units along with the mechanical power required by the compressor machines are shown in Figure 6.17. From Figure 6.17a, the controller saturation of the torque can be seen, especially during the first 98 s. The effect of the initial term $\omega_{0,j}$ is also seen when comparing the instantaneous application of the full torque for $j = \{2, 3\}$ at the start of the simulation versus the gradual torque increase for $j = 1$ starting at $t = 80$ s. It is noted that the torque outputs also show rapid changes in the transient regions with τ_3, e.g., changing from 40 N m to −28.4 N m at $t = 1529$ s before rising again to a maximum of 25.0 N m at $t = 1549$ s. The impact of such rapid changes in the torque necessitates similar rapid changes in the mechanical power $\tau_j\omega_j$ which the respective compressors require, as seen in Figure 6.17b. Continuing with the example for $j = 3$, an initial jump in the power from 20.1 kW to 56.8 kW is seen at $t = 1500$ s. The mechanical power then gradually increases to 83.0 kW at $t = 1529$ s before instantaneously dropping to −59.0 kW and subsequently increasing to 40.5 kW at $t = 1549$ s. Moreover, the steady-state power requirements for the periods where the compressors are active span a range of [20.1; 33.4] kW while the transient power demands span a range of [−97.5; 98.7] kW.

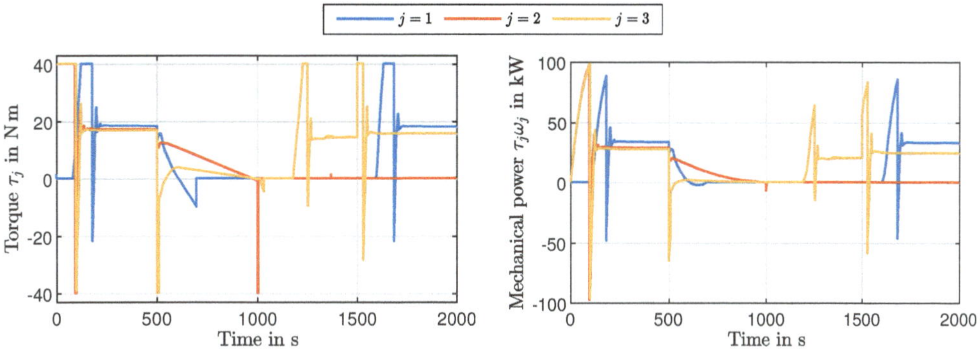

(a) Compressor input torque τ_j.

(b) Compressor mechanical power $\tau_j \omega_j$.

Figure 6.17: Compressor unit input torque (a) and mechanical power (b).

6.6 Discussion

The central work in this chapter involves achieving AS for the equilibria of a gas network in which a local oversupply of gas necessitates a bidirectional flow of gas to achieve pressure regulation in the network. This task is achieved by utilising parallels between the gas and electrical domains to inform the pipeline models proposed in Theorem 6.1 and through the use of EIP for the stability analysis in Theorem 6.8. It is also noteworthy that these parallels extend to the primary cause of instability in gas networks. Recall from Chapter 4 that loads with non-monotone incremental impedances represent the sources of instability in DC networks. Similarly, the non-monotone nature of the compressor characteristic $\Psi(\cdot)$ results in the unstable surge behaviour of compressors. Thus, the origin of instability in both networks can be traced to a simple sign change of the respective function gradients, with the local control action through the DGUs or the CCVs serving to counter these causes of instability. Furthermore, the gas network model allows for any EIP gas load to be connected at any node in the network. This mirrors the results from Chapters 4 and 5, where any EIP load functions may connect at any bus without impacting the verified AS of the steady states.

The gas pipeline model (Theorem 6.1), which uses the π-structure common in electrical networks, brings several advantages when dealing with gas networks and fulfils the requirements of RQ 3.1. The lumped-parameter and the simplified third-order models both exhibit a high simulation fidelity, comparing favourably with other discretised and commercial simulation models. While differences remain between the results obtained by using the proposed models and the state-of-the-art results, becoming more pronounced for larger elevation changes, the differences remain within 1 % in all cases, representing an acceptable trade-off w.r.t. to the analytical accessibility of the models. The accessibility for control engineers is demonstrated in Section 6.2, where the capacitive and inductive terms of the pipeline models are split apart to obtain a graph-based network similar to the network modelling in Chapter 4. Theorem 6.5 further demonstrates the analytical advantage by deriving an OS-EIP property for the inductive-like flow rate in the pipeline. This property, which holds across all friction types and accounts for the effect of gravity, is established by demonstrating the monotonicity of the friction factor w.r.t. the flow rate. As pointed out in Remark 6.14, this result directly implies the convexity of the steady-state pressure-difference-to-flow-rate relation for the nonlinear flow-rate equations, with potential applications in optimisation problems involving gas pipelines.

Although the pipeline model can be adapted for any homogenous fluid mixture (see Assumption 6.1 and Remark 6.5), the homogeneity poses a limitation for gas networks with hydrogen blending. Since the injection of different gas mixtures in a network can result in regions with different gas properties, the proposed model is only appropriate for the case where the injected gas matches the gas composition in the network.

The pipeline model has only been validated in a single benchmark test with high-pressures and long pipelines. Nevertheless, a similar fidelity is expected under different conditions provided that Assumption 6.1 holds. This expectation is substantiated by the models proposed in [Ke00; HG+09], which show similar levels of fidelity in low-, medium- and high-pressure setups, and against which the pipeline model in Theorem 6.1 shows a good fidelity (see Section 6.1.3).

Moving on to the control design, the simulation verifies the function of the controlled compressor units, where the backflow valve, the CCV, and the torque together achieve a pressure regulation on the inlet. The controllers together ensure a stable operation both inside and outside the surge region while yielding an EIP result for controlled compressor units. This EIP property allows these units to be included in gas networks or to be combined into various compressor station configurations while giving a direct pathway for establishing the AS of the equilibria of such compressor-equipped networks. The EIP result is obtained by designing the control gains such that four scalar inequalities hold simultaneously. Although the feasibility of this design process is not explicitly proven, the control gains which directly inject damping (see Remark 6.13) can be used to dominate the inequalities at the cost of a more aggressive control (in the case of $\beta_{\mathrm{B},j}$, $\beta_{\omega,j}$, β_{ε}, and β_{p}) or at the cost of a lower efficiency (arising due to a smaller $q_{\mathrm{F},j}$). Moreover, the EIP result in Theorem 6.7 can be made robust against parameter or model uncertainties by verifying that the inequalities hold in the presence of uncertain parameters or functions.

The EIP results of the pipelines and the controlled compressor units are key in deriving the AS equilibria result for an entire gas network. As demonstrated in Theorem 6.8, decentralised conditions for the compressor units and a connectivity result for the network topology are sufficient for AS. The AS is also robust against any piecewise-constant network changes which do not invalidate the requirements or Assumption 6.2. This robustness includes connectivity-preserving topology changes and parameter changes in the loads, nodes, lines, valves, or compressors (as long as the compressor units remain EIP). Moreover, by modelling the loads as EIP components, the AS property of the gas network holds for any dynamic loads, provided that a similar EIP property is maintained on their hydraulic ports. This again mirrors the DC network results (see Remark 4.14). The compressor controller together with the AS of the gas network thus provides an answer to RQ 3.2.

Finally, despite gas network pressures and flow rates showing settling times of several hundred seconds, the controller outputs exhibit fast transient behaviour in the range of seconds. Considering the case where the torque-dependent mechanical power needed for the compressors is supplied electrically via a time-varying load, the resulting electrical load can show transients on the same timescale as the DC network dynamics (see Chapter 5 in particular). This provides further motivation for RQ 4.2, which is considered in the next chapter.

6.7 Summary and Contributions

This chapter shows how pressure regulation and stabilisation can be achieved in a gas network with intermittent gas oversupply. The pressure regulation is achieved through feedback controllers for valve-equipped pipelines and compressor units, comprising valves and a centrifugal compressor. It is shown that these controllers produce the desired steady states and render the respective gas network components EIP. The controlled valve-equipped pipelines are inherently EIP, while a set of scalar inequalities must hold for the EIP of the compressor units. The unactuated gas network nodes and edges are shown to be EIP and OS-EIP, respectively, with the latter being derived by investigating the monotonicity of hydraulic friction. Finally, the component EIP properties are combined to obtain an AS result for the gas network equilibrium. This AS result is scalable for arbitrary network sizes and robust against parameter, load and topology changes.

The main contributions of this chapter are:

- A third-order lumped-parameter model for a gas pipeline using a π-structure with high simulation fidelity (Theorem 6.1);
- A proof for the OS-EIP of the derived gas pipeline model which derives from the monotonicity of the friction function (Theorem 6.5);
- A feedback controller for a compressor unit which regulates the inlet pressure to a desired setpoint while also yielding an EIP result for the closed-loop component (Theorem 6.7);
- A robust and scalable AS result for the gas network with top-down or bottom-up gas supplies (Theorem 6.8).

Together, these contributions meet the requirements of RQ 3.1 and RQ 3.2.

7 Stabilisation and Coordination in Coupled DC-Gas Networks

In the previous chapters, it has been shown how decentralised stabilisation and distributed coordination for DC networks (Chapters 4 and 5) and the decentralised stabilisation for gas networks (Chapter 6) can be achieved. While the results of these chapters are scalable, modular, and robust against various topology and parameter changes, the stability results established in Theorems 4.9, 5.9 and 6.8 do not consider the effects arising through the coupling with other networks. In this chapter, therefore, an NMES which consists of the DC and gas networks described in Chapters 4, 5 and 6 is considered. This is done to investigate the stability of the NMESs (RQ 4.1) as well as the effects of the dynamical coupling (RQ 4.2).

The chapter starts with **Section 7.1**, in which coupling components interconnecting DC and gas networks in an NMES are described and the NMES stability is investigated.

Next, in **Section 7.2**, an NMES comprising a DC and a gas network equipped with decentralised stabilisation controllers is presented. Thereafter, in **Section 7.3**, the same NMES is simulated for the case where the DC network is equipped with distributed coordination controllers. These simulations demonstrate the robustness of the NMES stability result.

The chapter concludes with a discussion of the methods and results in **Section 7.4** along with a summary of the work and the contributions in **Section 7.5**.

7.1 Interconnection and Stability of Networked Multi-Energy Systems

Before considering the stability of an NMES, a model for the NMES and its coupling components is required. For the NMES comprising DC and gas networks, the network models in Sections 4.1 and 6.2 are combined to obtain an NMES model. The coupling components interconnect the DC and gas networks via the previously modelled network components situated at the buses and nodes (i.e., ZIP loads, P sources and gas loads). E.g., a P2G component can be represented by a ZIP load and a negative gas load, while an external setpoint for the P2G component sets the parameters of the loads. The NMES model and coupling component descriptions are provided in **Section 7.1.1**.

Thereafter, in **Section 7.1.2**, stability conditions for an NMES comprising an arbitrary number of DC and gas networks are presented.

7.1.1 NMES and Coupling Component Descriptions

The NMES is described using a graph in which the vertices in a set $\mathcal{J}$ represent entire DC or gas networks which are connected via edges in a set $\mathcal{C}$ representing the coupling components. More formally, an NMES comprises a set $\mathcal{J}_{\mathrm{el}} = \{1, \ldots, |\mathcal{J}_{\mathrm{el}}|\}$ of DC networks and a set $\mathcal{J}_{\mathrm{g}} = \{|\mathcal{J}_{\mathrm{el}}| + 1, \ldots, |\mathcal{J}_{\mathrm{el}}| + |\mathcal{J}_{\mathrm{g}}|\}$ of gas networks, with $|\mathcal{J}_{\mathrm{el}}| \geq 0$, $|\mathcal{J}_{\mathrm{g}}| \geq 0$, and $\mathcal{J} = \mathcal{J}_{\mathrm{el}} \cup \mathcal{J}_{\mathrm{g}}$. The buses and nodes of all DC and gas networks are combined in the sets $\mathcal{N}_{\mathrm{el}} = \bigcup_{s \in \mathcal{J}_{\mathrm{el}}} \mathcal{N}_{\mathrm{el},s}$, $\mathcal{N}_{\mathrm{g}} = \bigcup_{s \in \mathcal{J}_{\mathrm{g}}} \mathcal{N}_{\mathrm{g},s}$, and $\mathcal{N} = \mathcal{N}_{\mathrm{el}} \cup \mathcal{N}_{\mathrm{g}} = \bigcup_{s \in \mathcal{J}} \mathcal{N}_s$. Let an edge $c \in \mathcal{C}$ with $|\mathcal{C}| \geq 0$ describe a coupling component that connects a source network $s_{c,+} \in \mathcal{J}$ to a sink network $s_{c,-} \in \mathcal{J}$. The directions of the edges in $\mathcal{C}$ indicate the directions of positive power flow. The NMES is thus described by the graph $\mathcal{G}_{\mathrm{NMES}} = (\mathcal{J}, \mathcal{C})$.

Corresponding to each coupling edge $c \in \mathcal{C}$ that describes the connections between the energy networks in $\mathcal{J}$, let there also be a set of coupling edges $\mathcal{E}_{\mathrm{Link}}$ that describes the interconnections between the buses and nodes in $\mathcal{N}$. Thus, an edge $c \in \mathcal{C}$ with the source and sink networks $s_{c,+}, s_{c,-} \in \mathcal{J}$ has a corresponding edge $o \in \mathcal{E}_{\mathrm{Link}}$ such that $n_{o,+} \in \mathcal{N}_{s_{c,+}}$ and $n_{o,-} \in \mathcal{N}_{s_{c,-}}$. While $\mathcal{G}_{\mathrm{NMES}} = (\mathcal{J}, \mathcal{C})$ describes the NMES at a network level, the graph $(\mathcal{N}, \mathcal{E}_{\mathrm{Link}})$ describes the same NMES at a bus and node level. An example of an NMES described by $(\mathcal{N}, \mathcal{E}_{\mathrm{Link}})$ is shown in Figure 7.1, whereas the NMES representation using $(\mathcal{J}, \mathcal{C})$ is shown in Figure 7.2.

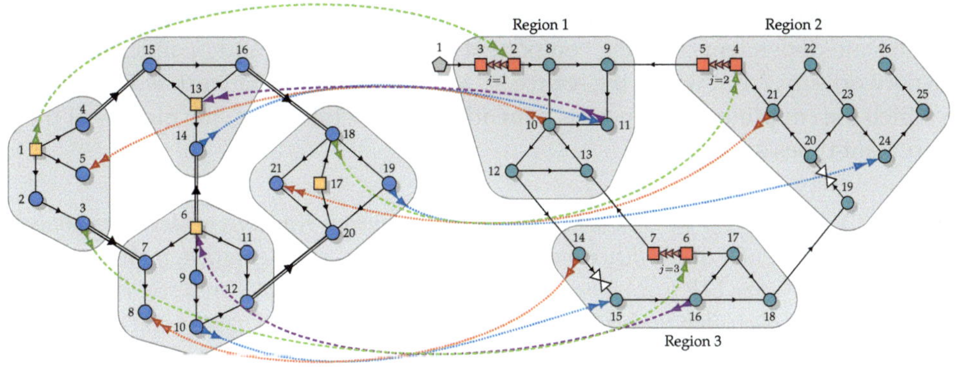

Figure 7.1: An NMES graph $(\mathcal{N}, \mathcal{E}_{\text{Link}})$ comprising the DC network from Figure 4.2 on the left and the gas network from Figure 6.7 on the right interconnected via coupling components in $\mathcal{E}_{\text{Link}}$ showing the direction of positive power flow. The set $\mathcal{E}_{\text{Link}}$ is divided into P2G components in $\mathcal{E}_{\text{P2G}}$, GFPPs in $\mathcal{E}_{\text{GFPP}}$, gas powered DGUs in $\mathcal{E}_{\text{gDGU}}$, and electric compressors in $\mathcal{E}_{\text{elC}}$.

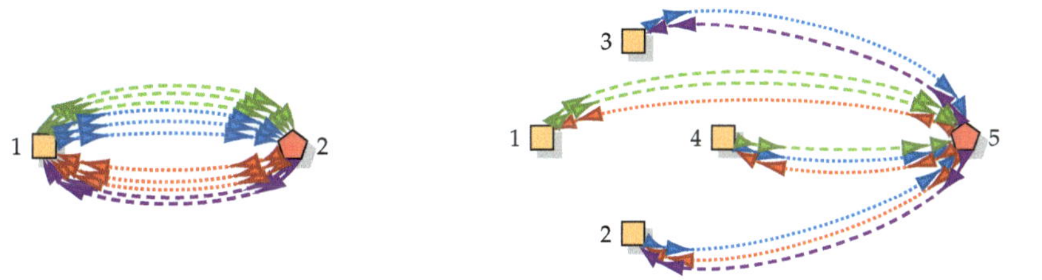

(a) Graph of Figure 7.1.

(b) Graph of Figure 7.1 with the electrical lines between clusters disconencted.

Figure 7.2: Graph $(\mathcal{J}, \mathcal{C})$ of an NMES comprising DC networks and gas networks connected via the coupling components in $\mathcal{C}$.

The coupling components are considered to be static functions which describe lossy conversions from one energy type to another. The coupling components are further considered to be either *setpoint dependent* or *state dependent*. In the case of setpoint-dependent coupling components, the power flowing through the edge in $\mathcal{E}_{\text{Link}}$ is dependent only on a piecewise-constant power setpoint. P2G components and GFPP belong to this category. For state-dependent coupling components, the power flowing through the edge in $\mathcal{E}_{\text{Link}}$ depends on state variables from networks of the source and/or sink nodes. Coupling components in this category include gas-powered DGUs and electric compressors. Since the state-dependent coupling components potentially allow transient behaviour to be transferred between networks, it is necessary to consider their directional causality. To this end, let there be an edge $d \in \mathcal{D}$ corresponding to each state-dependent coupling component $c \in \mathcal{C}$ such that d and c connect the same two networks in $\mathcal{J}$, but where the source $s_{d,+} \in \mathcal{J}$ denotes the network with the states on which the coupling component is dependent. In the following, the models of these component types are provided.

Remark 7.1. *Recall from Remark 4.14 that the instantaneous reactions of static components can be more detrimental to stability compared to a dynamical representation, if the dynamical model exhibits the same EIP port property and has an autonomous GAS property. Furthermore, for the purposes of analysing the NMES, elements such as pipes, tanks, valves, and compressors on the gas-network side of a coupling component may be considered as part of the gas network using the models in Section 6.2.*

Setpoint-Dependent P2G Components

Consider a generic P2G component described by an edge $o \in \mathcal{E}_{\mathrm{P2G}} \subseteq \mathcal{E}_{\mathrm{Link}}$ with a corresponding edge $c \in \mathcal{C}_{\mathrm{P2G}}$. The P2G component converts electrical power taken from a DC network bus $n_{o,+} \in \mathcal{N}_{\mathrm{el}}$ into gas injected at a gas network node $n_{o,-} \in \mathcal{N}_{\mathrm{g}}$. The component is controlled by its electrical power setpoint $p_o^\star$, while the losses are aggregated in an efficiency factor. In light of Remark 7.1 and recalling the thermal power associated with a gas flow rate (6.23), the P2G component may be represented by

$$P_{n_{o,+}} = p_o^\star, \qquad n_{o,+} \in \mathcal{N}_{\mathrm{el}}, \qquad s_{c,+} \in \mathcal{J}_{\mathrm{el}}, \qquad o \in \mathcal{E}_{\mathrm{P2G}}, \tag{7.1a}$$

$$\mathfrak{q}_{\mathfrak{l},n_{o,-}} = -\frac{\eta_{\mathrm{P2G},o} p_o^\star}{c_{\mathrm{cal}} \varrho_{\mathrm{n}}}, \qquad n_{o,-} \in \mathcal{N}_{\mathrm{g}}, \qquad s_{c,-} \in \mathcal{J}_{\mathrm{g}}, \qquad c \in \mathcal{C}_{\mathrm{P2G}}, \tag{7.1b}$$

where $P_{n_{o,+}}$ is the P component of a ZIP load (see (4.1)), $\mathfrak{q}_{\mathfrak{l},n_{o,-}}$ is the gas load (see (6.22)), $p_o^\star \in \mathbb{R}_{\geq 0}$ is the operating electrical power of the P2G component, and $\eta_{\mathrm{P2G},o} \in \mathbb{R}_{>0}$ with $\eta_{\mathrm{P2G},o} < 1$ is the aggregate efficiency factor for converting electrical power into gas. Note that the load components in (7.1) may simply be added to the ZIP and gas loads already present at the bus and node, respectively.

Remark 7.2. *Recalling the homogenous gas mixture requirement in Assumption 6.1 along with the discussion in Section 6.6, it is assumed that the gas injected by the P2G components has the same composition and temperature as the gas already in the network.*

Setpoint-Dependent GFPPs

Next, consider a generic GFPP described by an edge $o \in \mathcal{E}_{\mathrm{GFPP}} \subseteq \mathcal{E}_{\mathrm{Link}}$ with a corresponding edge $c \in \mathcal{C}_{\mathrm{GFPP}}$. The GFPP converts gas taken at a node $n_{o,+} \in \mathcal{N}_{\mathrm{g}}$

into electrical energy which is injected at a bus $n_{o,-} \in \mathcal{N}_{\mathrm{el}}$. Following the same argumentation as for the P2G components and again recalling the thermal power associated with a gas flow rate (6.23), a GFPP can be described by

$$\mathfrak{q}_{\mathfrak{l},n_{o,+}} = \frac{p_o^\star}{\eta_{\mathrm{GFPP},o} c_{\mathrm{cal}} \varrho_{\mathrm{n}}}, \qquad n_{o,+} \in \mathcal{N}_{\mathrm{g}}, \quad s_{c,+} \in \mathcal{J}_{\mathrm{g}}, \quad o \in \mathcal{E}_{\mathrm{GFPP}}, \tag{7.2a}$$

$$P_{\mathrm{ps},n_{o,-}} = -p_o^\star, \qquad n_{o,-} \in \mathcal{N}_{\mathrm{el}}, \quad s_{c,-} \in \mathcal{J}_{\mathrm{el}}, \quad c \in \mathcal{C}_{\mathrm{GFPP}}, \tag{7.2b}$$

where $p_o^\star \in \mathbb{R}_{\geq 0}$ is the electrical power at which the GFPP operates, $P_{\mathrm{ps},n_{o,+}}$ is the power parameter of the P source (see (4.5)), $\mathfrak{q}_{\mathfrak{l},n_{o,-}}$ is the gas load (see (6.22)), and $\eta_{\mathrm{GFPP},o} \in \mathbb{R}_{>0}$ with $\eta_{\mathrm{GFPP},o} < 1$ is the aggregate efficiency factor for converting gas into power.

State-Dependent Gas-Powered DGUs

In certain cases, instead of injecting power directly into the electrical network, a GFPP or a microturbine might be used to supply a DGU with power via the edge $o \in \mathcal{E}_{\mathrm{gDGU}}$ with a corresponding edge $c \in \mathcal{C}_{\mathrm{gDGU}}$. In such cases, gas might supply the full power demand $v_{\mathrm{b},n_{o,-}} i_{n_{o,-}} > 0$ of the DGU at a node $n_{o,-} \in \mathcal{N}_{\mathrm{el}}$, or it might only be used to cover the power demand extending past the base power $p_{\mathrm{base},o} \in \mathbb{R}_{\geq 0}$ supplied by local storage components or RESs. Recalling the DGU model in Figure 4.1, a gas-powered DGU may thus be represented by

$$\mathfrak{q}_{\mathfrak{l},n_{o,+}} = \frac{p_o^\star}{\eta_{\mathrm{gDGU},o} c_{\mathrm{cal}} \varrho_{\mathrm{n}}}, \qquad n_{o,+} \in \mathcal{N}_{\mathrm{g}}, \quad s_{c,+} \in \mathcal{J}_{\mathrm{g}}, \quad o \in \mathcal{E}_{\mathrm{gDGU}}, \tag{7.3a}$$

$$p_o^\star := \max(0, v_{\mathrm{b},n_{o,-}} i_{n_{o,-}} - p_{\mathrm{base},o}), \quad n_{o,-} \in \mathcal{N}_{\mathrm{el}}, \quad s_{c,-} \in \mathcal{J}_{\mathrm{el}}, \quad c \in \mathcal{C}_{\mathrm{gDGU}}, \tag{7.3b}$$

where $\mathfrak{q}_{\mathfrak{l},n_{o,+}}$ is the gas load (see (6.22)), $\eta_{\mathrm{gDGU},o} \in \mathbb{R}_{>0}$ with $\eta_{\mathrm{gDGU},o} < 1$ is the efficiency at which gas is converted into electrical power, and $p_o^\star \in \mathbb{R}_{\geq 0}$ is the electrical power required by the DGU from the GFPP. Note that the coupling component in (7.3) is dependent on the DC network, thus the edge $d \in \mathcal{D}$ corresponding to $c \in \mathcal{C}_{\mathrm{gDGU}}$ has its source in a DC network, i.e., $s_{d,+} = s_{c,-} \in \mathcal{J}_{\mathrm{el}}$.

State-Dependent Electric Compressors

Consider an electrically driven compressor $j \in \mathcal{E}_{\mathrm{Cs}}$ with its corresponding source node $n_{j,+} \in \mathcal{N}_{\mathrm{g}}$ which is powered via an edge $o \in \mathcal{E}_{\mathrm{elC}}$ with a corresponding edge $c \in \mathcal{C}_{\mathrm{elC}}$. The edge o thus describes the flow of power from the DC bus $n_{o,+} \in \mathcal{N}_{\mathrm{el}}$ to the gas node $n_{o,-} = n_{j,+}$. Recalling the compressor model in Figure 6.8 with the machine dynamics in (6.31b), the NMES coupling arising from an electric compressor can be described by

$$P_{n_{o,+}} = p_o^\star, \qquad n_{o,+} \in \mathcal{N}_{\text{el}}, \qquad s_{c,+} \in \mathcal{J}_{\text{el}}, \qquad o \in \mathcal{E}_{\text{elC}}, \tag{7.4a}$$

$$p_o^\star := \frac{\omega_j \tau_j}{\eta_{\text{elC},o}}, \qquad n_{j,+} = n_{o,-} \in \mathcal{N}_{\text{g}}, \qquad s_{c,-} \in \mathcal{J}_{\text{g}}, \qquad c \in \mathcal{C}_{\text{elC}}, \tag{7.4b}$$

where $P_{n_{o,+}}$ is the P component of a ZIP load (see (4.1)). The electrical power required by the compressor $p_o^\star \in \mathbb{R}$ is derived from the mechanical power $\omega_j \tau_j$ of the compressor j along with the power conversion efficiency factor $\eta_{\text{elC},o} \in \mathbb{R}_{>0}$. Notice that the description in (7.4) also allows for the generation of electrical power via regenerative braking of the compressor machine. Furthermore, the coupling component in (7.4) is dependent on the DC network, thus the edge $d \in \mathcal{D}$ corresponding to $c \in \mathcal{C}_{\text{elC}}$ has its source in a gas network, i.e., $s_{d,+} = s_{c,-} \in \mathcal{J}_{\text{g}}$.

7.1.2 Stability of NMESs

Building on the stability results in Chapters 4, 5 and 6, the AS of the equilibria of an NMES comprising DC and gas networks interconnected via the coupling components can now be considered. The AS result for the NMES is derived from the AS results of its constitutive networks under consideration for the effects of the coupling components. Since the coupling components are considered to be static functions, the piecewise-constant setpoint-dependent coupling components in $\mathcal{E}_{\text{P2G}} \cup \mathcal{E}_{\text{GFPP}}$ do not allow any transient behaviour to be transferred between the NMES networks. However, this is not the case for the state-dependent coupling components in $\mathcal{E}_{\text{gDGU}} \cup \mathcal{E}_{\text{elC}}$. Nevertheless, an AS result for the NMES can be established if the state-dependent coupling components form an acyclic interconnection of the DC and gas networks in the NMES.

Theorem 7.1 (AS of the NMES Equilibria)
Consider an NMES described by a graph $\mathcal{G}_{\mathrm{NMES}} = (\mathcal{J}, \mathcal{C})$. Let the equilibria of each network $s \in \mathcal{J}$ be AS in isolation. Consider now the coupling components between the networks in set $\mathcal{C}$, where the static edges in $\mathcal{C}_{\mathrm{P2G}} \cup \mathcal{C}_{\mathrm{GFPP}}$ have piecewise-constant power setpoints. Then, the equilibria of the NMES are AS, with a Lyapunov function comprising the sum of the Lyapunov functions of the networks $s \in \mathcal{J}$, if the following conditions hold simultaneously:

1. *The addition of the coupling elements in $\mathcal{C}$ does not violate the network feasibility assumptions for the respective networks;* [1]
2. *The addition of the constant P loads from the coupling components in $\mathcal{C}_{\mathrm{P2G}} \cup \mathcal{C}_{\mathrm{elC}}$ in the DC networks $\mathcal{J}_{\mathrm{el}}$ does not violate the AS conditions of the DC networks;* [2]
3. *The graph described by $(\mathcal{J}, \mathcal{D})$ with $\mathcal{D} = \mathcal{D}_{\mathrm{elC}} \cup \mathcal{D}_{\mathrm{gDGU}}$ is a directed acyclic graph.*

Proof:
The first condition ensures the feasibility of the networks. This is a necessary condition to ensure that their respective AS properties are retained when connecting the coupling components. For the second condition, recall that the coupling components in $\mathcal{E}_{\mathrm{Link}}$ interface with the bus and node dynamics through ZIP loads, P sources, and positive or negative gas loads. Any valid P source value yields an EIP property (see Remark 4.4) and the gas loads are independent of the pressure and thus also exhibit an EIP property irrespective of the load value. Thus, only changes to the ZIP loads resulting from the coupling components in $\mathcal{E}_{\mathrm{P2G}}$ (7.1) and $\mathcal{E}_{\mathrm{elC}}$ (7.4) have the potential to violate the AS conditions of the networks in $\mathcal{J}$ (see Remark 4.2).

For the third condition, recall that the state-dependent coupling components can allow transients to propagate from one network in $\mathcal{J}$ to another, whereas the powers flowing through the setpoint-dependent coupling components are piecewise constant. The graph $(\mathcal{J}, \mathcal{D})$ thus represents the directional causality of the networks arising from the state-dependent coupling components. Since the equilibria of each network $s \in \mathcal{J}$ are AS when the state-dependent coupling components are disregarded, any network $s \in \mathcal{J}$ which is not the sink node of some edge $d \in \mathcal{D}$ also has AS equilibria. Thus, the AS result of the network is retained since the network is independent of external time-varying parameters. Furthermore, the equilibria of any network $s \in \mathcal{J}$ in the graph $(\mathcal{J}, \mathcal{D})$ with incident edges in $\mathcal{D}$ that come only from networks in $\mathcal{J}$ that have AS equilibria are also AS. In this way, the AS property can be thought of as propagating from networks in $\mathcal{J}$ without any incident edges to $\mathcal{D}$ to other networks in $\mathcal{J}$. Moreover,

[1] See Assumptions 4.1, 5.1 and 6.2.
[2] See Theorems 4.9 and 5.9.

if $(\mathcal{J}, \mathcal{D})$ is a directed acyclic graph, then the propagation of the AS over the directional causality of the coupling components described by $\mathcal{D}$ ensures that every network $s \in \mathcal{J}$ has AS equilibria. ■

Through Theorem 7.1, the AS of the equilibria of an NMES comprising an arbitrary number of DC and gas networks can be investigated, thus providing a response to RQ 4.1. Note that Theorem 7.1 makes no distinction between the type of controllers with which the DC or gas networks are equipped and can thus be applied, e.g., for a DC network with a decentralised stabilisation or with a distributed coordination. The difference between these cases in the context of this work relates to Condition 2 of Theorem 7.1 where the ZIP loads in the DC networks with distributed coordination must have strictly monotone incremental impedances, while the DC networks with decentralised stabilisation allow non-monotone loads.

Remark 7.3. *Condition 3 in Theorem 7.1 is inherently fulfilled if* $\mathcal{C}_{\mathrm{gDGU}} = \emptyset$ *or* $\mathcal{C}_{\mathrm{elC}} = \emptyset$, *i.e., if there are no gas-powered DGUs or no electric compressors in the NMES. Considering the case where one of these sets is empty,* $(\mathcal{J}, \mathcal{D}_{\mathrm{elC}} \cup \mathcal{D}_{\mathrm{gDGU}})$ *describes a directed acyclic graph since the edges in* $\mathcal{D}_{\mathrm{elC}}$ *exclusively connect vertices in* $\mathcal{J}_{\mathrm{g}}$ *to vertices in* $\mathcal{J}_{\mathrm{el}}$ *and vice versa for* $\mathcal{D}_{\mathrm{gDGU}}$.

7.2 Simulated Networked Multi-Energy System with Stabilised DC Networks

In this section, a dynamical simulation is presented for an NMES comprising a gas network interconnected with a DC network equipped with controllers achieving decentralised stabilisation. This combines the networks and the results from Chapters 4 and 6 using all four coupling types described in Section 7.1.1. The simulation demonstrates the stability achieved by Theorem 7.1 in accordance with RQ 4.1, while the simulation event sequence is chosen such that the effects of the coupling components can be investigated as required by RQ 4.2. This section starts with a description of the simulation setup in **Section 7.2.1**. The results of the simulation are then presented in **Section 7.2.2**, which concludes with a summary of the main results.

7.2.1 Simulation Setup

The NMES in Figure 7.1 comprising the DC network from Figure 4.2 and the gas network from Figure 6.7 is simulated in MATLAB/SIMULINK using SIMSCAPE. The NMES serves as a synthetic test network (see [MTB20, Table 3]) to investigate the stability of the interconnected DC and gas networks. As indicated in Figure 7.1, the two networks are connected via three P2G components and three GFPP. Furthermore, two DGUs are partially powered with gas and all three compressors are powered electrically.

For the DC network, the network and control parameters in Table 4.2 are used, with the exception that the ZIP load and P source parameter restrictions are not applied to the coupling elements. The same line lengths as listed in Table B.3 in Appendix B.1 are used. Note that the DGU converter voltages again experience saturation to highlight the robustness properties of the controllers. The gas network setup is identical to Section 6.5.1, with the network and control parameters listed in Table 6.1 along with the gas properties (Table B.5) and the pipeline lengths (Table B.6) listed in Appendix B.3. The gas load values in Table B.7 are reused for different simulation time periods and the total gas load per region as given in Table 6.2 thus also remains applicable. The parameters for the coupling components are listed in Table 7.1. Note that only the power requirements of the gas-powered DGUs above the base power of 200 kW are supplied from the gas network, as per (7.3).

Table 7.1: NMES Coupling Component Parameters

Gas properties	$c_{\mathrm{cal}} = 47.1\,\mathrm{MJ/sm^3}$
	$\varrho_{\mathrm{n}} = 0.7179\,\mathrm{kg/m^3}$
P2G components (7.1), $o \in \mathcal{E}_{\mathrm{P2G}}$	$\eta_{\mathrm{P2G},o} = 0.38$
GFPPs (7.2), $o \in \mathcal{E}_{\mathrm{P2G}}$	$\eta_{\mathrm{GFPP},o} = 0.45$
Gas-powered DGUs (7.3),	$\eta_{\mathrm{gDGU},o} = 0.45$
$o \in \mathcal{E}_{\mathrm{gDGU}}$	$p_{\mathrm{base},o} = 200\,\mathrm{kW}$
Electric compressors (7.4), $o \in \mathcal{E}_{\mathrm{elC}}$	$\eta_{\mathrm{elC},o} = 0.92$

To investigate the dynamical coupling within the NMES, the simulation is run with the following sequence of events.

- $t = 0\,\mathrm{s}$: The DC network undergoes a black start with the ZIP load and P source parameters from Tables B.1 and B.2 at $t = 0\,\mathrm{s}$. The gas network starts with the node pressures initialised to the regional pressure setpoints in (6.7) and with the loads set to the valves in Table B.7 specified at $t = 500\,\mathrm{s}$. The pressure setpoints for Region 3 are stepped up to $\mathfrak{p}_n^\star = \mathfrak{p}_{\mathrm{n}} + 120\,\mathrm{mbar}$. The coupling components in $\mathcal{E}_{\mathrm{elC}}$ are enabled while those in $\mathcal{E}_{\mathrm{gDGU}}$ are disabled.
- $t = 500\,\mathrm{s}$: The ZIP load and P source parameters change to the values in Tables B.1 and B.2 specified at $t = 0.2\,\mathrm{s}$. The gas loads are set to the values in Table B.7 specified at $t = 0\,\mathrm{s}$.
- $t = 1000\,\mathrm{s}$: The lines interconnecting the clusters are disconnected, yielding four separated DC networks comprising one cluster each. The pressure setpoints for Region 3 are stepped down to $\mathfrak{p}_n^\star = \mathfrak{p}_{\mathrm{n}} + 80\,\mathrm{mbar}$, $n \in \mathcal{N}_{\mathrm{g},3}$. The coupling components in $\mathcal{E}_{\mathrm{gDGU}}$ are enabled.

- $t = 1500\,\mathrm{s}$: The DC lines are reconnected to form a single DC network.
- $1600\,\mathrm{s} \le t < 1700\,\mathrm{s}$: A sinusoidal disturbance with an amplitude of 30 kW and a period of 10 s is added to the power parameter P_{14} of the ZIP load at Bus 14 for 50 s. Thereafter, P_{14} is subjected to a random step disturbance in the range $[-30; 30]$ kW sampled from a uniform distribution every 5 s.
- $1800\,\mathrm{s} \le t < 1900\,\mathrm{s}$: A sinusoidal disturbance with an amplitude of $0.005\,\mathrm{sm}^3/\mathrm{s}$ and a period of 10 s is added to the gas load $\mathfrak{q}_{\mathfrak{l},17}$ of the at Node 17 for 50 s. Thereafter, $\mathfrak{q}_{\mathfrak{l},17}$ is subjected to a random step disturbance in the range $[-0.005; 0.005]\,\mathrm{sm}^3/\mathrm{s}$ sampled from a uniform distribution every 5 s.

The setpoints for the P2G and GFPP coupling components are listed in Table 7.2. Note that the ZIP load, P source and gas load values from Tables B.1, B.2 and B.7 in Appendix B do not include the effects of the coupling elements in $\mathcal{E}_{\mathrm{Link}}$.

Table 7.2: NMES Coupling Component Setpoints

		$\mathcal{E}_{\mathrm{P2G}}$			$\mathcal{E}_{\mathrm{GFPP}}$		
	o	1	2	3	4	5	6
	$n_{o,+}$	Bus 10	Bus 14	Bus 19	Node 10	Node 14	Node 21
	$n_{o,-}$	Node 15	Node 11	Node 24	Bus 5	Bus 8	Bus 21
$p_o^\star$ in kW	$t = 0\,\mathrm{s}$	0	0	0	80	80	80
	$t = 500\,\mathrm{s}$	80	80	80	0	0	0
	$t = 1000\,\mathrm{s}$	50	0	50	0	50	0

The DC and the gas networks are shown to have AS equilibria (see Sections 4.4.2 and 6.5.1), even for the case where the clusters of the DC network are disconnected. An AS for the equilibria of the NMES can thus be investigated via Theorem 7.1. For the period $0\,\mathrm{s} \le t < 1000\,\mathrm{s}$, the coupling components corresponding to the gas-powered DGUs are disabled and AS follows from Remark 7.3. For $1000\,\mathrm{s} \le t < 1500\,\mathrm{s}$, Theorem 7.1 is applied for the NMES depicted in Figure 7.2b and AS follows since the graph $(\mathcal{J}, \mathcal{D}_{\mathrm{elC}} \cup \mathcal{D}_{\mathrm{gDGU}})$ is a directed acyclic graph. After $t = 1500\,\mathrm{s}$, the coupling connections in $\mathcal{D}_{\mathrm{elC}} \cup \mathcal{D}_{\mathrm{gDGU}}$ result in a feedback cycle spanning the entire DC and gas networks. The stability in this period is not covered by Theorem 7.1 and is thus investigated purely in simulation.

7.2.2 Simulation Results

Starting with the bus voltages and injected powers for the DC network in the NMES in Figure 7.3, it can be seen that a stability according to Definition 2.1 is achieved throughout the simulation. A similar convergence rate is seen as in Section 4.4.3 (see Figure 4.10a), while notable transients occur for $500\,\mathrm{s} \le t \le 631\,\mathrm{s}$, and $1000\,\mathrm{s} \le t \le$

1146 s due to the influence of the coupling components in $\mathcal{E}_{\mathrm{gDGU}}$. Comparing these periods for the injected powers in Figure 7.3b, it can be seen that the voltage setting controllers inject power to counteract these effects. The effects isolating the clusters can also be seen for these time periods in Figures 7.3a and 7.3b with the transient behaviour being isolated to the buses in Cluster 1 for $1000\,\mathrm{s} \leq t \leq 1146\,\mathrm{s}$.

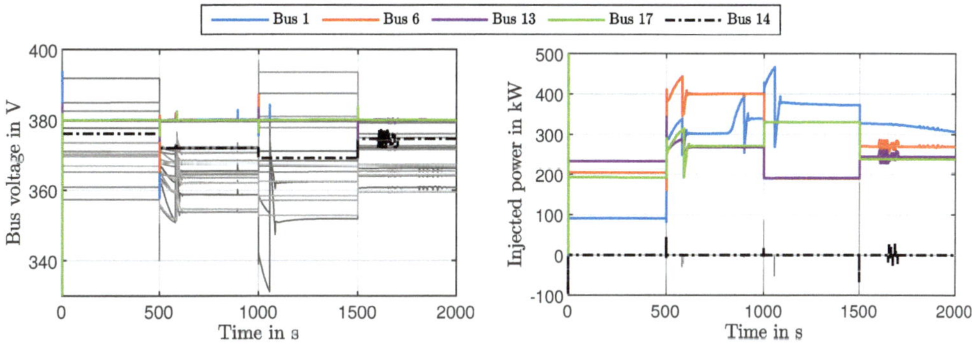

(a) Bus voltages v_n.

(b) Injected DGU power $v_n i_n$.

Figure 7.3: Simulated NMES with DC network stabilisation: Bus voltages (a) and injected powers (b). Voltage following buses are shown in shades of grey.

To analyse the DC network disturbance rejection, the bus voltages and injected powers are shown in Figure 7.4 for the load disturbance periods. Regarding the electrical disturbance, it can be seen that the sinusoidal disturbance is completely isolated to Bus 14. The step changes result in voltage deviations of less than $1.0\,\mathrm{V}$ for less than $20\,\mathrm{ms}$ at buses other than Bus 14. The injected power shows that only Buses 1, 13, and 14 react to the changes. This can be attributed to the fact that Bus 14 lies between the voltage setting Buses 1 and 13, which allows the disturbance to be isolated from the rest of the network. For the gas load disturbances during $1800\,\mathrm{s} \leq t \leq 1900\,\mathrm{s}$, voltage deviations no larger than $1.0\,\mathrm{V}$ are observed. The injected powers show transients no larger than $4\,\mathrm{kW}$.

The interactions between the DC and the gas networks are shown in Figure 7.5, where the load parameters arising from the coupling components are depicted. The P load transients during $500\,\mathrm{s} \leq t \leq 631\,\mathrm{s}$ and $1000\,\mathrm{s} \leq t \leq 1146\,\mathrm{s}$ confirm that the gas network dynamics cause the bus voltage transients in Figure 7.3a through the electrical compressors. Note that the P load variation at Bus 1 for $804\,\mathrm{s} \leq t \leq 918\,\mathrm{s}$ does not result in significant voltage deviations since the voltage setting controller at Bus 1 can sufficiently reject the disturbance (see the injected powers in Figure 7.3b).

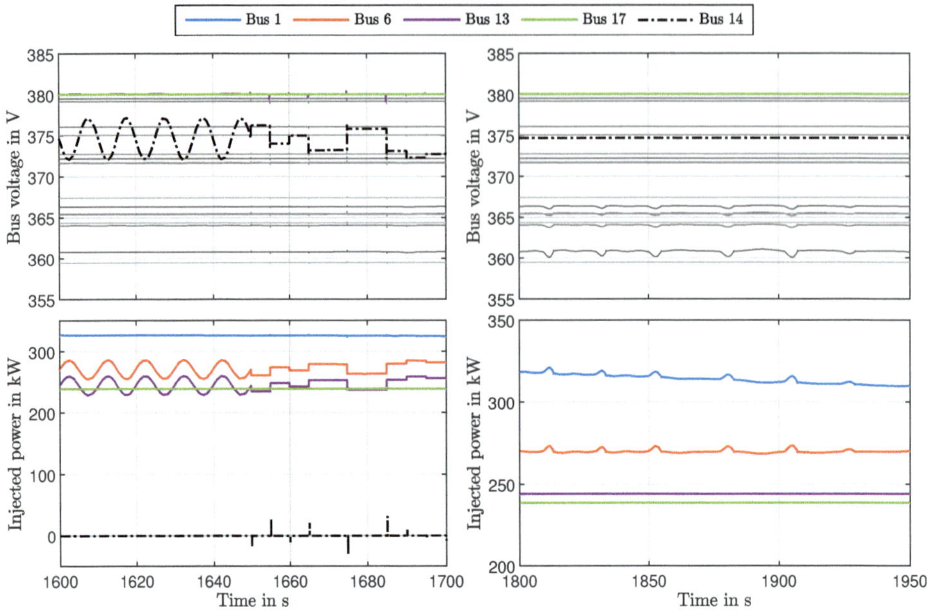

Figure 7.4: Simulated NMES with DC network stabilisation: DC network disturbance rejection showing the bus voltages (top) and injected powers (bottom).

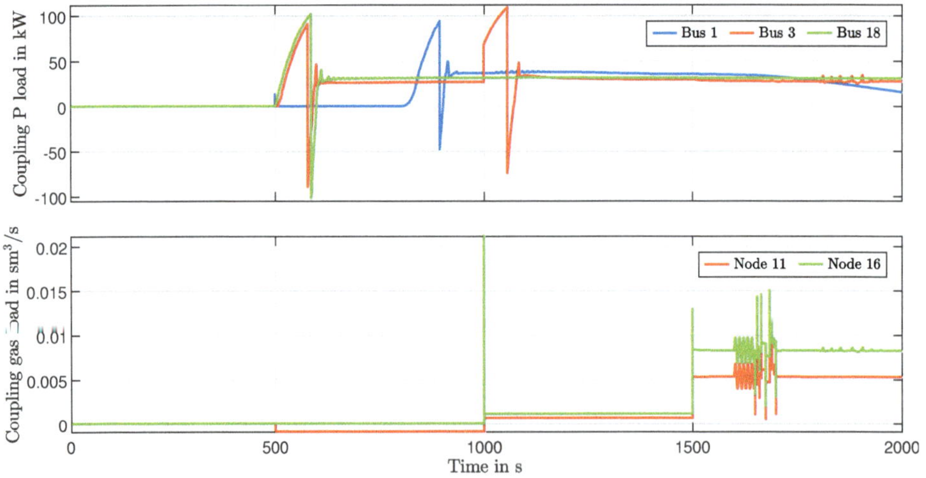

Figure 7.5: Simulated NMES with DC network stabilisation: Coupling effects of the P loads P_n (top) and the gas loads $q_{l,n}$ (bottom).

During the electrical disturbance ($1600\,\mathrm{s} \leq t \leq 1700\,\mathrm{s}$), the gas loads at Nodes 11 and 16 experience deviations of up to $1.43 \cdot 10^{-3}\,\mathrm{sm^3/s}$ from the steady state, apart from the peak values which last less than $2\,\mathrm{ms}$. In this period, the P loads show deviations of up to $1.31\,\mathrm{kW}$, indicating that the electrical disturbance propagates back to the DC network via the electrical compressors. During the gas disturbance ($1800\,\mathrm{s} \leq t \leq 1900\,\mathrm{s}$), the energy required by the electrical compressors results in the gas-powered DGUs drawing more power from the gas network. The largest of the transient peaks occurs at $t = 1905.2\,\mathrm{s}$. The delta power flows at various locations in the NMES between this peak and the nominal values at $t = 1907.3\,\mathrm{s}$ are shown in Figure 7.6. Notably, despite the gas-powered Bus 6 only covering $51\,\%$ of the electrical compressor load, the mechanical power delta of $6.7\,\mathrm{kW}$ requires $12.6\,\mathrm{kW}$ in gas. This increase arises due to losses in the DC network, DGUs, and the coupling components.

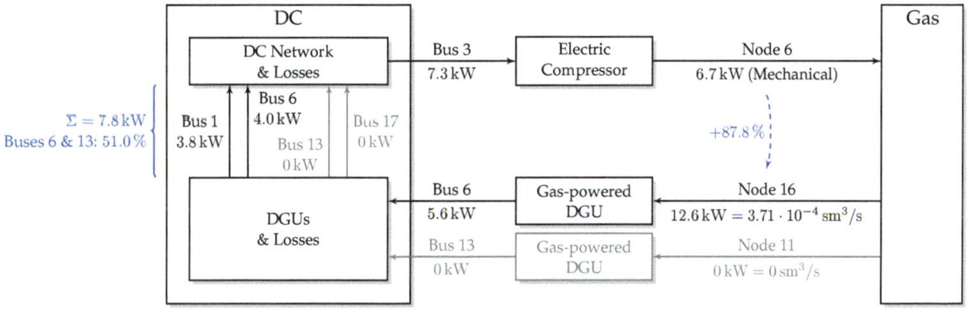

Figure 7.6: Simulated NMES with DC network stabilisation: Rounded power-flow deltas between $t = 1905.2\,\mathrm{s}$ and $t = 1907.3\,\mathrm{s}$ resulting from the gas load disturbances with arrows indicating the direction of power flow.

Moving on to the gas network results in Figure 7.7, the gas node pressures also show a stability according to Definition 2.2 in each simulation period. The events occurring before $t = 1500\,\mathrm{s}$ showcase how changes in the gas network propagate to the DC network, e.g., where the compressors are used to extract the excess gas at Nodes 4 and 6 for $500\,\mathrm{s} \leq t \leq 631\,\mathrm{s}$, and $1000\,\mathrm{s} \leq t \leq 1146\,\mathrm{s}$. At $t = 1500\,\mathrm{s}$, however, the DC clusters are again reconnected. Apart from the load disturbances, the changes during $1500\,\mathrm{s} \leq t \leq 2000\,\mathrm{s}$ and the new resulting steady state are thus a direct consequence of internal DC network changes.

The zoomed-in bus pressures for Region 3 of the gas network are shown for $1500\,\mathrm{s} \leq t \leq 2000\,\mathrm{s}$ in Figure 7.7e. The electrical and gas disturbances result in pressure deviations of up to $0.7\,\mathrm{mbar}$ and $1.0\,\mathrm{mbar}$ from the steady state, respectively. This indicates that a disturbance originating in either network can have a similar effect on the gas network.

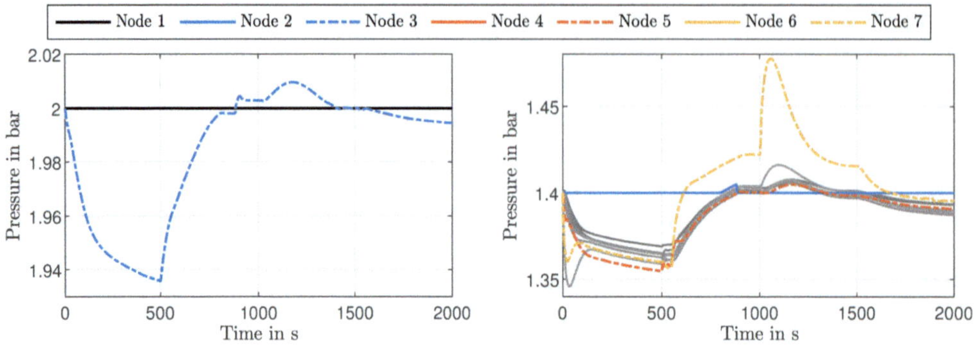

(a) Node pressures $\mathfrak{p}_n$ (high pressure).

(b) Node pressures $\mathfrak{p}_n$ (medium pressure).

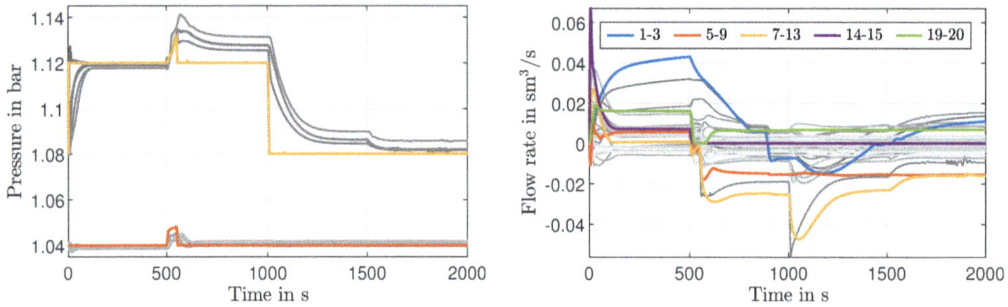

(c) Node pressures $\mathfrak{p}_n$ (low pressure).

(d) Pipe flow rates $\mathfrak{q}_{\mathrm{T},o}$ and $\mathfrak{q}_{v,o}$.

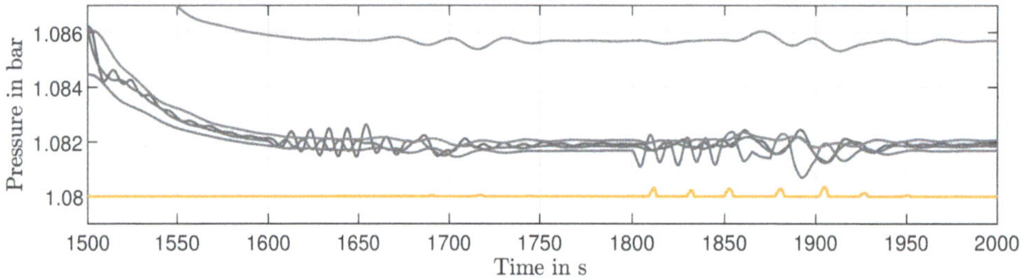

(e) Gas network disturbance rejection.

Figure 7.7: Simulated NMES with DC network stabilisation: Pressures and flow rates of the gas network split into high pressure (a), medium pressure (b), and low pressure (c) figures and the flow rates in the pipelines (d) along with a showcase of the disturbance rejection (e). Nodes and edges not denoted in the respective legends are shown in shades of grey.

In summary, the simulation in this section shows that:

- The NMES equilibria are AS in each time period, even when Theorem 7.1 does not apply after $t = 1500\,\mathrm{s}$.
- Gas network transients result in DC network transients.
- Disturbances in either network type can propagate to the other network type through the state-dependent coupling components.
- Gas disturbances can propagate through the DC network back to the gas network, but the slower gas network dynamics mean electrical disturbances do not propagate as readily through the gas network.

7.3 Simulated Networked Multi-Energy System with Coordinated DC Networks

The results in the previous section demonstrate the effects of the coupling components for a simulated NMES in which the DC network is equipped with a decentralised stabilisation scheme. Recalling the results from Chapters 4 and 5, replacing the decentralised stabilisation with a distributed coordination scheme yields proportional power sharing between the DGU at the cost of losing the OS-EIP property of the buses arising from the voltage setting controllers. In this section, the effects that this controller change in the DC network has on the NMES as a whole are investigated. The change supplements the investigation into dynamical coupling effects as per RQ 4.2 while also contrasting the qualities of the controllers from Chapters 4 and 5.

This section starts with a description of the simulation setup in **Section 7.3.1**, where the scenario is described and the simulation parameters are listed. The results of the simulation are then presented in **Section 7.3.2**.

7.3.1 Simulation Setup

The NMES in Figure 7.1 is again simulated in MATLAB/SIMULINK. The simulation setup is the same as in Section 7.2.1 and follows the same sequence of events, except for the following changes to the DC network. The voltage setting controllers are replaced with power regulators (see (5.2)) and the buses are equipped with a distributed power-sharing controller (see Figure 5.3). For the communication network, State A of Figure 5.6 is used, except for the period $1000\,\mathrm{s} \leq t < 1500\,\mathrm{s}$ where the DC network is split into four parts and during which State D is used. As in Section 5.3 (see Figure 5.13b) the effects of the leaky integrators are compensated by setting $v^\star = 390\,\mathrm{V}$ for the voltage setpoint of the four-stage controller in Figure 5.3.

Since the DC loads are not necessarily strictly monotone in the NMES, the AS result for the DC network equilibria cannot be verified using Theorem 5.9. By retaining the voltage following controllers of the DC network from the setup in Section 7.2, the simulation nevertheless can be used to demonstrate the AS of the NMES equilibria. Note that omitting the voltage following controllers from the simulation in this section results in instability.

7.3.2 Simulation Results

The results for the NMES with distributed coordination in the DC network mirror the results in Section 7.2.2. Starting with the bus voltages and injected powers in Figure 7.8, similar periods of transient behaviour are seen as in Figure 7.3. A notable difference to the DC stabilisation case is how pervasive the disturbances are, e.g., for $809\,\mathrm{s} \leq t \leq 920\,\mathrm{s}$ where the disturbance is no longer isolated to Bus 1 as in Figure 7.3. Moreover, the bus voltages experience greater extreme values of $212.4\,\mathrm{V}$ and $542.5\,\mathrm{V}$, indicating the lower damping of the network with distributed coordination. Nevertheless, despite the presence of non-monotone ZIP loads, the inclusion of the voltage following controllers allows for a stability in each time period.

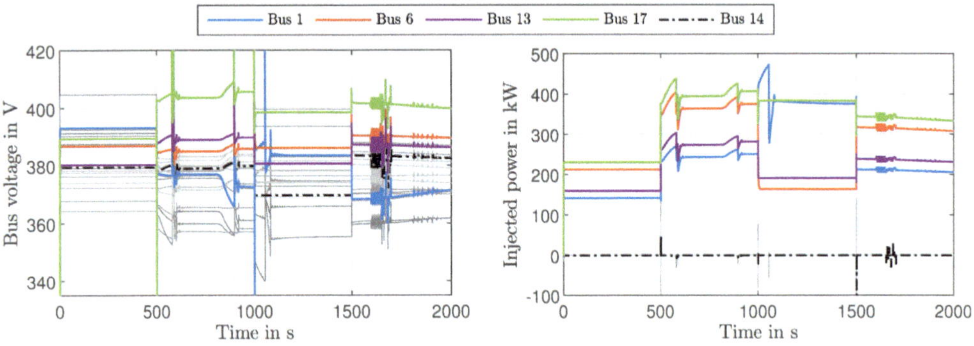

(a) Bus voltages v_n.

(b) Injected DGU power $v_n i_n$.

Figure 7.8: Simulated NMES with DC network coordination: Bus voltages (a) and injected powers (b). Voltage following buses are shown in shades of grey.

To analyse the DC network disturbance rejection, the bus voltages and injected powers are shown in Figure 7.9 for the load disturbance periods. This allows a direct comparison with the results in the previous section (see Figure 7.4). Specifically, it can be seen that the proportional power sharing comes at the cost of the disturbance affecting the entire network. During the electrical disturbance, buses other than Bus 14 see voltage disturbances of up to $8.6\,\mathrm{V}$. The gas load disturbance also causes voltage deviations from the steady state of up to $1.8\,\mathrm{V}$ while the transients of the injected power during $1800\,\mathrm{s} \leq t \leq 1900\,\mathrm{s}$ are no larger than $3.7\,\mathrm{kW}$.

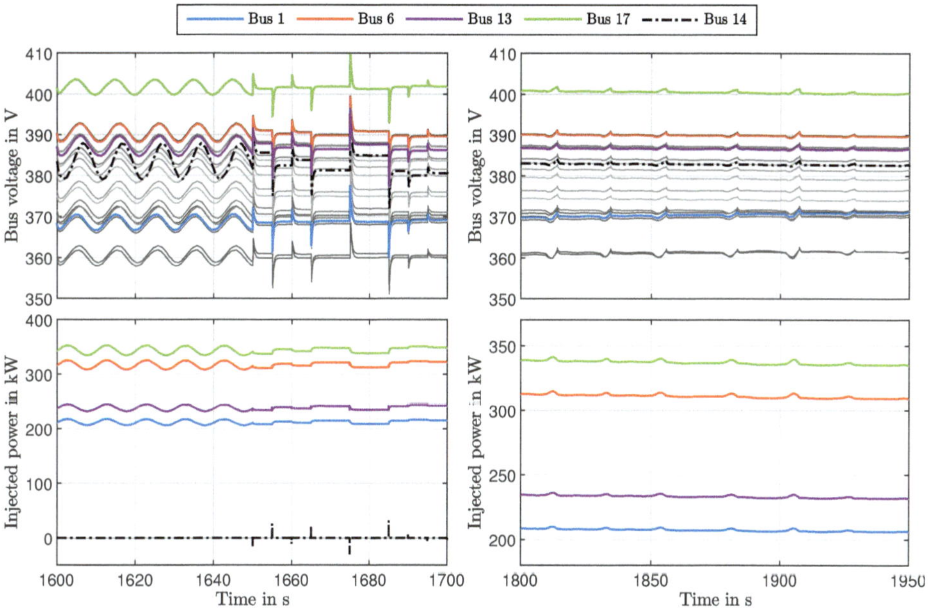

Figure 7.9: Simulated NMES with DC network coordination: DC network disturbance rejection showing the bus voltages (top) and injected powers (bottom).

The P load and gas load coupling effects are shown in Figure 7.10, where a significant difference is seen in the gas loads when compared to Figure 7.5. Specifically, the steady-state value of the load at Node 16 is 50.5 % ($4.17 \cdot 10^{-3}\,\mathrm{sm^3/s}$) greater when using distributed coordination in the DC network. In contrast, the oscillations arising from the P load disturbances are 42.0 % smaller than in the decentralised stabilisation case. During the gas disturbance ($1800\,\mathrm{s} < t < 1900\,\mathrm{s}$), the P load disturbance again propagates through the DC network to affect the gas loads through the coupling components. The largest peak of the occurs at $t = 1905.2\,\mathrm{s}$, with Figure 7.11 showing the power deltas at various locations in the NMES between the peak and the nominal values at $t = 1908.1\,\mathrm{s}$. Compared to Figure 7.6, both DGUs 6 and 13 require gas via the coupling components. These DGUs supply 49.9 % of the load, mirroring the results in Figure 7.6. Interestingly, the 9.6 kW mechanical power at Node 6 in Figure 7.11 draws 18.4 kW of gas, representing an increase of 91.2 %. The same relation in Figure 7.6 results in an increase of 87.8 %. When neglecting the slight percentage difference with which Buses 6 and 13 supply the disturbance, the distributed coordination requires 3.48 percentage points more power. This increase stems in part from the additional DC network losses which arise from the distributed nature of the disturbance rejection.

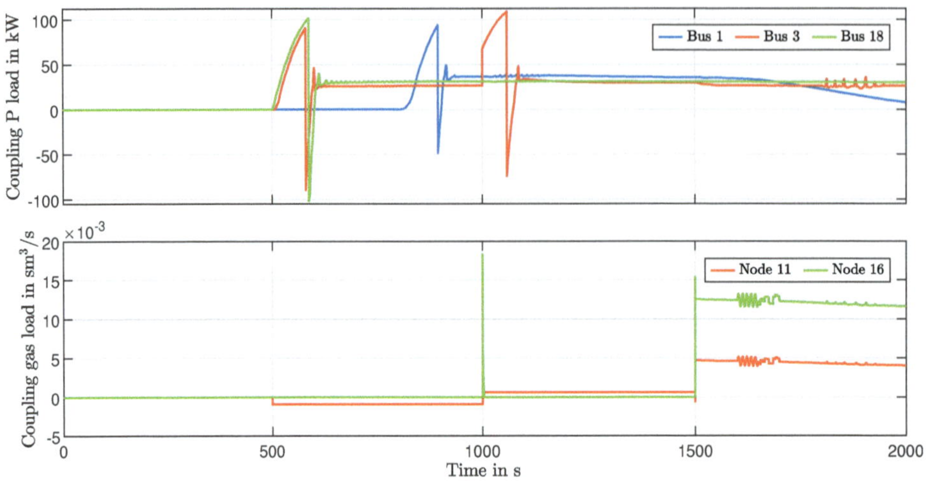

Figure 7.10: Simulated NMES with DC network coordination: Coupling effects of the P loads P_n (top) and the gas loads $q_{l,n}$ (bottom).

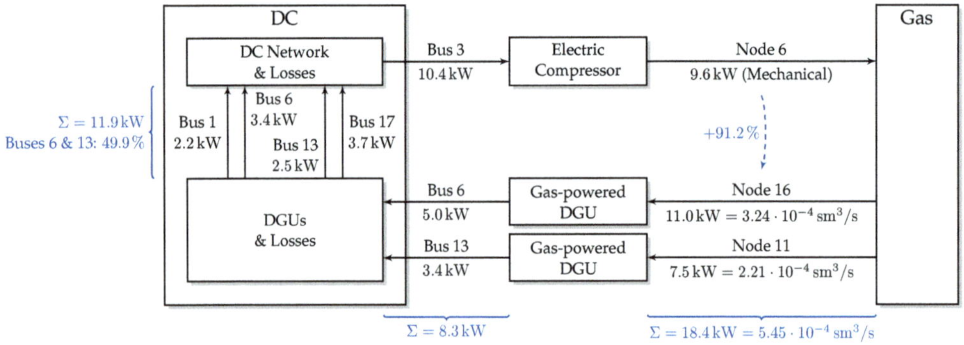

Figure 7.11: Simulated NMES with DC network coordination: Rounded power-flow deltas between $t = 1905.2$ s and $t = 1908.1$ s resulting from the gas load disturbances with arrows indicating the direction of power flow.

The simulated gas network pressures and flow rates have nearly identical trajectories to Figure 7.7, with the largest difference arising from the higher coupling-component gas load at Node 16 after $t = 1500$ s (see Figure 7.10). This results in the Region 3 pressures of the gas network being up to 1.2 mbar lower after $t = 1500$ s, as can be seen when comparing Figure 7.12 to Figure 7.7e. The electrical and gas disturbances result in similar steady-state deviations of up to 0.54 mbar and 1.1 mbar, respectively.

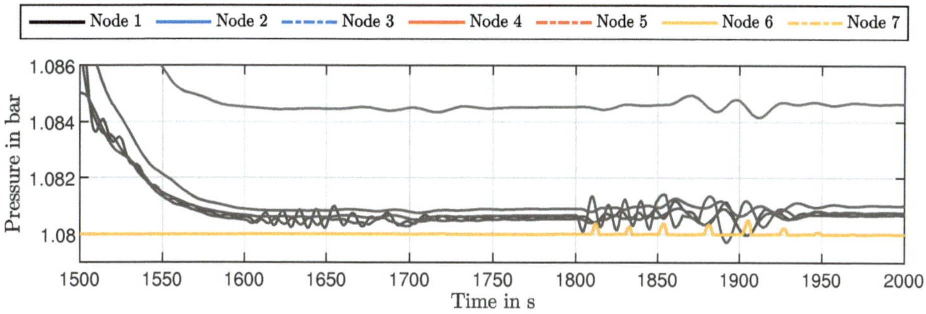

Figure 7.12: Simulated NMES with DC network stabilisation: Gas network disturbance rejection.

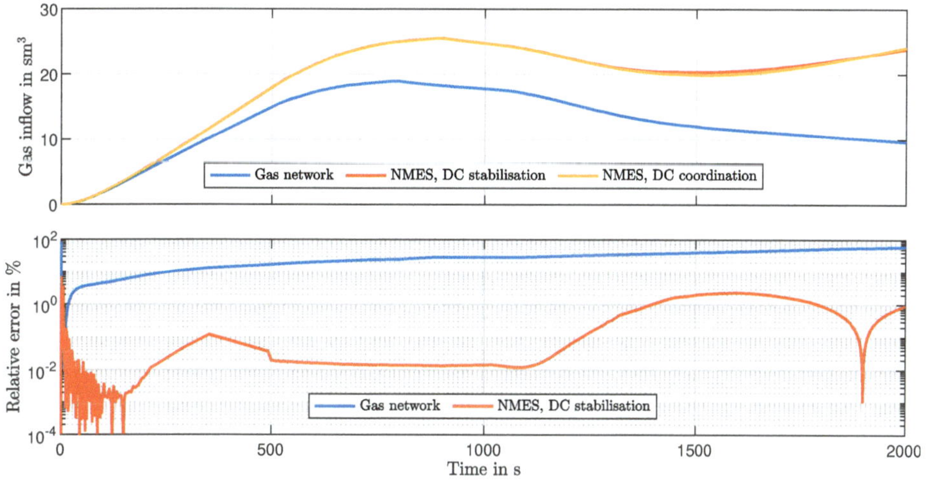

Figure 7.13: Comparison of the gas supplied by the gas network supply node for the uncoupled gas network as well as the NMESs with decentralised stabilisation and distributed coordination implemented for the DC network, showing the absolute value (above) and the error percentages compared to the distributed coordination case (below).

A final interesting comparison resulting from the coupling components and the differences between using decentralised stabilisation or distributed coordination in the DC network can be made through the accumulated gas supplied by the supply node of the gas network, as shown in Figure 7.13. Simulating the gas network without any of the coupling components leads to a difference of up to $14.5\,\mathrm{sm}^3$ of gas compared to the NMES cases by the end of the simulation. Before $t = 1000\,\mathrm{s}$ where the gas loads are only affected by the piecewise-constant setpoint-dependent coupling component, the relative error between the NMES cases remains below $0.12\,\%$. However, enabling the dynamic coupling from the gas-powered DGUs after $t = 1000\,\mathrm{s}$ leads to a divergence of the NMES cases in which an error of $2.0\,\%$ is reached within a further $500\,\mathrm{s}$. This error grows to $2.3\,\%$ at $1591\,\mathrm{s}$ before the error changes sign and again increases to $0.92\,\%$ at $t = 2000\,\mathrm{s}$. This demonstrates that the state-dependent coupling components together with the DC network controllers can have a significant impact on the operation of the gas network.

In summary, the simulation in this section shows that:

- The proportional power sharing of distributed coordination results in higher power usage and disturbances which propagate throughout the network compared to the decentralised stabilisation.
- Due to the power sharing, the distributed coordination allows a smaller portion of electrical disturbances to propagate to the gas network.
- The choice of DC network controllers has a measurable influence on the gas network once the DC-to-gas state-dependent coupling components are enabled.

7.4 Discussion

The work in this chapter presents sufficient conditions for the AS of the equilibria of an NMES comprising DC and gas networks while investigating the effects of the coupling components on the networks. This is achieved by letting the coupling components interface with the respective networks through the previously modelled loads and sources. This allows the theorems investigating the stability of the individual networks to retain their validity. By placing restrictions on the directions of causality for the coupling components, Theorem 7.1 derives an AS result for the NMES equilibria using the AS properties of the isolated NMES networks.

The novel NMES stability result in Theorem 7.1 continues the narrative of modularity and scalability seen in the results of the previous chapters. Additionally, the strength of this result lies in the generic requirements for the individual networks, i.e., Theorem 7.1 does not presuppose any specific topologies or the use of specific controllers for the constitutive NMES networks. Thus, in the context of this work, Theorem 7.1 can be applied for DC networks equipped with controllers yielding a decentralised stabilisation or a distributed coordination. Moreover, by describing coupling components interfacing with other types of energy networks, such as AC, water or district heating networks, Theorem 7.1 can be expanded to include other network types. Thus, Theorem 7.1 provides a novel result in response to RQ 4.1.

Despite the versatility of the stability result, Condition 3 in Theorem 7.1 presents a barrier to investigating the stability of arbitrarily interconnected NMESs since it prohibits feedback cycles between the networks. Conversely, this restriction means that state-dependent coupling components could lead to oscillating or unstable behaviour in an NMES. Examples of such oscillations propagating throughout the NMES have been seen in the simulations in this chapter, where the slower gas network dynamics provided sufficient damping for the disturbances. An analysis of the equilibria stability of an NMES with arbitrary state-dependent coupling components in the EIP framework requires establishing IFOF-EIP properties (or equilibrium-independent ISS properties, see Remark 3.3) on the ports for each type of component. However, this can be difficult for coupling components in which the power flow only indirectly depends on the network states, e.g., the mechanical power requirement of the electrical compressors.

The NMES in Figure 7.1 which is simulated in this chapter violates Condition 3 of Theorem 7.1 at the end of the simulations, thus preventing a stability result from being established via this theorem. Nevertheless, the simulation results for the strongly coupled NMES attest to a robust AS property. The lack of destabilising behaviour from the state-dependent coupling components in this chapter can be attributed qualitatively to the use of the electrical compressors to extract surplus gas from a region. If the required electricity is supplied by gas-powered DGUs, the compressor workload is lowered when the gas is taken from either the high- or low-pressure sides of the compressor. This therefore produces a negative feedback cycle which promotes stability. However, if a compressor is needed for supplying a network instead of extracting gas, a positive feedback cycle results.[3] This highlights the importance and necessity of a stability analysis of dynamical NMESs with state-dependent coupling components which extends beyond Theorem 7.1.

The results from the simulation studies in this chapter allow further conclusions on the dynamical effects of the coupling components to be drawn, as required by RQ 4.2. As per the summaries at the ends of Sections 7.2.2 and 7.3.2, transients in one network can propagate through state-dependent coupling components to cause transients in another network. Moreover, it is shown that disturbances in any network can propagate to other networks, but the slower dynamics of the gas networks limit the propagation of fast disturbances through the gas network. The common use of timescale separation for NMESs also remains valid, with the fast-acting bus and electrical line dynamics having no discernible impact on the NMES simulations. Nevertheless, any slow dynamical or static effects in the DC network have measurable impacts on the NMES as a whole. Specifically, this includes the control scheme implemented in the DC network, which affects both the NMES steady state and the disturbance propagation through state-dependent coupling components.

Even when the DC network dynamics are disregarded entirely, the slow or static behaviour of the DC networks can have a significant impact on the gas network, as demonstrated in Figure 7.13. In this comparison, and indeed by comparing the results in Sections 7.2 and 7.3 as a whole, it is shown that the same gas network and DC network with the same parameters and loads can exhibit significant transient differences when different control schemes for the DC network are used. While approximating an electrical network using the DC power flow equations might be a common method to ease the computational burden for higher-level controllers as discussed in Section 2.3,

3 In this case, gas used to power the DGUs results in a lower pressure. This in turn requires greater effort from the compressor.

the results in this chapter suggest the aforementioned approximation might not be appropriate if the control behaviour is not taken into account.[4]

Finally, the simulation results in Sections 7.2 and 7.3 present an interesting comparison between the proposed decentralised stabilisation and distributed cooperation schemes in the DC network w.r.t. their disturbance rejection properties (see Figures 7.4 and 7.9). Since the disturbances occur at a bus which exclusively connects to buses with steady-state power, the local voltage regulation from the voltage setting controllers seen in Figure 7.4 allows the noise to be isolated from the rest of the network at the cost of requiring larger power injections locally. This is in stark contrast to the four-stage distributed power-sharing controller used in Figure 7.9, where the disturbance is allowed to leak, affecting the entire rest of the network. The distributed nature of the power-sharing controllers also results in higher network losses and thus requires more power than the decentralised stabilisation approach when dealing with a disturbance. Nevertheless, the lower disturbance rejection capabilities of the distributed coordination also result in a smaller portion of the disturbance being propagated via the power-dependent gas-powered DGUs. These results reinforce the notion that the respective control schemes have distinct advantages and disadvantages and that their suitability is application and case specific.

7.5 Summary and Contributions

This chapter demonstrates the inclusion of coupling components in an NMES in a way that is compatible with the network descriptions of the previous chapters and which preserves of the AS results of the individual networks. An AS result for the NMES equilibria is derived using the AS results of the constitutive networks in isolation. The NMES stability result also requires that the feasibility assumptions of the networks are not invalidated when adding the coupling components while also restricting the interconnection of state-dependent coupling components. This theoretical result is followed by two simulation studies in which the work from Chapters 4, 5 and 6 is combined and compared, demonstrating the stable operation of an NMES comprising dynamical DC and gas networks. The simulations also highlight the coupling effects between the networks and show that changing the DGU controllers in the DC network can have a significant impact on the gas network dynamics.

---

4 Notice furthermore, that the distributed coordination scheme in Chapter 5 does not include or require a slack bus, complicating the application of the conventional DC power flow equations.

The main contributions of this chapter are:

- A representation of the static coupling components which is used to derive a scalable AS result for the equilibria of an NMES comprising DC and gas networks (Theorem 7.1);
- Simulation studies of a dynamical NMES along with an investigation of the effects that the coupling components have on the constitutive networks.

Together, these contributions provide responses to RQ 4.1 and RQ 4.2.

8 Conclusion

Modern energy systems are undergoing rapid change as countries seek to attain net-zero emissions through a greater reliance on renewable energy sources (RESs). At the same time, the development and adoption of new technologies result in previously independent networks becoming increasingly coupled. In such networked multi-energy systems (NMESs), the coupling can allow disturbances, e.g., from the volatility of RESs, to propagate throughout an entire NMES. However, the coupling of networks also has the potential to increase their resilience and the security of supply. Such beneficial coupling rests on the premise that the isolated networks are stable and that coupling components do not produce destabilising behaviour. Despite the extensive existing body of research on various energy networks, the state of the art does not sufficiently address the stability concerns of the individual networks or NMESs. This thesis contributes to the body of research by deriving controllers which ensure a decentralised stabilisation or a distributed coordination in DC networks, gas networks, and NMES.

In DC networks, controllers that ensure voltage regulation are designed for a decentralised stabilisation scheme. This extends the state of the art by allowing non-monotone loads anywhere in the network. Alternative controllers for a distributed coordination scheme are also designed. These novel controllers ensure a proportional power sharing and an average voltage regulation which incorporates measurements from all buses in a network with nonlinear loads. A direct comparison of these two approaches highlights their different steady-state and disturbance-rejection properties. This makes the appropriate control choice for DC networks application specific.

In the context of gas networks, stability risks arise from the integration of surge-capable compressors. Such compressors are necessary for the storage or redistribution of excess gas resulting from local gas production. Controllers are therefore designed for the compressors and valves to achieve pressure regulation in a decentralised stabilisation scheme. The resulting controlled compressor units improve upon existing approaches by allowing safe operation within the surge region while allowing for variable inlet and outlet pressures. The control design is complemented by a novel high-fidelity third-order gas pipeline model with a π-structure. By demonstrating the monotonicity of fluid friction, the model is shown to have equilibrium-independent passivity (EIP) properties that are useful for control and stability analysis purposes. This work contributes to the literature by allowing a stability proof for dynamical gas networks with and without surge-capable compressors.

For NMESs, a novel dynamical stability result which builds upon the DC and gas network results is presented. The NMESs are constructed by using the loads and sources of the respective networks as interfaces for setpoint- and state-dependent coupling components. By excluding feedback loops from state-dependent coupling components, the NMES stability can be derived directly from the stability of its constitutive networks in isolation. The simulation results also show quantitatively that stable behaviour is possible in the presence of state-dependent feedback loops. Moreover, the simulations demonstrate measurable transient and steady-state differences arising from coupling components between networks and the choice of controllers within networks. The presented work thus offers new insights into the effects of coupling components in NMESs while providing an extensible framework for investigating the dynamical stability of NMES.

For each network type and for the NMESs, the EIP framework is utilised to derive stability results. This thesis extends and formalises the theoretical framework to enable a constructive EIP-based analysis for the asymptotic stability (AS) of the equilibria of the interconnected subsystems. The constructive nature of EIP-based stability allows networks of arbitrary sizes to be analysed. At the same time, EIP yields modular results, since network components may be substituted with any other components that have the same EIP properties on the same ports. This modularity thus allows the presented results to retain their validity when the networks or NMESs are combined with components or controllers not explicitly considered in this thesis.

As a final conclusion, this thesis presents novel decentralised stabilisation and distributed coordination controllers for DC and gas networks, which yield scalable and modular dynamical stability results for the individual networks and for NMESs.

A Proofs

In this appendix, the full-length proofs for theorems and propositions with abbreviated proofs in the respective chapters are presented.

A.1 Proofs in Chapter 4

Proof of Theorem 4.2:
The OS-EIP of (4.16) is verified if $\dot{S}_s(\tilde{\boldsymbol{x}}_s) \leq \tilde{y}_s\tilde{u}_s - \rho\tilde{y}_s^2$ and if S_s is positive definite in $\tilde{\boldsymbol{x}}_s$. Starting with the first condition, calculate $\dot{S}_s$ while noting from (4.19) that $\boldsymbol{Q}_s\boldsymbol{P}_s = \boldsymbol{P}_s\boldsymbol{Q}_s$:

$$\dot{S}_s(\tilde{\boldsymbol{x}}_s) = \tfrac{1}{2}(\dot{\tilde{\boldsymbol{x}}}_s^T\boldsymbol{Q}_s\boldsymbol{P}_s\tilde{\boldsymbol{x}}_s + \tilde{\boldsymbol{x}}_s^T\boldsymbol{P}_s\boldsymbol{Q}_s\dot{\tilde{\boldsymbol{x}}}_s) = \tfrac{1}{2}(\tilde{\boldsymbol{x}}_s^T\boldsymbol{A}_s^T\boldsymbol{P}_s\tilde{\boldsymbol{x}}_s + \tilde{\boldsymbol{x}}_s^T\boldsymbol{P}_s\boldsymbol{A}_s\tilde{\boldsymbol{x}}_s) + \tilde{\boldsymbol{x}}_s^T\boldsymbol{b}_s u_s. \quad \text{(A.1)}$$

Thus, $\dot{S}_s(\boldsymbol{x}_s) \leq \tilde{y}_s\tilde{u}_s - \rho\tilde{y}_s^2 = \tilde{\boldsymbol{x}}_s^T\boldsymbol{b}_s u_s - \rho\tilde{\boldsymbol{x}}_s^T\boldsymbol{b}_s\boldsymbol{b}_s^T\tilde{\boldsymbol{x}}_s$ results in the LMI

$$-\boldsymbol{A}_s^T\boldsymbol{P}_s - \boldsymbol{P}_s\boldsymbol{A}_s - \rho\boldsymbol{b}_s\boldsymbol{b}_s^T = \begin{bmatrix} 2p_2 & p_3 + \kappa_i^\diamond p_2 & \kappa_v^\diamond p_2 - \kappa_\varepsilon^\diamond p_1 \\ p_3 + \kappa_i^\diamond p_2 & 2\kappa_i^\diamond p_3 & \kappa_v^\diamond p_3 - \kappa_\varepsilon^\diamond p_2 - 1 \\ \kappa_v^\diamond p_2 - \kappa_\varepsilon^\diamond p_1 & \kappa_v^\diamond p_3 - \kappa_\varepsilon^\diamond p_2 - 1 & 2(\delta - Y - \rho) \end{bmatrix} \succcurlyeq 0, \quad \text{(A.2)}$$

which can be verified using Sylvester's criterion [HJ12, Theorem 7.2.5]. This entails verifying $\det(\boldsymbol{A}_s^T\boldsymbol{P}_s + \boldsymbol{P}_s\boldsymbol{A}_s + \rho\boldsymbol{b}_s\boldsymbol{b}_s^T) = 0$ and that the last $N-1$ trailing principal minors are positive. The determinant of (A.2) is zero if the three minors obtained by removing the last row are zero. Starting with the first minor,

$$\det\begin{pmatrix} 2p_2 & p_3 + \kappa_i^\diamond p_2 \\ p_3 + \kappa_i^\diamond p_2 & 2\kappa_i^\diamond p_3 \end{pmatrix} = 0 \quad \Longrightarrow \quad 4\kappa_i^\diamond p_2 p_3 - (p_3 + \kappa_i^\diamond p_2)^2 = 0 \quad \text{(A.3a)}$$

$$\Longrightarrow \quad -(p_3 - \kappa_i^\diamond p_2)^2 = 0, \quad \text{(A.3b)}$$

which holds if p_3 is chosen as in (4.19b). For the second minor of the three minors, substitute p_3 in (4.19b) to obtain

$$\det\begin{pmatrix} 2p_2 & \kappa_v^\diamond p_2 - \kappa_\varepsilon^\diamond p_1 \\ p_3 + \kappa_i^\diamond p_2 & \kappa_v^\diamond p_3 - \kappa_\varepsilon^\diamond p_2 - 1 \end{pmatrix} = \det\begin{pmatrix} 2p_2 & \kappa_v^\diamond p_2 - \kappa_\varepsilon^\diamond p_1 \\ 2\kappa_i^\diamond p_2 & \kappa_i^\diamond\kappa_v^\diamond p_2 - \kappa_\varepsilon^\diamond p_2 - 1 \end{pmatrix} = 0 \quad \text{(A.4a)}$$

$$\Longrightarrow \quad 2p_2\left(\kappa_i^\diamond\kappa_v^\diamond p_2 - \kappa_\varepsilon^\diamond p_2 - 1\right) - 2\kappa_i^\diamond p_2(\kappa_v^\diamond p_2 - \kappa_\varepsilon^\diamond p_1) = 0 \quad \text{(A.4b)}$$

$$\Longrightarrow \quad -\kappa_\varepsilon^\diamond p_2 - 1 = -\kappa_i^\diamond\kappa_\varepsilon^\diamond p_1, \quad \text{(A.4c)}$$

from which the choice for p_1 in (4.19b) for $p_2 > 0$ stems. Substituting p_1 and p_3 into the last of the three minors and simplifying, it can be verified that

$$\det\begin{pmatrix} p_3 + \kappa_{\mathrm{i}}^{\diamond} p_2 & \kappa_{\mathrm{v}}^{\diamond} p_2 - \kappa_{\varepsilon}^{\diamond} p_1 \\ 2\kappa_{\mathrm{i}}^{\diamond} p_3 & \kappa_{\mathrm{v}}^{\diamond} p_3 - \kappa_{\varepsilon}^{\diamond} p_2 - 1 \end{pmatrix} = 0. \tag{A.5}$$

Thus, p_1 and p_3 in (4.19b) with $p_2 > 0$ guarantee $\det\big(\boldsymbol{A}_{\mathrm{s}}^T \boldsymbol{P}_{\mathrm{s}} + \boldsymbol{P}_{\mathrm{s}} \boldsymbol{A}_{\mathrm{s}}\big) = 0$. Next, conditions are formulated that ensure the last $N-1$ trailing principal minors of (A.2) are positive, i.e.,

$$2(\delta - Y - \rho) > 0, \tag{A.6}$$

$$\det\begin{pmatrix} 2\kappa_{\mathrm{i}}^{\diamond} p_3 & \kappa_{\mathrm{v}}^{\diamond} p_3 - \kappa_{\varepsilon}^{\diamond} p_2 - 1 \\ \kappa_{\mathrm{v}}^{\diamond} p_3 - \kappa_{\varepsilon}^{\diamond} p_2 - 1 & 2(\delta - Y - \rho) \end{pmatrix} > 0. \tag{A.7}$$

Calculating the determinant in (A.7) and substituting p_3 (4.19b) gives

$$\underbrace{4(\kappa_{\mathrm{i}}^{\diamond})^2(\delta - Y - \rho)}_{\mu_1} p_2 - \Big[\big(\underbrace{\kappa_{\varepsilon}^{\diamond} - \kappa_{\mathrm{i}}^{\diamond}\kappa_{\mathrm{v}}^{\diamond}}_{\mu_2}\big)p_2 + 1\Big]^2 > 0 \implies \mu_2^2 p_2^2 + (2\mu_2 - \mu_1)p_2 + 1 < 0, \tag{A.8}$$

which is a quadratic inequality in p_2. Since $p_2 > 0$, (A.8) is satisfied if the quadratic equation in p_2 has at least one positive root. This can be verified by ensuring that the turning point and the discriminant of the quadratic inequality are positive:

$$-\frac{(2\mu_2 - \mu_1)}{2\mu_2^2} > 0 \qquad \implies \mu_1 > 2\mu_2, \tag{A.9}$$

$$(2\mu_2 - \mu_1)^2 - 4\mu_2^2 > 0 \qquad \implies -4\mu_2\mu_1 + \mu_1^2 > 0 \qquad \implies \mu_1 > 4\mu_2, \tag{A.10}$$

Note that (A.6) ensures $\mu_1 > 0$, meaning (A.9) and (A.10) follow automatically if $\mu_2 \leq 0$. On the other hand, if $\mu_2 > 0$, (A.9) holds automatically if (A.10) is satisfied. Therefore, substituting μ_1 and μ_2 into (A.10) yields the sufficient condition

$$(\kappa_{\mathrm{i}}^{\diamond})^2(\delta - Y - \rho) > \kappa_{\varepsilon}^{\diamond} - \kappa_{\mathrm{i}}^{\diamond}\kappa_{\mathrm{v}}^{\diamond}, \tag{A.11}$$

which can be combined with (A.6) to obtain (4.18). Therefore, through (4.18), there is guaranteed to be a $p_2 > 0$ which, along with p_1 and p_3 in (4.19b), ensures that $\dot{S}_{\mathrm{s}}(\tilde{\boldsymbol{x}}_{\mathrm{s}}) \leq y_{\mathrm{s}} u_{\mathrm{s}} - \rho y_{\mathrm{s}}^2$.

Finally, it must be verified that S_{s} is positive definite in $\tilde{\boldsymbol{x}}_{\mathrm{s}}$. Since, from (4.19a), $\boldsymbol{Q}_{\mathrm{s}} \succ 0$, it is sufficient to ensure that $\boldsymbol{P}_{\mathrm{s}} \succ 0$. Again using Sylvester's criterion, $\boldsymbol{P}_{\mathrm{s}} \succ 0$ is guaranteed if its leading principal minors are positive, i.e.,

$$p_1 > 0, \tag{A.12}$$

$$\det\begin{pmatrix} p_1 & p_2 \\ p_2 & p_3 \end{pmatrix} = p_1 p_3 - p_2^2 > 0. \tag{A.13}$$

Since $\kappa_{\mathrm{i}}^{\diamond} > 0$ and $\kappa_{\varepsilon}^{\diamond} > 0$ (see (4.17)), the choices in (4.19b) guarantee that $p_1 > 0$ and $p_3 > 0$ for $p_2 > 0$. Thus (A.12) is met. Substituting p_1 and p_3 from (4.19b) into (A.13) yields

$$p_2^2 + \frac{1}{\kappa_{\varepsilon}^{\diamond}} p_2 - p_2^2 > 0 \implies \frac{1}{\kappa_{\varepsilon}^{\diamond}} p_2 > 0, \tag{A.14}$$

which automatically holds since $\kappa_{\varepsilon}^{\diamond} > 0$ (see (4.17)). ■

A.2 Proofs in Chapter 6

Proof of the derivation of (6.36) **and** (6.37)**:**
Starting with the backflow valve which sets $\mathfrak{q}_{\mathrm{B},j}$ in (6.31c), consider the dynamics of $\dot{\mathfrak{q}}_{\mathrm{B},j}$, which can be written similarly to (6.26). Applying the control for (6.34a) for a short pipe with $\mathfrak{L}_{\mathrm{F},j}\dot{\mathfrak{q}}_{\mathrm{F},j} = 0$, yields

$$\mathfrak{p}_{\mathrm{B},j} - \mathfrak{p}_k = \frac{\mathfrak{p}_{\mathrm{B},j}}{C_{v,j}^2 s_{\mathrm{B},j}^2} |\mathfrak{q}_{\mathrm{B},j}| \mathfrak{q}_{\mathrm{B},j}, \tag{A.15a}$$

$$C_{v,j}^2 s_{\mathrm{B},j}^2 |\mathfrak{p}_{\mathrm{B},j} - \mathfrak{p}_k| \operatorname{sign}(\mathfrak{p}_{\mathrm{B},j} - \mathfrak{p}_k) = \mathfrak{p}_{\mathrm{B},j} |\mathfrak{q}_{\mathrm{B},j}| \mathfrak{q}_{\mathrm{B},j}, \tag{A.15b}$$

$$C_{v,j}^2 \left[\beta_{\mathrm{B},j} (\mathfrak{p}_k - \mathfrak{p}_k^\star) + \frac{\mathfrak{q}_{\mathrm{B},j}}{C_{v,j}} \right]^2 = \mathfrak{q}_{\mathrm{B},j}^2, \tag{A.15c}$$

since $\operatorname{sign}(\mathfrak{p}_{\mathrm{B},j} - \mathfrak{p}_k) = \operatorname{sign}(\mathfrak{q}_{\mathrm{B},j})$. Eq. (A.15c) has the trivial solution $\mathfrak{p}_k = \mathfrak{p}_k^\star$ and $\mathfrak{q}_{\mathrm{B},j} \in \mathbb{R}$, meaning any flow rate for $\mathfrak{q}_{\mathrm{F},j}$ is allowed when the pressure is regulated to the desired setpoint. The second solution of (A.15c) which comprises the transient case is found by setting $\mathfrak{p}_k \neq \mathfrak{p}_k^\star$ and expanding, giving

$$C_{v,j}^2 \beta_{\mathrm{B},j}^2 (\mathfrak{p}_k - \mathfrak{p}_k^\star)^2 + 2 C_{v,j} \beta_{\mathrm{B},j} (\mathfrak{p}_k - \mathfrak{p}_k^\star) \mathfrak{q}_{\mathrm{B},j} = 0, \tag{A.15d}$$

$$-\frac{C_{v,j} \beta_{\mathrm{B},j}}{2} (\mathfrak{p}_k - \mathfrak{p}_k^\star) = \mathfrak{q}_{\mathrm{B},j}. \tag{A.15e}$$

The flow rate $\mathfrak{q}_{\mathrm{B},j}$ in (6.31c) is then substituted with (A.15e) to obtain the first state equation in (6.36a). The dynamics for $\mathfrak{p}_l$ in (6.36a) are taken directly from (6.31d). For the state $\mathfrak{q}_{\mathrm{F},j}$, the control law for τ_j (6.34b) is applied to (6.31a) and the nonlinear damping function $\mathfrak{R}_{\mathrm{F},j}(\cdot)$ is obtained as in (6.37a) through simplification. The torque control law (6.34c) is applied to the dynamics of ω (6.31b) and the substitution $\varepsilon_{\mathrm{C},j} = (z_j + J_j\omega_j)\beta_{\varepsilon,j}/\beta_{\mathfrak{p},j}$ is made according to the state transformation (6.35). This results in the functions $f_{\omega,j}(\cdot)$ (6.37b) and $f_{\mathrm{z},j}(\cdot)$ (6.37c). Finally, the dynamics of the augmented integral state in (6.34d) is included in $\dot{z}_j$, which is constructed from (6.35) as

$$\dot{z}_j = \frac{\beta_{\mathfrak{p},j}}{\beta_{\varepsilon,j}} \dot{\varepsilon}_{\mathrm{C},j} - J_j \dot{\omega}_j = \beta_{\mathfrak{p},j}(\mathfrak{p}_k - \mathfrak{p}_k^\star) - \frac{\beta_{\mathfrak{p},j}\beta_{\mathrm{B},j}}{\beta_{\varepsilon,j}} s_{\mathrm{B},j} - J_j \dot{\omega}_j = f_{\mathrm{z},j}(\cdot), \tag{A.16}$$

yielding the dynamics for the last state in (6.36a). ■

Proof of Proposition 6.6:
Applying Lemma 3.6 for the closed-loop valve-equipped pipeline and substituting in the control law (6.32) gives the nonlinear function

$$
\begin{aligned}
f_{v,o}(\mathfrak{q}_{v,o}) &= -\frac{\mathfrak{p}_{n_o,+}|\mathfrak{q}_{v,o}|\mathfrak{q}_{v,o}}{C_{v,o}^2 s_{v,o}^2}, && s_{v,o} > 0, && \text{(A.17a)}\\
&= -\frac{|\mathfrak{p}_{n_o,+} - \mathfrak{p}_{n_o,-}|\mathfrak{q}_{v,o}^2}{\left[C_{v,o}k_{v,o}(\mathfrak{p}^\star_{n_o,-} - \mathfrak{p}_{n_o,-}) + \mathfrak{q}_{\mathrm{T},o}\right]^2}, && \mathfrak{q}_{v,o} > 0, && \text{(A.17b)}
\end{aligned}
$$

where $-\nabla_{\mathfrak{q}_{v,o}} f_{v,o} > 0$ in $\mathfrak{q}_{v,o} > 0$ ensures the OS-EIP of the system with a $\rho > 0$. Recall that the case for $s_{v,o} = 0$ corresponds to a singularity in which $\mathfrak{q}_{v,o} = 0$ for as long as the valve remains closed, as discussed in Remark 6.8. Computing the derivative yields

$$
-\nabla_{\mathfrak{q}_{v,o}} f_{v,o} = \frac{\mu_1\mu_2\mathfrak{q}_{v,o}}{(\mu_2 + \mathfrak{q}_{v,o})^3}, \quad \mu_1 := 2|\mathfrak{p}_{n_o,+} - \mathfrak{p}_{n_o,-}|, \quad \mu_2 := C_{v,o}k_{v,o}(\mathfrak{p}^\star_{n_o,-} - \mathfrak{p}_{n_o,-}), \tag{A.18a}
$$

$$
-\nabla_{\mathfrak{q}_{v,o}} f_{v,o} > 0 \implies (\mu_1 > 0) \wedge (\mathfrak{q}_{v,o} > 0) \wedge \left[(\mu_2 > 0) \vee (\mu_2 < -\mathfrak{q}_{v,o})\right]. \tag{A.18b}
$$

Notice from its definition that $\mu_1 \geq 0$ and from the control law (6.32) that $\mu_1 = 0$ implies $s_{v,o} = 0$. Furthermore, writing the control law in terms of μ_1 and μ_2 gives

$$
s_{v,o} = C_{v,o}(\mu_2 + \mathfrak{q}_{v,o})\sqrt{\frac{2\mathfrak{p}_{n_o,+}}{\mu_1}}, \tag{A.19}
$$

from which it can be seen that $s_{v,o} > 0$ implies $[(\mu_2 > 0) \vee (\mu_2 < -\mathfrak{q}_{v,o})]$ if $\mathfrak{q}_{v,o} > 0$. Thus, the requirements for (A.18b) and thus of Lemma 3.6 are satisfied for $s_{v,o} > 0$, while the controlled system is static with $\mathfrak{q}_{v,o} = 0$ for $s_{v,o} = 0$. Finally, EIID follows directly from shifting the valve-equipped pipeline dynamics (6.27) w.r.t. an assumed equilibrium and setting $\tilde{\mathfrak{q}}_{v,o} = \tilde{y}_{v,o} \equiv 0$ to obtain $0 = \tilde{\mathfrak{p}}_{n_o,+} - \tilde{\mathfrak{p}}_{n_o,-}$. ∎

Proof of Theorem 6.7:
By applying Lemma 3.6 with $\rho = 0$, the EIP of the compressor unit can be verified by ensuring that $-\nabla \boldsymbol{f}_{\mathrm{C}}(\boldsymbol{x}_{\mathrm{C}}) \succeq 0$ for all $\boldsymbol{x}_{\mathrm{C}} \in \mathcal{X}_{\mathrm{C}}$, where

$$
\nabla \boldsymbol{f}_{\mathrm{C}} = \begin{bmatrix}
-\frac{1}{2}C_v\beta_{\mathrm{B}} & 0 & -1 & 0 & 0\\
0 & 0 & 1 & 0 & 0\\
\Psi & -1 & \mathfrak{p}_k\nabla_{\mathfrak{q}_{\mathrm{F}}}\Psi - \nabla_{\mathfrak{q}_{\mathrm{F}}}\mathfrak{R}_{\mathrm{F}} & \mathfrak{p}_k\nabla_\omega\Psi & 0\\
\nabla_{\mathfrak{p}_k} f_\omega & 0 & -\nabla_{\mathfrak{q}_{\mathrm{F}}} f_z & -\nabla_\omega f_z & -\nabla_z f_z\\
0 & 0 & \nabla_{\mathfrak{q}_{\mathrm{F}}} f_z & \nabla_\omega f_z & \nabla_z f_z
\end{bmatrix}. \tag{A.20}
$$

Since matrix definiteness is defined for symmetric matrices, $-\nabla \boldsymbol{f}_\mathrm{C}(\boldsymbol{x}_\mathrm{C}) \succcurlyeq 0$ becomes

$$-(\nabla \boldsymbol{f}_\mathrm{C} + \nabla \boldsymbol{f}_\mathrm{C}^T) = \begin{bmatrix} C_v\beta_\mathrm{B} & 0 & 1-\Psi & -\nabla_{\mathfrak{p}_k} f_\omega & 0 \\ 0 & 0 & 0 & 0 & 0 \\ 1-\Psi & 0 & 2(\nabla_{\mathfrak{q}_\mathrm{F}}\mathfrak{R}_\mathrm{F} - \mathfrak{p}_k\nabla_{\mathfrak{q}_\mathrm{F}}\Psi) & \nabla_{\mathfrak{q}_\mathrm{F}} f_\mathrm{z} - \mathfrak{p}_k\nabla_\omega\Psi & -\nabla_{\mathfrak{q}_\mathrm{F}} f_\mathrm{z} \\ -\nabla_{\mathfrak{p}_k} f_\omega & 0 & \nabla_{\mathfrak{q}_\mathrm{F}} f_\mathrm{z} - \mathfrak{p}_k\nabla_\omega\Psi & 2\nabla_\omega f_\mathrm{z} & \nabla_z f_\mathrm{z} - \nabla_\omega f_\mathrm{z} \\ 0 & 0 & -\nabla_{\mathfrak{q}_\mathrm{F}} f_\mathrm{z} & \nabla_z f_\mathrm{z} - \nabla_\omega f_\mathrm{z} & -2\nabla_z f_\mathrm{z} \end{bmatrix} \succcurlyeq 0, \tag{A.21}$$

which can be verified using the diagonal dominance of a matrix.[1] Investigating the diagonal dominance of (A.21) yields the following inequalities

$$0 \le C_v\beta_\mathrm{B} - |1-\Psi| - |\nabla_{\mathfrak{p}_k} f_\omega|, \tag{A.22a}$$
$$0 \le 2(\nabla_{\mathfrak{q}_\mathrm{F}}\mathfrak{R}_\mathrm{F} - \mathfrak{p}_k\nabla_{\mathfrak{q}_\mathrm{F}}\Psi) - |1-\Psi| - |\nabla_{\mathfrak{q}_\mathrm{F}} f_\mathrm{z} - \mathfrak{p}_k\nabla_\omega\Psi| - |\nabla_{\mathfrak{q}_\mathrm{F}} f_\mathrm{z}|, \tag{A.22b}$$
$$0 \le 2\nabla_\omega f_\mathrm{z} - |\nabla_{\mathfrak{p}_k} f_\omega| - |\nabla_{\mathfrak{q}_\mathrm{F}} f_\mathrm{z} - \mathfrak{p}_k\nabla_\omega\Psi| - |\nabla_z f_\mathrm{z} - \nabla_\omega f_\mathrm{z}|, \tag{A.22c}$$
$$0 \le -2\nabla_z f_\mathrm{z} - |\nabla_{\mathfrak{q}_\mathrm{F}} f_\mathrm{z}| - |\nabla_z f_\mathrm{z} - \nabla_\omega f_\mathrm{z}|. \tag{A.22d}$$

Thus, EIP is assured if (A.22) holds for all $\boldsymbol{x}_\mathrm{C} \in \mathcal{X}_\mathrm{C}$.

Next for the AS of the closed-loop equilibria with $\tilde{\boldsymbol{u}}_\mathrm{C} = \mathbf{0}$, consider the requirement

$$\exists \epsilon_\mathrm{C} > 0: \; -(\nabla \boldsymbol{f}_\mathrm{C} + \nabla \boldsymbol{f}_\mathrm{C}^T) \succcurlyeq \mathrm{diag}[(\epsilon_\mathrm{C}, 0, \epsilon_\mathrm{C}, \epsilon_\mathrm{C}, \epsilon_\mathrm{C})], \quad \forall \boldsymbol{x}_\mathrm{C} \in \mathcal{X}_\mathrm{C}, \tag{A.23}$$

which is equivalent to the inequalities in (A.22) being strict. In this case, S_C in (6.47) is a Lyapunov function with $\dot{S}_\mathrm{C} < 0$ being strict in all states except for $\tilde{\mathfrak{p}}_l$, i.e., the states $\tilde{\mathfrak{p}}_k$, $\tilde{\mathfrak{q}}_\mathrm{F}$, $\tilde{\omega}$, and $\tilde{z}$ converge to zero. Applying LaSalle's invariance principle, $\dot{S}_\mathrm{C} \equiv 0$ reduces the controlled compressor dynamics (6.36) shifted to the equilibrium to $\mathfrak{C}_{\mathrm{ef},k}\dot{\tilde{\mathfrak{p}}}_k = 0$. Moreover, due to the shift $\tilde{\mathfrak{p}}_k = \mathfrak{p}_k - \hat{\mathfrak{p}}_k$, it holds that $\tilde{\mathfrak{p}}_k = 0$ if $\dot{\tilde{\mathfrak{p}}}_k \equiv 0$. Thus, the closed-loop compressor unit with $\tilde{\boldsymbol{u}}_\mathrm{C} = \mathbf{0}$ is AS if (A.22) hold strictly. EISD is then also assured by the AS result, since $\tilde{\boldsymbol{u}}_\mathrm{C} = \mathbf{0} \implies \lim_{t\to\infty} \boldsymbol{x}_\mathrm{C} = \mathbf{0}$.

The inequalities in (A.22) can now be simplified as follows. Notice from (6.37b) that

$$\begin{aligned} |\nabla_{\mathfrak{p}_k} f_\omega| &= \beta_\mathfrak{p} \left| 1 - \frac{\beta_\mathrm{st}\beta_\mathrm{B}}{4\beta_\varepsilon} \sqrt{\frac{\mathfrak{p}_\mathrm{B}}{|\mathfrak{p}_\mathrm{B} - \mathfrak{p}_k|}} \left(2 + \frac{\mathfrak{p}_k - \mathfrak{p}_k^\star}{|\mathfrak{p}_\mathrm{B} - \mathfrak{p}_k|}\right) \right| \\ &= \beta_\mathfrak{p} \left| 1 - \frac{\beta_\mathrm{st}\beta_\mathrm{B}}{4\beta_c} \sqrt{\frac{\mathfrak{p}_\mathrm{B}}{|\mathfrak{p}_\mathrm{B} - \mathfrak{p}_k|}} \cdot \frac{2\mathfrak{p}_\mathrm{B} - \mathfrak{p}_k - \mathfrak{p}_k^\star}{|\mathfrak{p}_\mathrm{B} - \mathfrak{p}_k|} \right| \le \beta_\mathfrak{p}\varkappa_\mathrm{C}, \end{aligned} \tag{A.24}$$

and thus that the partial derivative is bounded by $\beta_\mathfrak{p}\varkappa_\mathrm{C}$ for all $\mathfrak{p}_k$ and $\mathfrak{p}_k^\star$. The upper bound $\varkappa_\mathrm{C}$ in (6.49e) is found by comparing the maximum of $\nabla_{\mathfrak{p}_k} f_\omega$ at ($\mathfrak{p}_k = 0$, $\mathfrak{p}_k^\star = \eta_\mathfrak{p}\mathfrak{p}_\mathrm{B}$) to its minimum at ($\mathfrak{p}_k = \eta_\mathfrak{p}\mathfrak{p}_\mathrm{B}$, $\mathfrak{p}_k^\star = 0$). Finally, substituting (A.24) along with the derivatives of f_ω and f_z into the strict versions of (A.22) and simplifying gives the conditions in (6.48). ■

[1] A matrix $\boldsymbol{A} = (a_{kl})$ is diagonally dominant if $a_{kk} \ge \sum_{l \ne k} |a_{kl}|$, $\forall k$, and strictly so if the inequality holds strictly. Furthermore, a (strictly) diagonally dominant matrix is (strictly) positive definite [Wil76; KB00].

B Supplementary Simulation Data

In this appendix, supplementary parameter data is provided for the simulations performed in the respective chapters.

B.1 Simulation Parameters in Chapter 4

Table B.1: Rounded ZIP Load Parameters for DC Stability

Bus n	(Z_n^{-1} in S; I_n in A; P_n in kW) $t = 0\,\text{s}$	$t = 0.2\,\text{s}$	$t = 0.4\,\text{s}$	$t = 0.8\,\text{s}$
1	0.159; 126.3; 40.33	0.402; 70.5; 17.57	0.402; 70.5; 17.57	0; 0; 17.12
2	−0.149; 88.8; −16.71	0.364; 68.0; 28.46	0.141; 90.7; 30.36	0; 0; −15.23
3	−0.006; 142.9; 1.12	−0.006; 142.9; 1.12	0.002; 77.7; 47.09	0; 0; 41.79
4	−0.129; 58.2; 45.59	−0.129; 58.2; 45.59	−0.129; 58.2; 45.59	0; 0; 45.59
5	−0.152; 97.8; −12.31	−0.152; 97.8; −12.31	0.086; 142.2; −27.37	0; 0; −28.66
6	0.304; 18.1; −5.56	0.185; −72.1; 2.34	0.185; −72.1; 2.34	0; 0; −13.48
7	−0.018; −74.7; −20.68	0.334; 90.6; 51.20	0.115; −18.6; 19.72	0; 0; 39.18
8	0.186; 20.6; 28.18	0.186; 20.6; 28.18	0.186; 20.6; 28.18	0; 0; 28.18
9	0.096; 125.3; −3.73	−0.163; −69.3; 6.87	−0.163; −69.3; 6.87	0; 0; 30.84
10	0.350; 17.0; 53.13	0.081; 36.4; 17.68	0.184; 63.6; −18.69	0; 0; −5.16
11	−0.041; −38.4; 5.84	0.232; 112.7; −1.70	0.148; −51.3; −3.00	0; 0; −26.06
12	0.042; 75.0; −14.09	0.058; −77.7; −20.63	0.058; −77.7; −20.63	0; 0; 37.46
13	0.320; 55.0; 51.86	0.320; 55.0; 51.86	0.218; −37.2; 0.03	0; 0; 29.00
14	0.064; 19.0; 35.92	0.026; 18.7; 5.74	0.026; 18.7; 5.74	0; 0; 4.74
15	0.410; 57.5; −7.79	−0.120; 154.1; −11.09	0.118; 99.3; 48.99	0; 0; 59.74
16	0.060; −60.9; −8.99	0.060; −60.9; −8.99	0.060; −60.9; −8.99	0; 0; 10.46
17	0.348; −44.7; 11.50	0.348; −44.7; 11.50	−0.099; −29.1; −29.86	0; 0; −25.98
18	0.374; 58.8; 29.31	0.374; 58.8; 29.31	0.220; 142.7; 30.40	0; 0; 30.40
19	0.299; −33.5; 5.80	0.299; −33.5; 5.80	0.151; 136.2; 15.13	0; 0; 10.47
20	0.350; −64.4; 49.30	0.330; 1.1; −17.73	0.330; 1.1; −17.73	0; 0; 8.27
21	0.147; 71.0; 19.21	0.147; 71.0; 19.21	0.147; 71.0; 19.21	0; 0; 19.21

Table B.2: Rounded P Source Parameters

Bus n	($P_{\mathrm{ps},n}$ in kW; $I_{\mathrm{ps,max},n}$ in A)				
	$t = 0\,\mathrm{s}$	$t = 0.2\,\mathrm{s}$	$t = 0.4\,\mathrm{s}$	$t = 0.6\,\mathrm{s}$	$t = 0.8\,\mathrm{s}$
1	0.00; −26.5	0.00; −26.5	0.00; −26.5	0.00; −26.5	−1.29; −20.5
2	−8.46; −27.4	−8.46; −27.4	0.00; −10.2	−3.84; −15.7	−3.84; −15.7
3	0.00; −10.1	−2.36; −28.0	0.00; −30.5	0.00; −30.5	0.00; −30.5
4	0.00; −30.9	0.00; −30.9	0.00; −30.9	0.00; −30.9	0.00; −10.3
5	0.00; −18.3	0.00; −18.3	−6.53; −22.5	0.00; −13.9	−1.85; −26.2
6	0.00; −18.4	0.00; −22.2	0.00; −19.8	0.00; −10.1	0.00; −34.0
7	0.00; −11.8	0.00; −28.6	0.00; −28.6	0.00; −30.1	−3.89; −34.3
8	−2.57; −31.7	0.00; −36.1	0.00; −36.1	0.00; −36.1	0.00; −8.8
9	0.00; −30.9	0.00; −30.9	−4.63; −21.8	−4.63; −21.8	−4.63; −21.8
10	−2.69; −24.6	−2.69; −24.6	0.00; −27.3	0.00; −27.3	0.00; −27.3
11	−5.47; −15.7	−5.47; −15.7	−5.47; −15.7	−6.82; −8.6	−6.82; −8.6
12	0.00; −14.3	0.00; −14.3	0.00; −14.3	0.00; −14.3	0.00; −14.3
13	0.00; −28.7	0.00; −36.5	−6.01; −23.1	−7.33; −27.5	−7.33; −27.5
14	−6.85; −21.5	−4.24; −8.9	−4.24; −8.9	0.00; −24.4	0.00; −24.4
15	−8.58; −16.9	0.00; −28.6	−9.16; −35.6	−9.16; −35.6	0.00; −31.6
16	0.00; −26.8	0.00; −17.0	0.00; −17.0	0.00; −33.0	−3.87; −16.7
17	0.00; −11.4	0.00; −11.4	−9.09; −16.3	−8.08; −15.1	−8.08; −15.1
18	0.00; −17.7	−6.97; −20.4	−7.39; −19.6	−7.39; −19.6	−6.91; −36.1
19	0.00; −24.8	0.00; −24.8	0.00; −24.8	0.00; −24.8	−1.26; −26.4
20	0.00; −19.7	0.00; −8.4	−8.05; −28.0	0.00; −12.7	0.00; −12.7
21	0.00; −17.1	−7.24; −31.9	−7.24; −31.9	−3.16; −31.3	0.00; −13.7

Table B.3: Rounded Transmission Line Lengths

Line	km	Line	km	Line	km	Line	km	Line	km	Line	km
1–2	1.68	1–4	4.65	1–5	4.03	2–3	2.78	3–7	2.75	4–15	1.67
6–7	1.03	6–9	4.56	6–11	4.47	6–14	4.37	7–8	3.02	9–10	0.50
10–12	3.56	11–12	0.64	11–20	3.31	13–14	4.34	13–15	4.64	13–15	4.18
15–16	4.89	16–18	0.97	17–18	3.92	17–20	2.80	18–19	4.11	18–21	2.89
19–20	4.01	20–21	0.25								

B.2 Simulation Parameters in Chapter 5

Table B.4: Rounded ZIP Load Parameters for DC Coordination

Bus n	(Z_n^{-1} in S; I_n in A; P_n in kW)				
	$t = 0$ s	$t = 10$ s	$t = 20$ s	$t = 30$ s	$t = 40$ s
1	0.331; 133.3; 11.26	0.408; 89.9; 1.14	0.408; 89.9; 1.14	0.319; −22.4; −8.30	0.115; −3.0; 0.94
2	0.196; 116.0; −14.09	0.392; 88.0; 5.98	0.297; 105.6; 6.82	0.297; 0.3; −13.44	0.297; 0.3; −13.44
3	0.169; 147.4; −6.17	0.169; 147.4; −6.17	0.307; 95.5; 14.26	0.188; 102.4; 7.07	0.267; 61.3; 11.91
4	0.264; 80.3; 13.59	0.264; 80.3; 13.59	0.264; 80.3; 13.59	0.264; 80.3; 13.59	0.264; 80.3; 13.59
5	0.096; 126.3; −12.14	0.096; 126.3; −12.14	0.284; 148.0; −18.83	0.284; 148.0; −18.83	0.168; 150.5; −19.41
6	0.370; 49.1; −9.14	0.305; −21.0; −5.63	0.305; −21.0; −5.63	0.289; 57.3; −12.66	0.289; 57.3; −12.66
7	0.367; −21.4; −15.86	0.398; 105.5; 16.09	0.253; 20.6; 2.10	0.189; 3.9; −2.83	0.259; 153.7; 10.75
8	0.312; 51.1; 5.86	0.312; 51.1; 5.86	0.312; 51.1; 5.86	0.312; 51.1; 5.86	0.312; 51.1; 5.86
9	0.228; 133.9; −8.32	0.076; 12.8; −3.61	0.076; 12.8; −3.61	0.393; 55.5; 19.02	0.409; 49.0; 7.04
10	0.402; 48.3; 10.95	0.228; 03.4; 1.19	0.377; 84.5; −14.08	0.302; 40.6; 12.17	0.136; 24.6; −8.96
11	0.174; 12.3; −4.07	0.308; 122.7; −7.42	0.259; −4.1; −8.00	0.259; −4.1; −8.00	0.290; 92.6; −18.25
12	0.197; 106.0; −12.93	0.262; 3.8; −15.83	0.262; 3.8; −15.83	0.392; 49.1; 9.98	0.392; 49.1; 9.98
13	0.395; 77.9; 16.38	0.395; 77.9; 16.38	0.300; 6.1; −6.65	0.300; 6.1; −6.65	0.254; 20.8; 6.22
14	0.283; 49.9; 9.30	0.187; 52.7; −4.11	0.187; 52.7; −4.11	0.313; 109.7; −4.56	0.313; 109.7; −4.56
15	0.412; 79.8; −10.13	0.101; 155.9; −11.59	0.343; 112.3; 15.11	0.415; 31.5; 19.89	0.415; 31.5; 19.89
16	0.207; 10.6; −10.66	0.207; 10.6; −10.66	0.207; 10.6; −10.66	0.348; 145.6; −2.02	0.348; 145.6; −2.02
17	0.376; 0.3; −1.55	0.376; 0.3; −1.55	0.147; 72.0; −19.94	0.147; 72.0; −19.94	0.303; 11.7; −18.21
18	0.397; 80.8; 6.36	0.397; 80.8; 6.36	0.331; 146.1; 6.84	0.331; 146.1; 6.84	0.331; 146.1; 6.84
19	0.347; 9.0; −4.09	0.347; 9.0; −4.09	0.261; 141.0; 0.06	0.261; 141.0; 0.06	0.256; 108.0; −2.02
20	0.400; −15.0; 15.24	0.365; 35.9; −14.55	0.365; 35.9; −14.55	0.365; 35.9; −14.55	0.201; −13.6; −2.99
21	0.269; 90.3; 1.87	0.269; 90.3; 1.87	0.269; 90.3; 1.87	0.269; 90.3; 1.87	0.269; 90.3; 1.87

B.3 Simulation Parameters in Chapter 6

Table B.5: Gas Pipeline Parameter Values

Parameter	Symbol	Value
Specific gas constant	R_{spec}	518.28 J/(kg K)
Dynamic viscosity	η_D	10^{-5} kg m/s
Critical pressure	p_{crit}	46.5 bar
Standard pressure	p_n	1.01325 bar = 1 atm
Critical temperature	T_{crit}	190.55 K
Standard temperature	T_n	273.15 K
Simulation temperature	T	278 K
Friction efficiency factor	ζ_{ef}	0.98
Pipe roughness	r_T	0.012 mm

Table B.6: Rounded Pipeline Lengths

Line	km	Line	km	Line	km	Line	km	Line	km	Line	km
1 3	5.00	2–8	2.85	4–21	1.66	5–9	2.33	6–17	1.04	7 13	3.30
8–9	3.25	8–10	1.08	9–10	3.39	10–11	1.34	10–12	1.71	10–13	2.52
12–13	3.36	12–14	3.13	15–16	1.01	16–17	2.30	16–18	2.38	17–18	2.21
18–19	2.92	20–21	1.40	20–23	2.91	21–22	1.05	22–23	1.34	23–24	1.29
24–25	1.77	25–26	2.68								

Table B.7: Rounded Gas Load Parameters

Node n	$q_{\mathrm{I},n}$ in $\mathrm{sm}^3/\mathrm{s}\ (\times 10^3)$				Node n	$q_{\mathrm{I},n}$ in $\mathrm{sm}^3/\mathrm{s}\ (\times 10^3)$			
	$t=0\,\mathrm{s}$	$t=500\,\mathrm{s}$	$t=1000\,\mathrm{s}$	$t=1500\,\mathrm{s}$		$t=0\,\mathrm{s}$	$t=500\,\mathrm{s}$	$t=1000\,\mathrm{s}$	$t=1500\,\mathrm{s}$
1	3.92	3.92	3.92	7.00	2	−8.63	5.51	5.51	5.51
3	5.92	5.92	−5.02	−5.02	4	−9.94	4.61	4.61	4.61
5	−2.77	−2.77	−5.78	−5.78	6	−4.12	−4.12	−4.12	−6.96
7	−2.31	−2.31	−2.31	−2.31	8	1.19	4.69	−1.42	−1.42
9	4.10	0.45	3.64	4.65	10	7.79	7.79	7.79	7.79
11	1.53	1.53	1.53	1.53	12	9.23	0.53	−3.06	−8.52
13	7.71	−6.81	6.81	2.96	14	5.52	2.40	2.40	2.40
15	−2.22	2.49	−6.39	−6.39	16	−5.32	−5.32	−4.93	−6.55
17	−4.57	5.89	5.89	5.89	18	−6.24	3.15	−0.87	−2.81
19	−6.06	−7.80	5.30	4.64	20	8.11	8.11	−2.43	−9.00
21	−4.08	4.53	4.53	1.80	22	−5.11	−5.11	−5.11	3.22
23	2.53	2.53	3.86	1.83	24	8.94	8.94	8.94	7.43
25	−1.56	−1.56	−1.56	−1.56	26	−4.14	−2.57	−7.24	−7.24

Abbreviations and Symbols

Abbreviations

AS	asymptotic stability/asymptotically stable
CCV	close-coupled valve
DDA	dynamic distributed averaging
DGU	distributed generation unit
EID	equilibrium-independent dissipative/dissipativity
EIID	equilibrium-independent-input detectable
EIO	equilibrium-independent observable
EIP	equilibrium-independent passive/passivity
EISD	equilibrium-independent-state detectable
EISO	equilibrium-independent-state observable
GAS	global asymptotic stability/globally asymptotically stable
GFPP	gas-fired power plant
IF-EIP	input-feedforward equilibrium-independent passive/passivity
IFOF-EIP	input-feedforward output-feedback equilibrium-independent passive/passivity
IFOFP	input-feedforward output-feedback passive/passivity
IFP	input-feedforward passive/passivity
IGV	inlet guide vanes
IS-EIP	input-strict(ly) equilibrium-independent passive/passivity
ISP	input-strict(ly) passive/passivity
ISS	input-to-state stable/stability
LMI	linear matrix inequality
MAS	multi-agent system
MPC	model predictive control
NMES	networked multi-energy system
OF-EIP	output-feedback equilibrium-independent passive/passivity
OFP	output-feedback passive/passivity
OS-EIP	output-strict(ly) equilibrium-independent passive/passivity
OSP	output-strict(ly) passive/passivity
P2G	power to gas
PDE	partial differential equation
PI	proportional-integral
PID	proportional-integral-derivative
RES	renewable energy source

RQ	research question
TCV	throttle control valve
w.r.t.	with respect to
ZIP	constant impedance (Z), constant current (I), constant power (P)

Latin Letters

Symbol	Description
A	cross-sectional pipe area, valve opening area
$\boldsymbol{A}$	linear state matrix
$\boldsymbol{b}, \boldsymbol{B}$	input vector, input matrix
c	speed of sound
c_{cal}	calorific value
$c_{\mathrm{p}}, c_{\mathrm{V}}$	(constant pressure, constant volume) specific heat
$C, \boldsymbol{C}$	capacitance (scalar, diagonal matrix)
C_{d}	valve discharge capacity
$\mathfrak{C}$	pipe capacitive volume
D	pipe diameter
$\boldsymbol{D}, d_{ij}$	auxiliary test matrix for inequalities (matrix, matrix element)
$\boldsymbol{e}_{\mathrm{row}}, \boldsymbol{e}_{\mathrm{col}}$	(row, column) vector of the incidence matrix
$\boldsymbol{E}$	incidence matrix of graph $\mathcal{G}$
$f(\cdot), \boldsymbol{f}(\cdot)$	auxiliary function / system dynamics (scalar, vector)
$\mathfrak{f}(\cdot)$	gas load function
g	gravitational constant
$g(\cdot)$	auxiliary / input function
$\mathfrak{g}(\cdot)$	gravitational pressure loss function
G	conductance
h	elevation difference between a pipe inlet and outlet
$h(\cdot), \boldsymbol{h}(\cdot)$	auxiliary / output function (scalar, vector)
$\boldsymbol{H}$	interconnection matrix
i	imaginary number
$\imath, \boldsymbol{\imath}$	current (scalar, vector)
$\imath_{\mathrm{l}}(\cdot), \boldsymbol{\imath}_{\mathrm{l}}(\cdot)$	load current function (scalar, vector)
I	constant current load parameter
$\boldsymbol{I}_a$	identity matrix of dimension a
J	rotational inertia
$k, \boldsymbol{K}$	controller gain (scalar, matrix)
K	isothermal bulk modulus
l	space variable
ℓ	pipe length
$L, \boldsymbol{L}$	inductance (scalar, diagonal matrix)

Symbol	Description
$\mathfrak{L}$	inductive volume
$m, \dot{m}$	mass, mass flow rate
$\mathfrak{m}$	dimension of the input variable
$\mathfrak{n}$	dimension of the state variable
p	electrical power
p_{th}	thermal power
$\mathfrak{p}, \boldsymbol{\mathfrak{p}}$	absolute pressure (scalar, vector)
P	constant power load parameter
$\boldsymbol{P}, p_{ij}$	symmetric storage weight (matrix, matrix element)
$\mathfrak{q}, \boldsymbol{\mathfrak{q}}$	volumetric flow rate (scalar, vector)
$\boldsymbol{Q}, q_{ij}$	diagonal system weight (matrix, matrix element)
r_1, r_2	centrifugal compressor radius (inlet, impeller)
r_{T}	pipe surface roughness
$R, \boldsymbol{R}$	electrical resistance (scalar, diagonal matrix)
Re	Reynolds number
R_{spec}	specific gas constant
$\mathfrak{R}(\cdot), \boldsymbol{\mathfrak{R}}(\cdot)$	nonlinear damping or friction function (scalar, matrix)
s	stem position
$S(\cdot)$	storage function
t	time variable
T	temperature
$\boldsymbol{T}$	transformation matrix
$u, \boldsymbol{u}$	input (scalar, vector)
$v, \boldsymbol{v}$	voltage (scalar, vector)
$\mathfrak{v}$	gas velocity
V	volume
$w(\cdot)$	supply rate function
$\boldsymbol{W}$	supply rate matrix
$x, \boldsymbol{x}$	state (scalar, vector)
$Y, \boldsymbol{Y}$	load admittance factor (scalar, diagonal matrix)
$y, \boldsymbol{y}$	output (scalar, vector)
$z, \boldsymbol{z}$	transformed state (scalar, vector)
Z	constant impedance load parameter / gas compressibility factor

Greek Letters

Symbol	Description
β	compressor control parameter
γ	system gain parameter
γ_{C}	zero incidence loss parameter of a centrifugal compressor
γ_{L_2}	L_2-gain of a system
$\delta, \boldsymbol{\delta}$	damping control parameter (scalar, diagonal matrix)
$\varepsilon, \boldsymbol{\varepsilon}$	control error variable (scalar, vector)
ζ	friction factor
ζ_{k}	fluid friction factor in a centrifugal compressor
η	auxiliary efficiency factor
η_{D}	dynamic viscosity of a fluid
η_{p}	pressure limit scaling factor
θ	pipeline inclination angle
$\kappa, \boldsymbol{\kappa}$	voltage setting control parameter (scalar, diagonal matrix)
$\varkappa$	auxiliary parameter
λ	eigenvalue of a matrix
$\lambda_2(\boldsymbol{\mathcal{L}})$	algebraic connectivity of a graph with a laplacian matrix $\boldsymbol{\mathcal{L}}$
$\boldsymbol{\Lambda}$	diagonal matrix of eigenvalues
μ	auxiliary parameter
ν	input passivity index
$\boldsymbol{\xi}$	right eigenvector of a matrix
$\boldsymbol{\Xi}$	matrix of right eigenvectors
ρ	output passivity index
ϱ	density
σ	auxiliary passivity index
ς	slip factor of a centrifugal compressor
τ	torque
ϕ	normalised electrical power
$\chi, \boldsymbol{\chi}$	auxiliary state (scalar, vector)
$\psi, \boldsymbol{\psi}$	voltage following control parameter (scalar, diagonal matrix)
$\Psi(\cdot)$	compressor characteristic function
ω	rotational velocity

Calligraphic and Blackboard Bold

Symbol	Description
$\mathcal{C}$	set of coupling component edges in a networked multi-energy system
$\mathcal{D}$	set of edges describing the directional causality of coupling components
$\mathcal{E}$	set of edges of a graph $\mathcal{G}$
$\mathcal{F}$	set of voltage following buses
$\mathcal{G}$	graph
$\mathcal{I}$	current state space
$\hat{\mathcal{I}}$	current equilibrium set
$\mathcal{J}$	set of systems or energy networks
$\mathcal{K}$	set of configurable systems
$\boldsymbol{\mathcal{L}}$	Laplacian matrix of a graph $\mathcal{G}$
$\mathcal{M}$	set of clusters
$\mathcal{N}$	set of buses or nodes
$\mathbb{N}$	set of natural numbers
$\mathbb{R}$	set of real numbers
$\mathcal{S}$	set of voltage setting buses
$\mathcal{T}$	set of external transmission lines
$\mathcal{U}$	input space
$\mathcal{V}$	voltage state space
$\mathcal{X}$	state space
$\hat{\mathcal{X}}$	state equilibrium set
$\mathcal{Y}$	output space
$\mathbb{1}_a$	vector of ones with dimension a

Indices and Exponents

Symbol	Description
$\square_{\mathrm{a}}$	corresponds to the dynamic distributed averaging controller(s)
$\square_{\mathrm{b}}$	corresponds to a buck converter
$\square_{\mathrm{base}}$	indicates the base version of a decomposed symbol
$\square_{\mathrm{B}}$	corresponds to the backward flow of gas
$\square_{c}$	index variable for coupling edges
$\square_{\mathrm{c}}$	corresponds to a capacitor
$\square_{\mathrm{cl}}$	corresponds to a cluster
$\square_{\mathrm{col}}$	corresponds to a column vector
$\square_{\mathrm{com}}$	corresponds to the communication network
$\square_{\mathrm{crit}}$	corresponds to the critical value
$\square_{\mathrm{C}}$	corresponds to a compressor
$\square_{\mathrm{Cs}}$	corresponds to the compressor set
$\square_{\mathfrak{c}}$	corresponds to the capacitive pipe volume
$\square_{d}$	index variable for edges showing the directional causality of coupling components
$\square_{\mathrm{d}}$	indicates the discharge factor (see C_{d})
$\square_{\mathrm{D}}$	indicates the dynamic factor (see η_{D})
$\square_{\mathrm{ef}}$	corresponds to the effective value
$\square_{\mathrm{el}}$	corresponds to the DC electrical network
$\square_{\mathrm{elC}}$	corresponds an electric compressor coupling component
$\square_{\mathrm{ex}}$	indicates to an external symbol
$\square_{\mathrm{f}}$	corresponds to a voltage following or an unactuated bus
$\square_{\mathrm{F}}$	corresponds to the forward flow of gas
$\square_{\mathrm{g}}$	corresponds to the gas network
$\square_{\mathrm{gDGU}}$	corresponds a gas-powered DGU coupling component
$\square_{\mathrm{GFPP}}$	corresponds to a gas-fired power plant
$\square_{\mathrm{i}}$	corresponds to the current
$\square_{\mathrm{in}}$	corresponds to an incoming value or a source
$\square_{\mathrm{int}}$	corresponds to an internal value
$\square^{\mathrm{I}}$	corresponds to an integral gain value
$\square_{\mathrm{Im}}$	corresponds to the image space
$\square_{j}$	index variable for voltage setting buses, compressor nodes, or subsystems
$\square_{k}$	index variable for voltage following buses, compressor souce nodes, or configurable subsystems
$\square_{l}$	index variable for compressor sink nodes
$\square_{\mathrm{l}}$	corresponds to an electrical load
$\square_{\mathfrak{l}}$	corresponds to a gas load
$\square_{\mathrm{L}}$	corresponds to the left section
$\square_{\mathrm{Lam}}$	indicates laminar friction

Symbol	Description
$\square_{\mathrm{Link}}$	corresponds to the linking/coupling components
$\square_{m}$	index variable for clusters
$\square^{\mathrm{m}}$	indicates a monotone function
$\square_{\mathrm{max}}$	corresponds to the maximum
$\square_{\mathrm{mean}}$	corresponds to the mean
$\square^{\mathrm{meas}}$	indicates the measureable value (e.g., of the volumetric flow rate)
$\square_{\mathrm{min}}$	corresponds to the minimum
$\square_{\mathrm{M}}$	corresponds to the middle section
$\square_{n}$	index variable for buses or nodes
$\square_{\mathrm{n}}$	indicates a variable at nominal conditions
$\square_{\mathrm{new}}$	indicates a new variable
$\square_{\mathrm{N}}$	corresponds to the set of normal nodes
$\square_{\mathrm{NMES}}$	corresponds to a networked multi-energy system
$\square_{o}$	index variable for edges or lines
$\square_{\mathrm{out}}$	corresponds to an outgoing value or a sink
$\square_{\mathrm{p}}$	corresponds to the power or a power controlled bus
$\square_{\mathrm{ps}}$	corresponds to a power source
$\square_{\mathfrak{p}}$	corresponds to the pressure
$\square_{\mathrm{P2G}}$	corresponds to a power-to-gas component
$\square^{\mathrm{P}}$	corresponds to a proportional gain value
$\square_{\mathfrak{q}}$	corresponds to the normalised volumetric flow rate
$\square_{\mathrm{r}}$	indicates the reduced version of a decomposed symbol
$\square_{\mathrm{ref}}$	corresponds to the reference value
$\square_{\mathrm{row}}$	corresponds to a row vector
$\square_{\mathrm{R}}$	corresponds to the right section
$\square_{\mathfrak{s}}$	index variable for subsystems or energy systems
$\square_{\mathrm{s}}$	corresponds to a voltage setting bus
$\square_{\mathrm{spec}}$	corresponds to the specific value (see R_{spec})
$\square_{\mathrm{st}}$	corresponds to the valve stem position
$\square_{\mathrm{S}}$	corresponds to a supply node
$\square_{\mathrm{t}}$	corresponds to an electrical transmission line
$\square_{\mathrm{th}}$	indicates the thermal component
$\square_{\mathrm{T}}$	corresponds to a gas pipeline
$\square_{\mathrm{Tur}}$	indicates turbulent friction
$\square^{T}$	matrix or vector transpose
$\square_{\mathrm{u}}$	corresponds to the Stage 3 leaky PI controllers
$\square_{\mathrm{v}}$	corresponds to the voltage
$\square_{\upsilon}$	corresponds to the valve
$\square_{\mathrm{w}}$	corresponds to the weight function
$\square_{\mathrm{z}}$	corresponds to the transformed variable
$\square_{\varepsilon}$	corresponds to the error
$\square_{\Sigma}$	corresponds to the overall system

Symbol	Description
$\square_\tau$	corresponds to the torque
$\square_v$	corresponds to a valve
$\square_\omega$	corresponds to the rotational velocity
$\square_{\geq 0}$	set of positive real or natural numbers
$\square_{>0}$	set of strictly positive real or natural numbers
$\square_{\leq 0}$	set of negative real or natural numbers
$\square_{<0}$	set of strictly negative real or natural numbers
$\square^{-1}$	inverse of a nonsingular matrix
$\square^{+}$	sink set
$\square^{-}$	source set
$\square^{\star}$	desired setpoint or vector of setpoints to be established in steady state
$\square^{\diamond}$	modified parameter shorthand
$\square'$	per-length parameter value
$\dot{\square}$	time derivative
$\hat{\square}$	equilibrium of a variable
$\tilde{\square}$	error of a variable with respect to its equilibrium $\hat{\square}$
$\overline{\square}$	indicates the upper limit
$\underline{\square}$	indicates the lower limit

List of Figures

List of Tables

References

Public References

[Abb+21] ABBASI, M. H.; ABDULLAH, B.; AHMAD, M. W.; ROSTAMI, A.; CULLEN, J.: Heat transition in the European building sector: Overview of the heat decarbonisation practices through heat pump technology. In: *Sustainable Energy Technologies and Assessments* 48 (2021), p. 101630. DOI: 10.1016/j.seta.2021.101630.

[ABD22] ABDOLMALEKI, B.; BERGNA-DIAZ, G.: Distributed Control and Optimization of DC Microgrids: A Port-Hamiltonian Approach. In: *IEEE Access* 10 (2022), pp. 64222–64233. DOI: 10.1109/ACCESS.2022.3183209.

[ABNG12] ALAMIAN, R.; BEHBAHANI-NEJAD, M.; GHANBARZADEH, A.: A state space model for transient flow simulation in natural gas pipelines. In: *Journal of Natural Gas Science and Engineering* 9 (2012), pp. 51–59. DOI: 10.1016/j.jngse.2012.05.013.

[AI21] AL-ISMAIL, F. S.: DC Microgrid Planning, Operation, and Control: A Comprehensive Review. In: *IEEE Access* 9 (2021), pp. 36154–36172. DOI: 10.1109/ACCESS.2021.3062840.

[Ala+15] ALABDULWAHAB, A.; ABUSORRAH, A.; ZHANG, X.; SHAHIDEHPOUR, M.: Coordination of Interdependent Natural Gas and Electricity Infrastructures for Firming the Variability of Wind Energy in Stochastic Day-Ahead Scheduling. In: *IEEE Transactions on Sustainable Energy* 6.2 (2015), pp. 606–615. DOI: 10.1109/TSTE.2015.2399855.

[Ama06] AMATO, F.: *Robust Control of Linear Systems Subject to Uncertain Time-Varying Parameters*. Vol. 325. Lecture Notes in Control and Information Sciences. Springer Berlin Heidelberg, 2006. DOI: 10.1007/3-540-33276-6.

[AMP16] ARCAK, M.; MEISSEN, C.; PACKARD, A.: *Networks of Dissipative Systems: Compositional Certification of Stability, Performance, and Safety*. SpringerBriefs in Control, Automation and Robotics. Cham: Springer, 2016. DOI: 10.1007/978-3-319-29928-0.

[AN+17] AL-NUSSAIRI, M. K.; BAYINDIR, R.; PADMANABAN, S.; MIHET-POPA, L.; SIANO, P.: Constant Power Loads (CPL) with Microgrids: Problem Definition, Stability Analysis and Compensation Techniques. In: *Energies* 10.10 (2017), p. 1656. DOI: 10.3390/en10101656.

[Arm+22] ARMSTRONG MCKAY, D. I.; STAAL, A.; ABRAMS, J. F.; WINKELMANN, R.; SAKSCHEWSKI, B.; LORIANI, S.; FETZER, I.; CORNELL, S. E.; ROCKSTRÖM, J.; LENTON, T. M.: Exceeding 1.5°C global warming could trigger multiple climate tipping points. In: *Science* 377.6611 (2022), eabn7950. DOI: 10.1126/science.abn7950.

[BB12] BUNDESNETZAGENTUR FÜR ELEKTRIZITÄT, GAS, TELEKOMMUNIKATION, POST UND EISENBAHNEN; BUNDESKARTELLAMT: *Monitoringbericht 2012 von Bundesnetzagentur und Bundeskartellamt [Monitoring report 2012 from the Federal Network Agency and the Federal Cartel Office]*. Tech. rep. 2012.

[BB23] BUNDESNETZAGENTUR FÜR ELEKTRIZITÄT, GAS, TELEKOMMUNIKATION, POST UND EISENBAHNEN; BUNDESKARTELLAMT: *Monitoringbericht 2023 von Bundesnetzagentur und Bundeskartellamt [Monitoring report 2023 from the Federal Network Agency and the Federal Cartel Office]*. Tech. rep. 2023.

[Bel+20] BELDERBOS, A.; VALKAERT, T.; BRUNINX, K.; DELARUE, E.; D'HAESELEER, W.: Facilitating renewables and power-to-gas via integrated electrical power-gas system scheduling. In: *Applied Energy* 275 (2020), p. 115082. DOI: 10.1016/j.apenergy.2020.115082.

[Ben+02] BENENSON, W.; HARRIS, J. W.; STOCKER, H.; LUTZ, H.: *Handbook of Physics*. New York, NY: Springer New York, 2002. DOI: 10.1007/0-387-21632-4.

[Ben+15] BENTALEB, T.; CACITTI, A.; FRANCISCIS, S. DE; GARULLI, A.: Model Predictive Control for pressure regulation and surge prevention in centrifugal compressors. In: *2015 European Control Conference (ECC)*. IEEE, 2015, pp. 3346–3351. DOI: 10.1109/ECC.2015.7331051.

[BG05] BØHAGEN, B.; GRAVDAHL, J. T.: Active control of compression systems using drive torque; a backstepping approach. In: *44th IEEE Conference on Decision and Control (CDC)*. IEEE, 2005, pp. 2493–2498. DOI: 10.1109/CDC.2005.1582537.

[BG06] BØHAGEN, B.; GRAVDAHL, J. T.: Active surge control using drive torque: dynamic control laws. In: *45th IEEE Conference on Decision and Control*. IEEE, 2006, pp. 6437–6444. DOI: 10.1109/CDC.2006.376722.

[BG08] BØHAGEN, B.; GRAVDAHL, J. T.: Active surge control of compression system using drive torque. In: *Automatica* 44.4 (2008), pp. 1135–1140. DOI: 10.1016/j.automatica.2007.11.002.

[BGL05] BENZI, M.; GOLUB, G. H.; LIESEN, J.: Numerical solution of saddle point problems. In: *Acta Numerica* 14 (2005), pp. 1–137. DOI: 10.1017/S0962492904000212.

[BGS16] Backi, C. J.; Gravdahl, J. T.; Skogestad, S.: Robust control of a two-state Greitzer compressor model by state-feedback linearization. In: *2016 IEEE Conference on Control Applications (CCA)*. IEEE, 2016, pp. 1226–1231. doi: 10.1109/CCA.2016.7587974.

[BH12] Brouwer, A. E.; Haemers, W. H.: *Spectra of Graphs*. New York, NY: Springer, 2012. doi: 10.1007/978-1-4614-1939-6.

[Bil+24] Billerbeck, A.; Kiefer, C. P.; Winkler, J.; Bernath, C.; Sensfuss, F.; Kranzl, L.; Müller, A.; Ragwitz, M.: The race between hydrogen and heat pumps for space and water heating: A model-based scenario analysis. In: *Energy Conversion and Management* 299 (2024), p. 117850. doi: 10.1016/j.enconman.2023.117850.

[BJ21] Bothe, D.; Janssen, M.: *The value of hydrogen in the heating market: An analysis of different heating technologies with a focus on hydrogen condensing boilers and electric heat pumps: Study for FNB Gas (convenience translation – original study in German)*. Frontier Economics. 2021. url: https://www.frontier-economics.com/media/byojuyut/the-value-of-hydrogen-in-the-heating-market.pdf.

[BL07] Bao, J.; Lee, P. L.: *Process control: The passive systems approach*. 1st ed. Advances in industrial control. London: Springer, 2007. doi: 10.1007/978-1-84628-893-7.

[Bro+20] Brogliato, B.; Lozano, R.; Maschke, B.; Egeland, O.: *Dissipative systems analysis and control: Theory and applications*. 3rd ed. Communications and control engineering. Cham, Switzerland: Springer, 2020. doi: 10.1007/978-3-030-19420-8.

[BS18] Buttler, A.; Spliethoff, H.: Current status of water electrolysis for energy storage, grid balancing and sector coupling via power-to-gas and power-to-liquids: A review. In: *Renewable and Sustainable Energy Reviews* 82 (2018), pp. 2440–2454. doi: 10.1016/j.rser.2017.09.003.

[BS19] Bermúdez, A.; Shabani, M.: Finite element solution of isothermal gas flow in a network. In: *Journal of Computational Physics* 396 (2019), pp. 616–652. doi: 10.1016/j.jcp.2019.06.063.

[BV04] Boyd, S. P.; Vandenberghe, L.: *Convex optimization*. Cambridge: Cambridge University Press, 2004. doi: 10.1017/CBO9780511804441.

[Car+22] Carvalho, E. L.; Blinov, A.; Chub, A.; Emiliani, P.; Carne, G. de; Vinnikov, D.: Grid Integration of DC Buildings: Standards, Requirements and Power Converter Topologies. In: *IEEE Open Journal of Power Electronics* 3 (2022), pp. 798–823. doi: 10.1109/OJPEL.2022.3217741.

[Cer+18] Cerbus, R. T.; Liu, C.-C.; Gioia, G.; Chakraborty, P.: Laws of Resistance in Transitional Pipe Flows. In: *Physical review letters* 120.5 (2018), p. 054502. doi: 10.1103/PhysRevLett.120.054502.

[CGW99] CALCEV, G.; GOREZ, R.; WERTZ, V.: Passivity and fuzzy control of singularly perturbed systems. In: *38th IEEE Conference on Decision and Control (CDC)*. IEEE, 1999, pp. 4364–4367. DOI: 10.1109/CDC.1999.833229.

[Cha+18] CHACZYKOWSKI, M.; SUND, F.; ZARODKIEWICZ, P.; HOPE, S. M.: Gas composition tracking in transient pipeline flow. In: *Journal of Natural Gas Science and Engineering* 55 (2018), pp. 321–330. DOI: 10.1016/j.jngse.2018.03.014.

[Che+21] CHEN, W.; WANG, D.; LIU, J.; CHEN, Y.; KHONG, S. Z.; BASAR, T.; JOHANSSON, K. H.; QIU, L.: On Spectral Properties of Signed Laplacians With Connections to Eventual Positivity. In: *IEEE Transactions on Automatic Control* 66.5 (2021), pp. 2177–2190. DOI: 10.1109/TAC.2020.3008300.

[CK06] CELLIER, F. E.; KOFMAN, E.: *Continuous system simulation*. 1st ed. New York, NY: Springer, 2006. DOI: 10.1007/0-387-30260-3.

[CKS07] CUCUZZELLA, M.; KOSARAJU, K. C.; SCHERPEN, J. M. A.: Distributed Passivity-Based Control of DC Microgrids. In: *2019 American Control Conference (ACC)*. IEEE, 10/07/2019 - 12/07/2019, pp. 652–657. DOI: 10.23919/ACC.2019.8814756.

[CKS23] CUCUZZELLA, M.; KOSARAJU, K. C.; SCHERPEN, J. M. A.: Voltage Control of DC Microgrids: Robustness for Unknown ZIP-Loads. In: *IEEE Control Systems Letters* 7 (2023), pp. 139–144. DOI: 10.1109/LCSYS.2022.3187925.

[CM16] CLEGG, S.; MANCARELLA, P.: Storing renewables in the gas network: modelling of power-to-gas seasonal storage flexibility in low-carbon power systems. In: *IET Generation, Transmission & Distribution* 10.3 (2016), pp. 566–575. DOI: 10.1049/iet-gtd.2015.0439.

[Cor+15] CORTINOVIS, A.; FERREAU, H. J.; LEWANDOWSKI, D.; MERCANGÖZ, M.: Experimental evaluation of MPC-based anti-surge and process control for electric driven centrifugal gas compressors. In: *Journal of Process Control* 34 (2015), pp. 13–25. DOI: 10.1016/j.jprocont.2015.07.001.

[CPSM15] CORREA-POSADA, C. M.; SÁNCHEZ-MARTIN, P.: Integrated Power and Natural Gas Model for Energy Adequacy in Short-Term Operation. In: *IEEE Transactions on Power Systems* 30.6 (2015), pp. 3347–3355. DOI: 10.1109/TPWRS.2014.2372013.

[Cuc+19] CUCUZZELLA, M.; TRIP, S.; PERSIS, C. DE; CHENG, X.; FERRARA, A.; VAN DER SCHAFT, A.: A Robust Consensus Algorithm for Current Sharing and Voltage Regulation in DC Microgrids. In: *IEEE Transactions on Control Systems Technology* 27.4 (2019), pp. 1583–1595. DOI: 10.1109/TCST.2018.2834878.

[Cvi+17] CVIJIC, S.; LANG, J.; ILIC, M.; BABAEI, S.; STEFOPOULOS, G.: Reliable adaptive optimization demonstration using big data. In: *2017 IEEE Power & Energy Society General Meeting*. IEEE, 2017, pp. 1–5. DOI: 10.1109/PESGM.2017.8274271.

[dAb07] DE ABREU, N. M. M.: Old and new results on algebraic connectivity of graphs. In: *Linear Algebra and its Applications* 423.1 (2007), pp. 53–73. DOI: 10.1016/j.laa.2006.08.017.

[DB13] DÖRFLER, F.; BULLO, F.: Kron Reduction of Graphs With Applications to Electrical Networks. In: *IEEE Transactions on Circuits and Systems I: Regular Papers* 60.1 (2013), pp. 150–163. DOI: 10.1109/TCSI.2012.2215780.

[DF11] DORAO, C. A.; FERNANDINO, M.: Simulation of transients in natural gas pipelines. In: *Journal of Natural Gas Science and Engineering* 3.1 (2011), pp. 349–355. DOI: 10.1016/j.jngse.2011.01.004.

[DL18] DAM, D.-H.; LEE, H.-H.: A Power Distributed Control Method for Proportional Load Power Sharing and Bus Voltage Restoration in a DC Microgrid. In: *IEEE Transactions on Industry Applications* 54.4 (2018), pp. 3616–3625. DOI: 10.1109/TIA.2018.2815661.

[Dom+11] DOMSCHKE, P.; GEISSLER, B.; KOLB, O.; LANG, J.; MARTIN, A.; MORSI, A.: Combination of Nonlinear and Linear Optimization of Transient Gas Networks. In: *INFORMS Journal on Computing* 23.4 (2011), pp. 605–617. DOI: 10.1287/ijoc.1100.0429.

[Dom+21] DOMSCHKE, P.; HILLER, B.; LANG, J.; MEHRMANN, V.; MORANDIN, R.; TISCHENDORF, C.: *Gas Network Modeling: An Overview*. preprint. 2021. URL: https://opus4.kobv.de/opus4-trr154/frontdoor/index/index/docId/411.

[Dou+22] DOU, Y.; CHI, M.; LIU, Z.-W.; WEN, G.; SUN, Q.: Distributed Secondary Control for Voltage Regulation and Optimal Power Sharing in DC Microgrids. In: *IEEE Transactions on Control Systems Technology* 30.6 (2022), pp. 2561–2572. DOI: 10.1109/TCST.2022.3156391.

[Dra+16a] DRAGIČEVIĆ, T.; LU, X.; VASQUEZ, J. C.; GUERRERO, J. M.: DC Microgrids—Part I: A Review of Control Strategies and Stabilization Techniques. In: *IEEE Transactions on Power Electronics* 31.7 (2016), pp. 4876–4891. DOI: 10.1109/TPEL.2015.2478859.

[Dra+16b] DRAGIČEVIĆ, T.; LU, X.; VASQUEZ, J. C.; GUERRERO, J. M.: DC Microgrids—Part II: A Review of Power Architectures, Applications, and Standardization Issues. In: *IEEE Transactions on Power Electronics* 31.5 (2016), pp. 3528–3549. DOI: 10.1109/TPEL.2015.2464277.

[Dui+09] DUINDAM, V.; MACCHELLI, A.; STRAMIGIOLI, S.; BRUYNINCKX, H.: *Modeling and Control of Complex Physical Systems: The Port-Hamiltonian Approach*. Berlin: Springer, 2009. DOI: 10.1007/978-3-642-03196-0.

[Dvi+15] DVIJOTHAM, K.; VUFFRAY, M.; MISRA, S.; CHERTKOV, M.: *Natural Gas Flow Solutions with Guarantees: A Monotone Operator Theory Approach.* 2015. DOI: 10.48550/arXiv.1506.06075.

[Dvo+22] DVORKIN, V.; MALLAPRAGADA, D.; BOTTERUD, A.; KAZEMPOUR, J.; PINSON, P.: Multi-stage linear decision rules for stochastic control of natural gas networks with linepack. In: *Electric Power Systems Research* 212 (2022), p. 108388. DOI: 10.1016/j.epsr.2022.108388.

[DWD18] DE PERSIS, C.; WEITENBERG, E. R.; DÖRFLER, F.: A power consensus algorithm for DC microgrids. In: *Automatica* 89 (2018), pp. 364–375. DOI: 10.1016/j.automatica.2017.12.026.

[Ekh+19] EKHTIARI, A.; DASSIOS, I.; LIU, M.; SYRON, E.: A Novel Approach to Model a Gas Network. In: *Applied Sciences* 9.6 (2019), p. 1047. DOI: 10.3390/app9061047.

[EKK23] ENERGY INSTITUTE; KPMG; KEARNEY: *Statistical Review of World Energy.* Tech. rep. Energy Institute, 2023. URL: https://www.energyinst.org/statistical-review/home.

[Eme23] EMERSON: *Control Valve Handbook.* 2023. URL: https://www.emerson.com/documents/automation/control-valve-handbook-en-3661206.pdf.

[Erd+14] ERDENER, B. C.; PAMBOUR, K. A.; LAVIN, R. B.; DENGIZ, B.: An integrated simulation model for analysing electricity and gas systems. In: *International Journal of Electrical Power & Energy Systems* 61 (2014), pp. 410–420. DOI: 10.1016/j.ijepes.2014.03.052.

[Eur12] EUROPEAN TELECOMMUNICATIONS STANDARDS INSTITUTE: *ETSI EN 300 132-3-1: Environmental Engineering (EE); Power supply interface at the input to telecommunications and datacom (ICT) equipment; Part 3: Operated by rectified current source, alternating current source or direct current source up to 400 V.* 2012. URL: https://www.etsi.org/deliver/etsi_en/300100_300199/3001320301/02.01.01_60/en_3001320301v020101p.pdf.

[Eur+20] EUROPEAN COMMISSION; DIRECTORATE-GENERAL FOR ENERGY; SARDI, K.; DE VITA, A.; CAPROS, P.: *The role of gas DSOs and distribution networks in the context of the energy transition.* Publications Office, 2020. DOI: 10.2833/53338.

[Eur22] EUROPEAN COMMISSION: REPowerEU Plan. In: *Communication from the Commission to the European Parliament, the Council, the European Economic and Social Committee and Committee of the Regions.* 2022, pp. 1–21.

[Eur23] EUROPEAN COMMISSION: A Green Deal Industrial Plan for the Net-Zero Age. In: *Communication from the Commission to the European Parliament, the Council, the European Economic and Social Committee and Committee of the Regions.* 2023, pp. 1–21.

[Fan+18] FANG, J.; ZENG, Q.; AI, X.; CHEN, Z.; WEN, J.: Dynamic Optimal Energy Flow in the Integrated Natural Gas and Electrical Power Systems. In: *IEEE Transactions on Sustainable Energy* 9.1 (2018), pp. 188–198. DOI: 10.1109/TSTE.2017.2717600.

[Fan+20] FAN, B.; GUO, S.; PENG, J.; YANG, Q.; LIU, W.; LIU, L.: A Consensus-Based Algorithm for Power Sharing and Voltage Regulation in DC Microgrids. In: *IEEE Transactions on Industrial Informatics* 16.6 (2020), pp. 3987–3996. DOI: 10.1109/TII.2019.2941268.

[Far+20] FARROKHABADI, M.; CAÑIZARES, C. A.; SIMPSON-PORCO, J. W.; NASR, E.; FAN, L.; MENDOZA-ARAYA, P. A.; TONKOSKI, R.; TAMRAKAR, U.; HATZIARGYRIOU, N.; LAGOS, D.; WIES, R. W.; PAOLONE, M.; LISERRE, M.; MEEGAHAPOLA, L.; KABALAN, M.; HAJIMIRAGHA, A. H.; PERALTA, D.; ELIZONDO, M. A.; SCHNEIDER, K. P.; TUFFNER, F. K.; REILLY, J.: Microgrid Stability Definitions, Analysis, and Examples. In: *IEEE Transactions on Power Systems* 35.1 (2020), pp. 13–29. DOI: 10.1109/TPWRS.2019.2925703.

[FCS21] FERGUSON, J.; CUCUZZELLA, M.; SCHERPEN, J. M. A.: Exponential Stability and Local ISS for DC Networks. In: *IEEE Control Systems Letters* 5.3 (2021), pp. 893–898. DOI: 10.1109/LCSYS.2020.3007222.

[FCS23] FERGUSON, J.; CUCUZZELLA, M.; SCHERPEN, J. M. A.: Increasing the region of attraction in DC microgrids. In: *Automatica* 151 (2023), p. 110883. DOI: 10.1016/j.automatica.2023.110883.

[Fie73] FIEDLER, M.: Algebraic connectivity of graphs. In: *Czechoslovak Mathematical Journal* 23.2 (1973), pp. 298–305. DOI: 10.21136/CMJ.1973.101168.

[Fra23] FRAUNHOFER INSTITUTE FOR SOLAR ENERGY SYSTEMS: *Electricity generation in Germany in 2023*. Tech. rep. Fraunhofer ISE, 2023. URL: https://www.energy-charts.info/downloads/electricity_generation_germany_2023.pdf.

[FS23] FOERSTER, A.; SPENCER, M.: Corporate net zero pledges: a triumph of private climate regulation or more greenwash? In: *Griffith Law Review* 32.1 (2023), pp. 110–142. DOI: 10.1080/10383441.2023.2210450.

[FYL06] FREEMAN, R. A.; YANG, P.; LYNCH, K. M.: Stability and Convergence Properties of Dynamic Average Consensus Estimators. In: *45th IEEE Conference on Decision and Control (CDC)*. IEEE, 2006, pp. 338–343. DOI: 10.1109/CDC.2006.377078.

[GAH16] GLASGO, B.; AZEVEDO, I. L.; HENDRICKSON, C.: How much electricity can we save by using direct current circuits in homes? Understanding the potential for electricity savings and assessing feasibility of a transition towards DC powered buildings. In: *Applied Energy* 180 (2016), pp. 66–75. DOI: 10.1016/j.apenergy.2016.07.036.

[Gao+19] GAO, F.; KANG, R.; CAO, J.; YANG, T.: Primary and secondary control in DC microgrids: a review. In: *Journal of Modern Power Systems and Clean Energy* 7.2 (2019), pp. 227–242. DOI: 10.1007/s40565-018-0466-5.

[GasNZV10] *Verordnung über den Zugang zu Gasversorgungsnetzen (Gasnetzzugangsverordnung - GasNZV) [Ordinance on access to gas supply networks (Gas Network Access Ordinance - GasNZV)]*. Version released 3 September 2010. 2010. URL: http://bundesrecht.juris.de/bundesrecht/gasnzv_2010/gesamt.pdf.

[GCC17] GUANDALINI, G.; COLBERTALDO, P.; CAMPANARI, S.: Dynamic modeling of natural gas quality within transport pipelines in presence of hydrogen injections. In: *Applied Energy* 185 (2017), pp. 1712–1723. DOI: 10.1016/j.apenergy.2016.03.006.

[GE99a] GRAVDAHL, J. T.; EGELAND, O.: Centrifugal compressor surge and speed control. In: *IEEE Transactions on Control Systems Technology* 7.5 (1999), pp. 567–579. DOI: 10.1109/87.784420.

[GE99b] GRAVDAHL, J. T.; EGELAND, O.: *Compressor Surge and Rotating Stall*. London: Springer, 1999. DOI: 10.1007/978-1-4471-0827-6.

[Ger+18] GERBER, D. L.; VOSSOS, V.; FENG, W.; MARNAY, C.; NORDMAN, B.; BROWN, R.: A simulation-based efficiency comparison of AC and DC power distribution networks in commercial buildings. In: *Applied Energy* 210 (2018), pp. 1167–1187. DOI: 10.1016/j.apenergy.2017.05.179.

[GEV02] GRAVDAHL, J. T.; EGELAND, O.; VATLAND, S. O.: Drive torque actuation in active surge control of centrifugal compressors. In: *Automatica* 38.11 (2002), pp. 1881–1893. DOI: 10.1016/S0005-1098(02)00113-9.

[Gha+18] GHANAVATI, M.; SALAHSHOOR, K.; MOTLAGH, M. R. J.; RAMAZANI, A.; MOAREFIANPOUR, A.: A novel combined approach for gas compressors surge suppression based on robust adaptive control and backstepping. In: *Journal of Mechanical Science and Technology* 32.2 (2018), pp. 823–833. DOI: 10.1007/s12206-018-0133-1.

[Gra+00] GRAVDAHL, J. T.; WILLEMS, F.; JAGER, B. DE; EGELAND, O.: Modeling for surge control of centrifugal compressors: comparison with experiment. In: *39th IEEE Conference on Decision and Control (CDC)*. IEEE, 2000, pp. 1341–1346. DOI: 10.1109/CDC.2000.912043.

[Gra+04] GRAVDAHL, J. T.; WILLEMS, F.; JAGER, B. DE; EGELAND, O.: Modeling of Surge in Free-Spool Centrifugal Compressors: Experimental Validation. In: *Journal of Propulsion and Power* 20.5 (2004), pp. 849–857. DOI: 10.2514/1.10052.

[Gue+19] GUELPA, E.; BISCHI, A.; VERDA, V.; CHERTKOV, M.; LUND, H.: Towards future infrastructures for sustainable multi-energy systems: A review. In: *Energy* 184 (2019), pp. 2–21. DOI: 10.1016/j.energy.2019.05.057.

[Han+18] HAN, R.; MENG, L.; GUERRERO, J. M.; VASQUEZ, J. C.: Distributed Nonlinear Control With Event-Triggered Communication to Achieve Current-Sharing and Voltage Regulation in DC Microgrids. In: *IEEE Transactions on Power Electronics* 33.7 (2018), pp. 6416–6433. DOI: 10.1109/TPEL.2017.2749518.

[Han+19] HAN, R.; TUCCI, M.; MARTINELLI, A.; GUERRERO, J. M.; FERRARI-TRECATE, G.: Stability Analysis of Primary Plug-and-Play and Secondary Leader-Based Controllers for DC Microgrid Clusters. In: *IEEE Transactions on Power Systems* 34.3 (2019), pp. 1780–1800. DOI: 10.1109/TPWRS.2018.2884876.

[HAP11] HINES, G. H.; ARCAK, M.; PACKARD, A. K.: Equilibrium-independent passivity: A new definition and numerical certification. In: *Automatica* 47.9 (2011), pp. 1949–1956. DOI: 10.1016/j.automatica.2011.05.011.

[Hat+21] HATZIARGYRIOU, N.; MILANOVIC, J.; RAHMANN, C.; AJJARAPU, V.; CANIZARES, C.; ERLICH, I.; HILL, D.; HISKENS, I.; KAMWA, I.; PAL, B.; POURBEIK, P.; SANCHEZ-GASCA, J.; STANKOVIC, A.; VAN CUTSEM, T.; VITTAL, V.; VOURNAS, C.: Definition and Classification of Power System Stability – Revisited & Extended. In: *IEEE Transactions on Power Systems* 36.4 (2021), pp. 3271–3281. DOI: 10.1109/TPWRS.2020.3041774.

[Hau+20] HAUSCHILD, S.-A.; MARHEINEKE, N.; MEHRMANN, V.; MOHRING, J.; BADLYAN, A. M.; REIN, M.; SCHMIDT, M.: Port-Hamiltonian Modeling of District Heating Networks. In: *Progress in Differential-Algebraic Equations II*. Ed. by REIS, T.; GRUNDEL, S.; SCHÖPS, S. Differential-Algebraic Equations Forum. Cham: Springer International Publishing, 2020, pp. 333–355. DOI: 10.1007/978-3-030-53905-4_11.

[HB93] HADDAD, W. M.; BERNSTEIN, D. S.: Explicit construction of quadratic Lyapunov functions for the small gain, positivity, circle, and Popov theorems and their application to robust stability. Part I: Continuous-time theory. In: *International Journal of Robust and Nonlinear Control* 3.4 (1993), pp. 313–339. DOI: 10.1002/rnc.4590030403.

[He+17] HE, C.; WU, L.; LIU, T.; SHAHIDEHPOUR, M.: Robust Co-Optimization Scheduling of Electricity and Natural Gas Systems via ADMM. In: *IEEE Transactions on Sustainable Energy* 8.2 (2017), pp. 658–670. DOI: 10.1109/TSTE.2016.2615104.

[He+18] HE, C.; ZHANG, X.; LIU, T.; WU, L.; SHAHIDEHPOUR, M.: Coordination of Interdependent Electricity Grid and Natural Gas Network—a Review. In: *Current Sustainable/Renewable Energy Reports* 5.1 (2018), pp. 23–36. DOI: 10.1007/s40518-018-0093-9.

[HG+09] HERRÁN-GONZÁLEZ, A.; LA CRUZ, J. M. DE; ANDRÉS-TORO, B. DE; RISCO-MARTÍN, J. L.: Modeling and simulation of a gas distribution pipeline network. In: *Applied Mathematical Modelling* 33.3 (2009), pp. 1584–1600. DOI: 10.1016/j.apm.2008.02.012.

[HHY08] HUANG, B.-N.; HWANG, M. J.; YANG, C. W.: Causal relationship between energy consumption and GDP growth revisited: A dynamic panel data approach. In: *Ecological Economics* 67.1 (2008), pp. 41–54. DOI: 10.1016/j.ecolecon.2007.11.006.

[HJ12] HORN, R. A.; JOHNSON, C. R.: *Matrix analysis*. 2nd ed. Cambridge: Cambridge University Press, 2012. DOI: 10.1017/CBO9781139020411.

[HJ91] HORN, R. A.; JOHNSON, C. R.: *Topics in Matrix Analysis*. Cambridge University Press, 1991. DOI: 10.1017/CBO9780511840371.

[HMJ22] HASSAN, M. S.; MAHMOOD, H.; JAVAID, A.: The impact of electric power consumption on economic growth: a case study of Portugal, France, and Finland. In: *Environmental Science and Pollution Research* 29.30 (2022), pp. 45204–45220. DOI: 10.1007/s11356-022-19097-y.

[HMS10] HERTY, M.; MOHRING, J.; SACHERS, V.: A new model for gas flow in pipe networks. In: *Mathematical Methods in the Applied Sciences* 33.7 (2010), pp. 845–855. DOI: 10.1002/mma.1197.

[Höh+21] HÖHNE, N.; GIDDEN, M. J.; ELZEN, M. DEN; HANS, F.; FYSON, C.; GEIGES, A.; JEFFERY, M. L.; GONZALES-ZUÑIGA, S.; MOOLDIJK, S.; HARE, W.; ROGELJ, J.: Wave of net zero emission targets opens window to meeting the Paris Agreement. In: *Nature Climate Change* 11.10 (2021), pp. 820–822. DOI: 10.1038/s41558-021-01142-2.

[IEEE22] IEEE Guide for Load Modeling and Simulations for Power Systems. In: *IEEE Std 2781-2022* (2022), pp. 1–88. DOI: 10.1109/IEEESTD.2022.9905546.

[Int23a] INTERNATIONAL ENERGY AGENCY: *The Oil and Gas Industry in Net Zero Transitions*. Tech. rep. Paris, 2023. URL: https://www.iea.org/reports/the-oil-and-gas-industry-in-net-zero-transitions.

[Int23b] INTERNATIONAL ENERGY AGENCY: *World Energy Balances 2023 Highlights*. Tech. rep. Paris, 2023. URL: https://www.iea.org/data-and-statistics/data-product/world-energy-statistics-and-balances.

[Iov+18] IOVINE, A.; DAMM, G.; SANTIS, E. DE; DI BENEDETTO, M. D.; GALAI-DOL, L.; PEPE, P.: Voltage Stabilization in a DC MicroGrid by an ISS-like Lyapunov Function implementing Droop Control. In: *2018 European Control Conference (ECC)*. IEEE, 2018, pp. 1130–1135. DOI: 10.23919/ECC.2018.8550492.

[Jay+07] JAYAWARDHANA, B.; ORTEGA, R.; GARCÍA-CANSECO, E.; CASTAÑOS, F.: Passivity of nonlinear incremental systems: Application to PI stabilization of nonlinear RLC circuits. In: *Systems & Control Letters* 56.9-10 (2007), pp. 618–622. DOI: 10.1016/j.sysconle.2007.03.011.

[Jus04] JUSOH, A. B.: The instability effect of constant power loads. In: *PECon 2004. Proceedings. National Power and Energy Conference, 2004.* IEEE, 2004, pp. 175–179. DOI: 10.1109/PECON.2004.1461638.

[Jus+13] JUSTO, J. J.; MWASILU, F.; LEE, J.; JUNG, J.-W.: AC-microgrids versus DC-microgrids with distributed energy resources: A review. In: *Renewable and Sustainable Energy Reviews* 24 (2013), pp. 387–405. DOI: 10.1016/j.rser.2013.03.067.

[JXG13] JIANG, Q.; XUE, M.; GENG, G.: Energy Management of Microgrid in Grid-Connected and Stand-Alone Modes. In: *IEEE Transactions on Power Systems* 28.3 (2013), pp. 3380–3389. DOI: 10.1109/TPWRS.2013.2244104.

[KAA19] KUMAR, J.; AGARWAL, A.; AGARWAL, V.: A review on overall control of DC microgrids. In: *Journal of Energy Storage* 21 (2019), pp. 113–138. DOI: 10.1016/j.est.2018.11.013.

[Kar+21] KARAMI, Z.; SHAFIEE, Q.; KHAYAT, Y.; YARIBEYGI, M.; DRAGICEVIC, T.; BEVRANI, H.: Decentralized Model Predictive Control of DC Microgrids With Constant Power Load. In: *IEEE Journal of Emerging and Selected Topics in Power Electronics* 9.1 (2021), pp. 451–460. DOI: 10.1109/JESTPE.2019.2957231.

[KB00] KASZKUREWICZ, E.; BHAYA, A.: *Matrix Diagonal Stability in Systems and Computation*. Boston, MA: Birkhäuser Boston, 2000. DOI: 10.1007/978-1-4612-1346-8.

[Ke00] KE, S.: Transient analysis of isothermal gas flow in pipeline network. In: *Chemical Engineering Journal* 76.2 (2000), pp. 169–177. DOI: 10.1016/S1385-8947(99)00122-9.

[KH21] KRISTOFFERSEN, T. T.; HOLDEN, C.: Modeling and Control of a Wet-Gas Centrifugal Compressor. In: *IEEE Transactions on Control Systems Technology* 29.3 (2021), pp. 1175–1190. DOI: 10.1109/TCST.2020.2993224.

[Kha02] KHALIL, H. K.: *Nonlinear Systems*. 3rd. Upper Saddle River, NJ: Prentice Hall, 2002.

[KKO99] KOKOTOVIĆ, P.; KHALIL, H. K.; O'REILLY, J.: *Singular Perturbation Methods in Control: Analysis and Design*. Society for Industrial and Applied Mathematics, 1999. DOI: 10.1137/1.9781611971118.

[KKS19] KOSARAJU, K. C.; KAWANO, Y.; SCHERPEN, J. M.: Krasovskii's Passivity. In: *IFAC-PapersOnLine* 52.16 (2019), pp. 466–471. DOI: 10.1016/j.ifacol.2019.12.005.

[KKS21] Kawano, Y.; Kosaraju, K. C.; Scherpen, J. M. A.: Krasovskii and Shifted Passivity-Based Control. In: *IEEE Transactions on Automatic Control* 66.10 (2021), pp. 4926–4932. doi: 10.1109/TAC.2020.3040252.

[Koc+15] Koch, T.; Hiller, B.; Pfetsch, M. E.; Schewe, L.: *Evaluating Gas Network Capacities*. Philadelphia, PA: Society for Industrial and Applied Mathematics, 2015. doi: 10.1137/1.9781611973693.

[Kos+21] Kosaraju, K. C.; Cucuzzella, M.; Scherpen, J. M. A.; Pasumarthy, R.: Differentiation and Passivity for Control of Brayton–Moser Systems. In: *IEEE Transactions on Automatic Control* 66.3 (2021), pp. 1087–1101. doi: 10.1109/TAC.2020.2994317.

[Kun+04] Kundur, P.; Paserba, J.; Ajjarapu, V.; Andersson, G.; Bose, A.; Canizares, C.; Hatziargyriou, N.; Hill, D.; Stankovic, A.; Taylor, C.; Van Cutsem, T.; Vittal, V.: Definition and Classification of Power System Stability IEEE/CIGRE Joint Task Force on Stability Terms and Definitions. In: *IEEE Transactions on Power Systems* 19.3 (2004), pp. 1387–1401. doi: 10.1109/TPWRS.2004.825981.

[Kun94] Kundur, P.: *Power System Stability and Control*. New York: McGraw-Hill, 1994.

[Las01] Lasseter, B.: Microgrids [distributed power generation]. In: *2001 IEEE Power Engineering Society Winter Meeting*. Vol. 1. 2001, pp. 146–149. doi: 10.1109/PESW.2001.917020.

[Lee+23] Lee, H. et al.: *IPCC, 2023: Climate Change 2023: Synthesis Report. Contribution of Working Groups I, II and III to the Sixth Assessment Report of the Intergovernmental Panel on Climate Change [Core Writing Team, H. Lee and J. Romero (eds.)]. IPCC, Geneva, Switzerland*. Tech. rep. 2023. doi: 10.59327/IPCC/AR6-9789291691647.

[LES08] Li, T.; Eremia, M.; Shahidehpour, M.: Interdependency of Natural Gas Network and Power System Security. In: *IEEE Transactions on Power Systems* 23.4 (2008), pp. 1817–1824. doi: 10.1109/TPWRS.2008.2004739.

[Li+17] Li, C.; Bosio, F. de; Chen, F.; Chaudhary, S. K.; Vasquez, J. C.; Guerrero, J. M.: Economic Dispatch for Operating Cost Minimization Under Real-Time Pricing in Droop-Controlled DC Microgrid. In: *IEEE Journal of Emerging and Selected Topics in Power Electronics* 5.1 (2017), pp. 587–595. doi: 10.1109/JESTPE.2016.2634026.

[Li+18] Li, J.; Liu, F.; Wang, Z.; Low, S. H.; Mei, S.: Optimal Power Flow in Stand-Alone DC Microgrids. In: *IEEE Transactions on Power Systems* 33.5 (2018), pp. 5496–5506. doi: 10.1109/TPWRS.2018.2801280.

[Liu+09] Liu, C.; Shahidehpour, M.; Fu, Y.; Li, Z.: Security-Constrained Unit Commitment With Natural Gas Transmission Constraints. In: *IEEE Transactions on Power Systems* 24.3 (2009), pp. 1523–1536. doi: 10.1109/T PWRS.2009.2023262.

[Liu+21] LIU, Z.; MU, Y.; GUO, X.; JIA, H.; WU, G.; ZU, G.: Static Stability Analysis of Integrated Electricity and Gas System Considering Voltage and Pressure. In: *2021 IEEE 5th Conference on Energy Internet and Energy System Integration (EI2)*. IEEE, 2021, pp. 1750–1754. DOI: 10.1109/EI252483.2021.9712954.

[LLC21] LONGORIA, G.; LYNCH, M.; CURTIS, J.: Green hydrogen for heating and its impact on the power system. In: *International Journal of Hydrogen Energy* 46.53 (2021), pp. 26725–26740. DOI: 10.1016/j.ijhydene.2021.05.171.

[Loh+14] LOHMANN, B.; DEROO, F.; HIRCHE, S.; LUNZE, J.; MEINEL, M.; PHILIPP, P.; ULBRICH, M.: 'Distributed Estimation and Control'. In: *Control Theory of Digitally Networked Dynamic Systems*. Ed. by LUNZE, J. Heidelberg: Springer International Publishing, 2014, pp. 81–110. DOI: 10.1007/978-3-319-01131-8_3.

[Lun19] LUNZE, J.: *Networked control of multi-agent systems: Consensus and synchronisation, communication structure design, self-organisation in networked systems, event-triggered control*. Rotterdam: Bookmundo, 2019.

[Lur08] LURIE, M. V.: *Modeling of Oil Product and Gas Pipeline Transportation*. Wiley, 2008. DOI: 10.1002/9783527626199.

[Mac+21] MACHADO, J. E.; ORTEGA, R.; ASTOLFI, A.; AROCAS-PÉREZ, J.; PYRKIN, A.; BOBTSOV, A. A.; GRIÑÓ, R.: An Adaptive Observer-Based Controller Design for Active Damping of a DC Network With a Constant Power Load. In: *IEEE Transactions on Control Systems Technology* 29.6 (2021), pp. 2312–2324. DOI: 10.1109/TCST.2020.3037859.

[Man14] MANCARELLA, P.: MES (multi-energy systems): An overview of concepts and evaluation models. In: *Energy* 65 (2014), pp. 1–17. DOI: 10.1016/j.energy.2013.10.041.

[MB+15] MOHAMADI-BAGHMOLAEI, M.; AZIN, R.; OSFOURI, S.; MOHAMADI-BAGHMOLAEI, R.; ZAREI, Z.: Prediction of gas compressibility factor using intelligent models. In: *Natural Gas Industry B* 2.4 (2015), pp. 283–294. DOI: 10.1016/j.ngib.2015.09.001.

[MBB08] MACHOWSKI, J.; BIALEK, J. W.; BUMBY, J. R.: *Power System Dynamics: Stability and Control*. 2nd. Chichester, United Kingdom: John Wiley & Sons, Ltd., 2008.

[Mei+22] MEIRA, R. L.; MARTINS, M. A.; KALID, R. A.; COSTA, G. M.: Implementable MPC-based surge avoidance nonlinear control strategies for non-ideally modeled natural gas compression systems. In: *Journal of Natural Gas Science and Engineering* 102 (2022), p. 104573. DOI: 10.1016/j.jngse.2022.104573.

[Men05] MENON, E. S.: *Gas Pipeline Hydraulics*. CRC Press, 2005. DOI: 10.1201/9781420038224.

[Men+17] MENG, L.; SHAFIEE, Q.; TRECATE, G. F.; KARIMI, H.; FULWANI, D.; LU, X.; GUERRERO, J. M.: Review on Control of DC Microgrids and Multiple Microgrid Clusters. In: *IEEE Journal of Emerging and Selected Topics in Power Electronics* 5.3 (2017), pp. 928–948. DOI: 10.1109/JESTPE.2017.2690219.

[MF19] MANRING, N. D.; FALES, R. C.: *Hydraulic Control Systems*. Wiley, 2019. DOI: 10.1002/9781119418528.

[Mil+20] MILOSAVLJEVIC, P.; MARCHETTI, A. G.; CORTINOVIS, A.; FAULWASSER, T.; MERCANGÖZ, M.; BONVIN, D.: Real-time optimization of load sharing for gas compressors in the presence of uncertainty. In: *Applied Energy* 272 (2020), p. 114883. DOI: 10.1016/j.apenergy.2020.114883.

[MKC09] MOORE, R. E.; KEARFOTT, R. B.; CLOUD, M. J.: *Introduction to Interval Analysis*. Society for Industrial and Applied Mathematics, 2009. DOI: 10.1137/1.9780898717716.

[Mod+23] MODU, B.; ABDULLAH, M. P.; SANUSI, M. A.; HAMZA, M. F.: DC-based microgrid: Topologies, control schemes, and implementations. In: *Alexandria Engineering Journal* 70 (2023), pp. 61–92. DOI: 10.1016/j.aej.2023.02.021.

[Mon+19] MONSHIZADEH, N.; MONSHIZADEH, P.; ORTEGA, R.; VAN DER SCHAFT, A.: Conditions on shifted passivity of port-Hamiltonian systems. In: *Systems & Control Letters* 123 (2019), pp. 55–61. DOI: 10.1016/j.sysconle.2018.10.010.

[MTB20] MEINECKE, S.; THURNER, L.; BRAUN, M.: Review of Steady-State Electric Power Distribution System Datasets. In: *Energies* 13.18 (2020), p. 4826. DOI: 10.3390/en13184826.

[MVZ20] MISRA, S.; VUFFRAY, M.; ZLOTNIK, A.: Monotonicity Properties of Physical Network Flows and Application to Robust Optimal Allocation. In: *Proceedings of the IEEE* 108.9 (2020), pp. 1558–1579. DOI: 10.1109/JPROC.2020.3014069.

[Nah+20] NAHATA, P.; SOLOPERTO, R.; TUCCI, M.; MARTINELLI, A.; FERRARI-TRECATE, G.: A Passivity-Based Approach to Voltage Stabilization in DC Microgrids with ZIP Loads. In: *Automatica* 113 (2020), p. 108770. DOI: 10.1016/j.automatica.2019.108770.

[Nas+15] NASIRIAN, V.; MOAYEDI, S.; DAVOUDI, A.; LEWIS, F. L.: Distributed Cooperative Control of DC Microgrids. In: *IEEE Transactions on Power Electronics* 30.4 (2015), pp. 2288–2303. DOI: 10.1109/TPEL.2014.2324579.

[NFT20] NAHATA, P.; FERRARI-TRECATE, G.: On Existence of Equilibria, Voltage Balancing, and Current Sharing in Consensus-Based DC Microgrids. In: *2020 European Control Conference (ECC)*. IEEE, 2020, pp. 1216–1223. DOI: 10.23919/ECC51009.2020.9143766.

[NTFT22] NAHATA, P.; TURAN, M. S.; FERRARI-TRECATE, G.: Consensus-Based Current Sharing and Voltage Balancing in DC Microgrids With Exponential Loads. In: *IEEE Transactions on Control Systems Technology* 30.4 (2022), pp. 1668–1680. DOI: 10.1109/TCST.2021.3120321.

[NTG18] NOROOZI, N.; TRIP, S.; GEISELHART, R.: Model predictive control of DC microgrids: current sharing and voltage regulation. In: *IFAC-PapersOnLine* 51.23 (2018), pp. 124–129. DOI: 10.1016/j.ifacol.2018.12.022.

[OC20] OSIADACZ, A. J.; CHACZYKOWSKI, M.: Modeling and Simulation of Gas Distribution Networks in a Multienergy System Environment. In: *Proceedings of the IEEE* 108.9 (2020), pp. 1580–1595. DOI: 10.1109/JPROC.2020.2989114.

[OCK11] OLIVARES, D. E.; CAÑIZARES, C. A.; KAZERANI, M.: A centralized optimal energy management system for microgrids. In: *2011 IEEE Power and Energy Society General Meeting*. IEEE, 2011, pp. 1–6. DOI: 10.1109/PES.2011.6039527.

[Oli+14] OLIVARES, D. E.; MEHRIZI-SANI, A.; ETEMADI, A. H.; CAÑIZARES, C. A.; IRAVANI, R.; KAZERANI, M.; HAJIMIRAGHA, A. H.; GOMIS-BELLMUNT, O.; SAEEDIFARD, M.; PALMA-BEHNKE, R.; JIMEÉNEZ-ESTÉVEZ, G. A.; HATZIARGYRIOU, N. D.: Trends in Microgrid Control. In: *IEEE Transactions on Smart Grid* 5.4 (2014), pp. 1905–1919. DOI: 10.1109/TSG.2013.2295514.

[OMa21] O'MALLEY, C.: *Coordination of Gas-Electric Networks: Modeling, Optimization and Uncertainty*, phdthesis. ETH Zurich, 2021. DOI: 10.3929/ethz-b-000510208.

[OW15] OVERBYE, T. J.; WEBER, J.: Smart Grid Wide-Area Transmission System Visualization. In: *Engineering* 1.4 (2015), pp. 466–474. DOI: 10.15302/J-ENG-2015098.

[Ozc+23] OZCAN, H.; EL-EMAM, R. S.; CELIK, S.; AMINI HORRI, B.: Recent advances, challenges, and prospects of electrochemical water-splitting technologies for net-zero transition. In: *Cleaner Chemical Engineering* 8 (2023), p. 100115. DOI: 10.1016/j.clce.2023.100115.

[Pam+17] PAMBOUR, K. A.; CAKIR ERDENER, B.; BOLADO-LAVIN, R.; DIJKEMA, G. P.: Development of a Simulation Framework for Analyzing Security of Supply in Integrated Gas and Electric Power Systems. In: *Applied Sciences* 7.1 (2017), p. 47. DOI: 10.3390/app7010047.

[Pam18] PAMBOUR, K. A.: *Modelling, simulation and analysis of security of supply scenarios in integrated gas and electricity transmission networks*, phdthesis. University of Groningen, 2018. URL: https://hdl.handle.net/11370/e2b138e4-c4fb-4596-810d-0d25e94bff8d.

[Pap21] PAPAILIOU, K. O.: *Springer Handbook of Power Systems*. Singapore: Springer, 2021. DOI: 10.1007/978-981-32-9938-2.

[PBLD16] PAMBOUR, K. A.; BOLADO-LAVIN, R.; DIJKEMA, G. P.: An integrated transient model for simulating the operation of natural gas transport systems. In: *Journal of Natural Gas Science and Engineering* 28 (2016), pp. 672–690. DOI: 10.1016/j.jngse.2015.11.036.

[Per+18] PEREIRINHA, P. G.; GONZÁLEZ, M.; CARRILERO, I.; ANSEÁN, D.; ALONSO, J.; VIERA, J. C.: Main Trends and Challenges in Road Transportation Electrification. In: *Transportation Research Procedia* 33 (2018), pp. 235–242. DOI: 10.1016/j.trpro.2018.10.096.

[Per+20] PEREZ, F.; IOVINE, A.; DAMM, G.; GALAI-DOL, L.; RIBEIRO, P. F.: Stability Analysis of a DC MicroGrid for a Smart Railway Station Integrating Renewable Sources. In: *IEEE Transactions on Control Systems Technology* 28.5 (2020), pp. 1802–1816. DOI: 10.1109/TCST.2019.2924615.

[Pla+15] PLANAS, E.; ANDREU, J.; GÁRATE, J. I.; MARTÍNEZ DE ALEGRÍA, I.; IBARRA, E.: AC and DC technology in microgrids: A review. In: *Renewable and Sustainable Energy Reviews* 43 (2015), pp. 726–749. DOI: 10.1016/j.rser.2014.11.067.

[PPC23] PIRES, V. F.; PIRES, A.; CORDEIRO, A.: DC Microgrids: Benefits, Architectures, Perspectives and Challenges. In: *Energies* 16.3 (2023), p. 1217. DOI: 10.3390/en16031217.

[Qad+19] QADRDAN, M.; FAZELI, R.; JENKINS, N.; STRBAC, G.; SANSOM, R.: Gas and electricity supply implications of decarbonising heat sector in GB. In: *Energy* 169 (2019), pp. 50–60. DOI: 10.1016/j.energy.2018.11.066.

[Qad+20] QADRDAN, M.; ABEYSEKERA, M.; WU, J.; JENKINS, N.; WINTER, B.: *The Future of Gas Networks*. SpringerBriefs in Energy. Cham: Springer, 2020. DOI: 10.1007/978-3-319-66784-3.

[RFFT13] RIVERSO, S.; FARINA, M.; FERRARI-TRECATE, G.: Plug-and-Play Decentralized Model Predictive Control for Linear Systems. In: *IEEE Transactions on Automatic Control* 58.10 (2013), pp. 2608–2614. DOI: 10.1109/TAC.2013.2254641.

[RMBS15] RÍOS-MERCADO, R. Z.; BORRAZ-SÁNCHEZ, C.: Optimization problems in natural gas transportation systems: A state-of-the-art review. In: *Applied Energy* 147 (2015), pp. 536–555. DOI: 10.1016/j.apenergy.2015.03.017.

[RW98] ROCKAFELLAR, R. T.; WETS, R. J.-B.: *Variational analysis*. Berlin: Springer, 1998. DOI: 10.1007/978-3-642-02431-3.

[SAA09] SHEHATA, R. S.; ABDULLAH, H. A.; AREED, F. F.: Variable structure surge control for constant speed centrifugal compressors. In: *Control Engineering Practice* 17.7 (2009), pp. 815–833. DOI: 10.1016/j.conengprac.2009.02.002.

[Sas+21] SASANPOUR, S.; CAO, K.-K.; GILS, H. C.; JOCHEM, P.: Strategic policy targets and the contribution of hydrogen in a 100% renewable European power system. In: *Energy Reports* 7 (2021), pp. 4595–4608. DOI: 10.1016/j.egyr.2021.07.005.

[Sch15] SCHIFFER, J.: *Stability and power sharing in microgrids*, Doctoral Thesis. Berlin: Technische Universität Berlin, 2015. DOI: 10.14279/depositonce-4581.

[Sch+17] SCHMIDT, M.; ASSMANN, D.; BURLACU, R.; HUMPOLA, J.; JOORMANN, I.; KANELAKIS, N.; KOCH, T.; OUCHERIF, D.; PFETSCH, M. E.; SCHEWE, L.; SCHWARZ, R.; SIRVENT, M.: GasLib—A Library of Gas Network Instances. In: *Data* 2.4 (2017), p. 40. DOI: 10.3390/data2040040.

[Sch+22] SCHÄFER, B.; PESCH, T.; MANIK, D.; GOLLENSTEDE, J.; LIN, G.; BECK, H.-P.; WITTHAUT, D.; TIMME, M.: Understanding Braess' Paradox in power grids. In: *Nature Communications* 13.1 (2022), p. 5396. DOI: 10.1038/s41467-022-32917-6.

[SGF17] SINGH, S.; GAUTAM, A. R.; FULWANI, D.: Constant power loads and their effects in DC distributed power systems: A review. In: *Renewable and Sustainable Energy Reviews* 72 (2017), pp. 407–421. DOI: 10.1016/j.rser.2017.01.027.

[Sil+22] SILANI, A.; CUCUZZELLA, M.; SCHERPEN, J. M.; YAZDANPANAH, M. J.: Robust output regulation for voltage control in DC networks with time-varying loads. In: *Automatica* 135 (2022), p. 109997. DOI: 10.1016/j.automatica.2021.109997.

[SJK97] SEPULCHRE, R.; JANKOVIĆ, M.; KOKOTOVIĆ, P. V.: *Constructive Nonlinear Control*. London: Springer, 1997. DOI: 10.1007/978-1-4471-0967-9.

[Smi21] SMIL, V.: *Grand Transitions: How the Modern World Was Made*. Oxford University Press, Apr. 2021. DOI: 10.1093/oso/9780190060664.001.0001.

[Son08] SONTAG, E. D.: 'Input to State Stability: Basic Concepts and Results'. In: *Nonlinear and Optimal Control Theory*. Ed. by NISTRI, P.; STEFANI, G. Lecture Notes in Mathematics. Berlin, Heidelberg: Springer, 2008, pp. 163–220. DOI: 10.1007/978-3-540-77653-6_3.

[SP19] SIMPSON-PORCO, J. W.: Equilibrium-Independent Dissipativity With Quadratic Supply Rates. In: *IEEE Transactions on Automatic Control* 64.4 (2019), pp. 1440–1455. DOI: 10.1109/TAC.2018.2838664.

[SS17] SUREK, D.; STEMPIN, S.: *Technische Strömungsmechanik*. Wiesbaden: Springer, 2017. DOI: 10.1007/978-3-658-18757-6.

[SSB21] SADABADI, M. S.; SAHOO, S.; BLAABJERG, F.: Stability Oriented Design of Cyber Attack Resilient Controllers for Cooperative DC Microgrids. In: *IEEE Transactions on Power Electronics* (2021), p. 1. DOI: 10.1109/TPEL.2021.3104721.

[SSK18] SADABADI, M. S.; SHAFIEE, Q.; KARIMI, A.: Plug-and-Play Robust Voltage Control of DC Microgrids. In: *IEEE Transactions on Smart Grid* 9.6 (2018), pp. 6886–6896. DOI: 10.1109/TSG.2017.2728319.

[SSW15] SCHMIDT, M.; STEINBACH, M. C.; WILLERT, B. M.: High detail stationary optimization models for gas networks. In: *Optimization and Engineering* 16.1 (2015), pp. 131–164. DOI: 10.1007/s11081-014-9246-x.

[Sto09] STOUSTRUP, J.: Plug & Play Control: Control Technology Towards New Challenges. In: *European Journal of Control* 15.3-4 (2009), pp. 311–330. DOI: 10.3166/ejc.15.311-330.

[Str24] STREHLE, F.: *A Framework for Decentralized Stabilization in Networked Energy Systems: A Passivity-Based Approach*, phdthesis. Karlsruhe Institute for Technology (KIT), 2024. DOI: 10.5445/IR/1000167942.

[Tay+22] TAYLOR, P. C.; ABEYSEKERA, M.; BIAN, Y.; ĆETENOVIĆ, D.; DEAKIN, M.; EHSAN, A.; LEVI, V.; LI, F.; ODURO, R.; PREECE, R.; TAYLOR, P. G.; TERZIJA, V.; WALKER, S. L.; WU, J.: An interdisciplinary research perspective on the future of multi-vector energy networks. In: *International Journal of Electrical Power & Energy Systems* 135 (2022), p. 107492. DOI: 10.1016/j.ijepes.2021.107492.

[THM17] TAHERINEJAD, M.; HOSSEINALIPOUR, S. M.; MADOLIAT, R.: Dynamic simulation of gas pipeline networks with electrical analogy. In: *Journal of the Brazilian Society of Mechanical Sciences and Engineering* 39.11 (2017), pp. 4431–4441. DOI: 10.1007/s40430-017-0821-x.

[Tor+17] TORRISI, G.; GRAMMATICO, S.; CORTINOVIS, A.; MERCANGOZ, M.; MORARI, M.; SMITH, R. S.: Model Predictive Approaches for Active Surge Control in Centrifugal Compressors. In: *IEEE Transactions on Control Systems Technology* 25.6 (2017), pp. 1947–1960. DOI: 10.1109/TCST.2016.2636027.

[Tra+18] TRAN, T. H.; FRENCH, S.; ASHMAN, R.; KENT, E.: Linepack planning models for gas transmission network under uncertainty. In: *European Journal of Operational Research* 268.2 (2018), pp. 688–702. DOI: 10.1016/j.ejor.2018.01.033.

[TRFT18] TUCCI, M.; RIVERSO, S.; FERRARI TRECATE, G.: Line-Independent Plug-and-Play Controllers for Voltage Stabilization in DC Microgrids. In: *IEEE Transactions on Control Systems Technology* 26.3 (2018), pp. 1115–1123. DOI: 10.1109/TCST.2017.2695167.

[Tri+18] TRIP, S.; HAN, R.; CUCUZZELLA, M.; CHENG, X.; SCHERPEN, J.; GUERRERO, J.: Distributed Averaging Control for Voltage Regulation and Current Sharing in DC Microgrids: Modelling and Experimental Validation. In: *IFAC-PapersOnLine* 51.23 (2018), pp. 242–247. DOI: 10.1016/j.ifacol.2018.12.042.

[Tri+19] TRIP, S.; CUCUZZELLA, M.; CHENG, X.; SCHERPEN, J.: Distributed Averaging Control for Voltage Regulation and Current Sharing in DC Microgrids. In: *IEEE Control Systems Letters* 3.1 (2019), pp. 174–179. DOI: 10.1109/LCSYS.2018.2857559.

[Tuc+16] TUCCI, M.; RIVERSO, S.; VASQUEZ, J. C.; GUERRERO, J. M.; FERRARI-TRECATE, G.: A Decentralized Scalable Approach to Voltage Control of DC Islanded Microgrids. In: *IEEE Transactions on Control Systems Technology* 24.6 (2016), pp. 1965–1979. DOI: 10.1109/TCST.2016.2525001.

[Tuc+18] TUCCI, M.; MENG, L.; GUERRERO, J. M.; FERRARI-TRECATE, G.: Stable current sharing and voltage balancing in DC microgrids: A consensus-based secondary control layer. In: *Automatica* 95 (2018), pp. 1–13. DOI: 10.1016/j.automatica.2018.04.017.

[TWZ19] TAN, Y.; WANG, X.; ZHENG, Y.: A new modeling and solution method for optimal energy flow in electricity-gas integrated energy system. In: *International Journal of Energy Research* 43.9 (2019), pp. 4322–4343. DOI: 10.1002/er.4558.

[UMR14] ULABY, F. T.; MICHIELSSEN, E.; RAVAIOLI, U.: *Fundamentals of applied electromagnetics*. 6th ed. Harlow: Pearson Education, 2014.

[vdS17] VAN DER SCHAFT, A.: *L2-Gain and Passivity Techniques in Nonlinear Control*. 3rd ed. Communications and control engineering. Cham: Springer, 2017. DOI: 10.1007/978-3-319-49992-5.

[Vis06] VISIOLI, A.: *Practical PID Control*. 1st ed. Advances in industrial control. London: Springer, 2006. DOI: 10.1007/1-84628-586-0.

[vJ14] VAN DER SCHAFT, A. J.; JELTSEMA, D.: Port-Hamiltonian Systems Theory: An Introductory Overview. In: *Foundations and Trends in Systems and Control* 1.2–3 (2014), pp. 173–378. DOI: 10.1561/2600000002.

[VMC12] VUFFRAY, M.; MISRA, S.; CHERTKOV, M.: Monotonicity of dissipative flow networks renders robust maximum profit problem tractable: General analysis and application to natural gas flows. In: *54th IEEE Conference on Decision and Control (CDC)*. IEEE, 15.12.2015 - 18.12.2015, pp. 4571–4578. DOI: 10.1109/CDC.2015.7402933.

[vRc20] VAN RENSSEN, S.: The hydrogen solution? In: *Nature Climate Change* 10.9 (2020), pp. 799–801. DOI: 10.1038/s41558-020-0891-0.

[Vu+17] VU, T. V.; PARAN, S.; DIAZ-FRANCO, F.; EL-MEZYANI, T.; EDRINGTON, C. S.: An Alternative Distributed Control Architecture for Improvement in the Transient Response of DC Microgrids. In: *IEEE Transactions on Industrial Electronics* 64.1 (2017), pp. 574–584. DOI: 10.1109/TIE.2016.2607681.

[vV98] VAN CUTSEM, T.; VOURNAS, C.: *Voltage Stability of Electric Power Systems*. New York, NY: Springer, 1998. DOI: 10.1007/978-0-387-75536-6.

[VvH18] VANDERMEULEN, A.; VAN DER HEIJDE, B.; HELSEN, L.: Controlling district heating and cooling networks to unlock flexibility: A review. In: *Energy* 151 (2018), pp. 103–115. DOI: 10.1016/j.energy.2018.03.034.

[WC13] WINNING, H. K.; COOLE, T.: Explicit Friction Factor Accuracy and Computational Efficiency for Turbulent Flow in Pipes. In: *Flow, Turbulence and Combustion* 90.1 (2013), pp. 1–27. DOI: 10.1007/s10494-012-9419-7.

[Wei+18] WEITENBERG, E.; JIANG, Y.; ZHAO, C.; MALLADA, E.; DÖRFLER, F.; PERSIS, C. DE: Robust decentralized frequency control: A leaky integrator approach. In: *2018 European Control Conference (ECC)*. IEEE, 2018, pp. 764–769. DOI: 10.23919/ECC.2018.8550060.

[Wen+18] WEN, K.; XIA, Z.; YU, W.; GONG, J.: A New Lumped Parameter Model for Natural Gas Pipelines in State Space. In: *Energies* 11.8 (2018), p. 1971. DOI: 10.3390/en11081971.

[Wer+18] WERTH, A.; ANDRE, A.; KAWAMOTO, D.; MORITA, T.; TAJIMA, S.; TOKORO, M.; YANAGIDAIRA, D.; TANAKA, K.: Peer-to-Peer Control System for DC Microgrids. In: *IEEE Transactions on Smart Grid* 9.4 (2018), pp. 3667–3675. DOI: 10.1109/TSG.2016.2638462.

[Wey12] WEYMOUTH, T. R.: Problems in Natural Gas Engineering. In: *Transactions of the American Society of Mechanical Engineers* 34 (1912), pp. 185–234. URL: https://archive.org/details/sim_american-society-of-mechanical-engineers-transactions_1912_34/page/185.

[Wil72] WILLEMS, J. C.: Dissipative dynamical systems part I: General theory. In: *Archive for Rational Mechanics and Analysis* 45.5 (1972), pp. 321–351. DOI: 10.1007/BF00276493.

[Wil76] WILLEMS, J. C.: Lyapunov functions for diagonally dominant systems. In: *Automatica* 12.5 (1976), pp. 519–523. DOI: 10.1016/0005-1098(76)90011-X.

[WLC20] WIID, A. J.; LE ROUX, J. D.; CRAIG, I. K.: Modelling of methane-rich gas pipeline networks for simulation and control. In: *Journal of Process Control* 92 (2020), pp. 234–245. DOI: 10.1016/j.jprocont.2020.06.010.

[WM11] WOLDEYOHANNES, A. D.; MAJID, M. A. A.: Simulation model for natural gas transmission pipeline network system. In: *Simulation Modelling Practice and Theory* 19.1 (2011), pp. 196–212. DOI: 10.1016/j.simpat.2010.06.006.

[WR19] WILSON, G.; ROWLEY, P.: *Flexibility in Great Britain's gas networks: analysis of linepack and linepack flexibility using hourly data*. 2019. URL: https://ukerc.ac.uk/publications/linepack/.

[WYW22] WANG, Y.; YE, J.; WU, S.: An accurate correlation for calculating natural gas compressibility factors under a wide range of pressure conditions. In: *Energy Reports* 8 (2022), pp. 130–137. DOI: 10.1016/j.egyr.2021.11.029.

[Xu+20] XU, D.; WU, Q.; ZHOU, B.; LI, C.; BAI, L.; HUANG, S.: Distributed Multi-Energy Operation of Coupled Electricity, Heating, and Natural Gas Networks. In: *IEEE Transactions on Sustainable Energy* 11.4 (2020), pp. 2457–2469. DOI: 10.1109/TSTE.2019.2961432.

[Yan+20] YANG, J.; ZHANG, N.; BOTTERUD, A.; KANG, C.: Situation awareness of electricity-gas coupled systems with a multi-port equivalent gas network model. In: *Applied Energy* 258 (2020), p. 114029. DOI: 10.1016/j.apenergy.2019.114029.

[Zag+23] ZAGOROWSKA, M.; DEGNER, M.; ORTMANN, L.; AHMED, A.; BOLOGNANI, S.; DEL RIO CHANONA, E. A.; MERCANGÖZ, M.: Online Feedback Optimization of Compressor Stations with Model Adaptation using Gaussian Process Regression. In: *Journal of Process Control* 121 (2023), pp. 119–133. DOI: 10.1016/j.jprocont.2022.12.001.

[ZD15] ZHAO, J.; DÖRFLER, F.: Distributed control and optimization in DC microgrids. In: *Automatica* 61 (2015), pp. 18–26. DOI: 10.1016/j.automatica.2015.07.015.

[Zha+17] ZHANG, C.; ZHOU, K.; YANG, S.; SHAO, Z.: On electricity consumption and economic growth in China. In: *Renewable and Sustainable Energy Reviews* 76 (2017), pp. 353–368. DOI: 10.1016/j.rser.2017.03.071.

[Zha+20] ZHANG, H.; ZHANG, H.; SONG, L.; LI, Y.; HAN, Z.; POOR, H. V.: Peer-to-Peer Energy Trading in DC Packetized Power Microgrids. In: *IEEE Journal on Selected Areas in Communications* 38.1 (2020), pp. 17–30. DOI: 10.1109/JSAC.2019.2951991.

[Zha+23] ZHANG, S.; GU, W.; ZHANG, X.-P.; LU, H.; YU, R.; QIU, H.; LU, S.: Dynamic Modeling and Simulation of Integrated Electricity and Gas Systems. In: *IEEE Transactions on Smart Grid* 14.2 (2023), pp. 1011–1026. DOI: 10.1109/TSG.2022.3203485.

[Zho+17] ZHOU, Y.; GU, C.; WU, H.; SONG, Y.: An Equivalent Model of Gas Networks for Dynamic Analysis of Gas-Electricity Systems. In: *IEEE Transactions on Power Systems* 32.6 (2017), pp. 4255–4264. DOI: 10.1109/TPWRS.2017.2661762.

[Zlo+06] ZLOTNIK, A.; MISRA, S.; VUFFRAY, M.; CHERTKOV, M.: Monotonicity of actuated flows on dissipative transport networks. In: *2016 European Control Conference (ECC)*. IEEE, 29/06/2016 - 01/07/2016, pp. 831–836. DOI: 10.1109/ECC.2016.7810392.

[ZM20] ZHANG, N.; MALEKGOUDARZI, M.: Compressor surge control using a new robust adaptive method in the presence of uncertainty and unmatched disturbance. In: *Systems Science & Control Engineering* 8.1 (2020), pp. 405–412. DOI: 10.1080/21642583.2020.1785970.

[ZRA19] Zeghadnia, L.; Robert, J. L.; Achour, B.: Explicit solutions for turbulent flow friction factor: A review, assessment and approaches classification. In: *Ain Shams Engineering Journal* 10.1 (2019), pp. 243–252. doi: 10.1016/j.asej.2018.10.007.

[ZST20] Zagorowska, M.; Skourup, C.; Thornhill, N. F.: Influence of compressor degradation on optimal operation of a compressor station. In: *Computers & Chemical Engineering* 143 (2020), p. 107104. doi: 10.1016/j.compchemeng.2020.107104.

Own Publications and Conference Proceedings

[Gie+23] Giessler, A.; Jané-Soneira, P.; **Malan, A. J.**; Hohmann, S.: Economic Dispatch for DC Microgrids: An Optimal Power Sharing Approach with Batteries. In: *2023 62nd IEEE Conference on Decision and Control (CDC)*. 2023, pp. 1555–1562. doi: 10.1109/CDC49753.2023.10383211.

[JS+23a] Jané-Soneira, P.; **Malan, A. J.**; Prodan, I.; Hohmann, S.: Passivity-Based Economic Ports for Optimal Operation of Networked DC Microgrids. In: *2023 62nd IEEE Conference on Decision and Control (CDC)*. 2023, pp. 1549–1554. doi: 10.1109/CDC49753.2023.10383694.

[JS+23b] Jané-Soneira, P.; Prodan, I.; **Malan, A. J.**; Hohmann, S.: On MPC-based Strategies for Optimal Voltage References in DC Microgrids. In: *2023 European Control Conference (ECC)*. 2023, pp. 1–6. doi: 10.23919/ECC57647.2023.10178379.

[Köl+23] Kölsch, L.; Jané-Soneira, P.; **Malan, A. J.**; Hohmann, S.: Learning feedback Nash strategies for nonlinear port-Hamiltonian systems. In: *International Journal of Control* 96.1 (2023), pp. 201–213. doi: 10.1080/00207179.2021.1986233.

[Mal+23] **Malan, A. J.**; Rausche, L.; Strehle, F.; Hohmann, S.: Port-Hamiltonian Modelling for Analysis and Control of Gas Networks. In: *IFAC-PapersOnLine* 56.2 (2023). 22nd IFAC World Congress, pp. 5431–5437. doi: 10.1016/j.ifacol.2023.10.193.

[Mal+24a] **Malan, A. J.**; Giessler, A.; Strehle, F.; Hohmann, S.: Passivity-based pressure control for grid-forming compressors in gas networks. In: *2024 European Control Conference (ECC)*. 2024, pp. 1097–1104. doi: 10.23919/ECC64448.2024.10590807.

[Mal+24b] **Malan, A. J.**; Jané-Soneira, P.; Strehle, F.; Hohmann, S.: Passivity-Based Power Sharing and Voltage Regulation in DC Microgrids With Unactuated Buses. In: *IEEE Transactions on Control Systems Technology* 32.4 (2024), pp. 1410–1425. doi: 10.1109/TCST.2024.3372308.

[Mal+25] **MALAN, A. J.**; FERGUSON, J.; CUCUZZELLA, M.; HOHMANN, S.; SCHERPEN, J. M. A.: Passivation of Clustered DC Microgrids with Non-Monotone Loads. In: *IEEE Transactions on Control Systems Technology* 33.3 (2025), pp. 1069–1084. DOI: 10.1109/TCST.2025.3537861.

[Mau+21] MAURER, J.; RATZEL, O. M.; **MALAN, A. J.**; HOHMANN, S.: Comparison of discrete dynamic pipeline models for operational optimization of District Heating Networks. In: *Energy Reports* 7 (2021). The 17th International Symposium on District Heating and Cooling, pp. 244–253. DOI: 10.1016/j.egyr.2021.08.150.

[MJSH22] **MALAN, A. J.**; JANÉ-SONEIRA, P.; HOHMANN, S.: Constructive Analysis and Design of Interconnected Krasovskii Passive and Quadratic Dissipative Systems. In: *2022 IEEE 61st Conference on Decision and Control (CDC)*. 2022, pp. 7059–7065. DOI: 10.1109/CDC51059.2022.9992956.

[MPH22] **MALAN, A. J.**; PFEIFER, M.; HOHMANN, S.: Distributed coordination of physically-interconnected multi-agent systems with actuated and unactuated agents. In: *European Journal of Control* (2022), p. 100673. DOI: 10.1016/j.ejcon.2022.100673.

[Sta+22] STARK, O.; ECKERT, M.; **MALAN, A. J.**; HOHMANN, S.: Fractional Systems' Identification Based on Implicit Modulating Functions. In: *Mathematics* 10.21 (2022). DOI: 10.3390/math10214106.

[Ste+21] STEINLE, S.; RUF, J.; VAYAS, L.; ISIK, V.; MOHAN, J.; JANÉ-SONEIRA, P.; **MALAN, A. J.**; KÜCHLIN, R.; SAUERSCHELL, S.; HEROLD, L.; WALTER, J.; SURIYAH, M. R.; BAJOHR, S.; KERN, T.; KOLB, T.; HOHMANN, S.; KÖPPEL, W.; LEIBFRIED, T.: Das Verbundprojekt RegEnZell: Zellenübergreifende Regionalisierung der Energieversorgung durch betriebs- optimierte Sektorenkopplung. German. In: *ETG-Fb. 163: ETG-Kongress 2021: Das Gesamtsystem im Fokus der Energiewende, Online*. ETG-Kongress (Online). VDE Verlag, 2021, pp. 124–129.

[Str+19] STREHLE, F.; **MALAN, A. J.**; KREBS, S.; HOHMANN, S.: A Port-Hamiltonian Approach to Plug-and-Play Voltage and Frequency Control in Islanded Inverter-Based AC Microgrids. In: *2019 IEEE 58th Conference on Decision and Control (CDC)*. 2019, pp. 4648–4655. DOI: 10.1109/CDC40024.2019.9029272.

[Str+20a] STREHLE, F.; **MALAN, A. J.**; KREBS, S.; HOHMANN, S.: Passivity Conditions for Plug-and-Play Operation of Nonlinear Static AC Loads. In: *IFAC-PapersOnLine* 53.2 (2020). 21st IFAC World Congress, pp. 12237–12243. DOI: 10.1016/j.ifacol.2020.12.1128.

[Str+20b] STREHLE, F.; PFEIFER, M.; **MALAN, A. J.**; KREBS, S.; HOHMANN, S.: A Scalable Port-Hamiltonian Approach to Plug-and-Play Voltage Stabilization in DC Microgrids. In: *2020 IEEE Conference on Control Technology *and Applications (CCTA)*. 2020, pp. 787–794. DOI: 10.1109/CCTA41146.2020.9206323.

[Str+22a] STREHLE, F.; MACHADO, J. E.; CUCUZZELLA, M.; **MALAN, A. J.**; SCHERPEN, J. M.; HOHMANN, S.: Port-Hamiltonian Modeling of Hydraulics in 4th Generation District Heating Networks. In: *2022 IEEE 61st Conference on Decision and Control (CDC)*. 2022, pp. 1182–1189. DOI: 10.1109/CDC51059.2022.9992887.

[Str+22b] STREHLE, F.; NAHATA, P.; **MALAN, A. J.**; HOHMANN, S.; FERRARI-TRECATE, G.: A Unified Passivity-Based Framework for Control of Modular Islanded AC Microgrids. In: *IEEE Transactions on Control Systems Technology* 30.5 (2022), pp. 1960–1976. DOI: 10.1109/TCST.2021.3136489.

[Str+24] STREHLE, F.; MACHADO, J. E.; CUCUZZELLA, M.; **MALAN, A. J.**; HOHMANN, S.; SCHERPEN, J. M. A.: A Unifying Passivity-Based Framework for Pressure and Volume Flow Rate Control in District Heating Networks. In: *IEEE Transactions on Control Systems Technology* 32.4 (2024), pp. 1323–1340. DOI: 10.1109/TCST.2024.3365250.

Supervised Student Theses

[Blö21] BLÖTSCHER, M. G.: *Modellierung von Gasleitungen mit und ohne Brennwertdynamik*, Master's thesis. Institute of Control Systems (IRS), Karlsruhe Institute of Technology (KIT), 2021.

[Bre25] BREISACHER, P.: *Dynamic Simulation of Networked Multi-Energy Systems*, Master's thesis. Institute of Control Systems (IRS), Karlsruhe Institute of Technology (KIT), 2025.

[Gra23] GRAJCAR, J.: *Modular compensation of arbitrary DC loads through control design and passivity analysis*, Bachelor's Thesis. Institute of Control Systems (IRS), Karlsruhe Institute of Technology (KIT), 2023.

[Hor21] HORNEFF, N.: *Implementierung und Validierung von Port-Hamiltonischen Wärmenetzmodellen*, Bachelor's thesis. Institute of Product Engineering (IPEK), Karlsruhe Institute of Technology (KIT), 2021.

[Hu21] HU, X.: *Control of Coupling Components Between Microgrids*, Master's thesis. Institute of Control Systems (IRS), Karlsruhe Institute of Technology (KIT), 2021.

[Kar21] KARG, W.: *Verification and Improvement of a Standardized Method for the Passivity Analysis of Interconnected Systems*, Bachelor's thesis. Institute of Control Systems (IRS), Karlsruhe Institute of Technology (KIT), 2021.

[Mer21] MERZ, L.: *Modellierung von Kopplungselementen in Sektorübergreifenden Energiesystemen*, Master's thesis. Institute for Applied Thermofluidics (IATF), Karlsruhe Institute of Technology (KIT), 2021.

[Qia20] QIAN, Y.: *The Application of Multi-agent Systems Theory in Microgrid Control*, Bachelor's thesis. Institute of Control Systems (IRS), Karlsruhe Institute of Technology (KIT), 2020.

[Rat23] RATZEL, O. M.: *Verteilte Koordinierung von leistungsgeregelten Knoten in Drehstrom-Microgrids*, Master's thesis. Institute of Control Systems (IRS), Karlsruhe Institute of Technology (KIT), 2023.

[Rei21] REINHARD, T. F.: *Modellierung und leistungsbasierte Regelung von Stromspeicher für Stromnetzsimulationen*, Bachelor's thesis. Institute of Control Systems (IRS), Karlsruhe Institute of Technology (KIT), 2021.

[Sch22] SCHMIEDER, F. I.: *Anwendung einer koordinierten Regelung für physikalisch vernetzte Multi-Agenten-Systeme auf das Gasnetz*, Bachelor's thesis. Institute of Control Systems (IRS), Karlsruhe Institute of Technology (KIT), 2022.

[Stä23] STÄUDLE, S.: *Entwurf und Vergleich robuster Regelungen für ein Ball-auf-Platte-System*, Master's thesis. Institute of Control Systems (IRS), Karlsruhe Institute of Technology (KIT), 2023.

Karlsruher Beiträge zur Regelungs- und Steuerungstechnik (ISSN 2511-6312)
Institut für Regelungs- und Steuerungssysteme

Band 01
Diehm, Gunter
Identifikation des menschlichen Bewegungsverhaltens auf der Basis von Primitiven.
ISBN 978-3-7315-0608-9

Band 02
Flad, Michael
Kooperative Regelungskonzepte auf Basis der Spieltheorie und deren Anwendung auf Fahrerassistenzsysteme.
ISBN 978-3-7315-0610-2

Band 03
Eckert, Marius
Modellbasierte Identifikation fraktionaler Systeme und ihre Anwendung auf die Lithium-Ionen-Zelle.
ISBN 978-3-7315-0690-4

Band 04
Krebs, Stefan
Intervallbeobachter für lineare parametervariante Systeme und deren Anwendung auf die Asynchronmaschine.
ISBN 978-3-7315-0857-1

Band 05
Kaspar, Stephan
Fahrdynamikuntersuchungen eines Elektrofahrzeugs mit Einzelrad-Hinterradantrieb.
ISBN 978-3-7315-0916-5

Band 06
Sauter, Patrick S.
Modellierung und zentrale prädiktive Regelung von multimodalen Energieverteilnetzen.
ISBN 978-3-7315-0963-9

Band 07
Kupper, Martin
Verteilte Zustandsschätzung fraktionaler Systeme und ihre Anwendung auf Lithium-Ionen-Batteriesysteme.
ISBN 978-3-7315-0971-4

Band 08
Merkert, Lennart
Optimal Scheduling of Combined Heat and Power Generation Considering Heating Grid Dynamics.
ISBN 978-3-7315-1056-7

Die Bände sind unter www.ksp.kit.edu als PDF frei verfügbar oder als Druckausgabe bestellbar.

Band 09 Ludwig, Julian
Automatisierte kooperative Transition einer Regelungsaufgabe zwischen Mensch und Maschine am Beispiel des hochautomatisierten Fahrens.
ISBN 978-3-7315-1069-7

Band 10 Inga Charaja, Juan Jairo
Inverse Dynamic Game Methods for Identification of Cooperative System Behavior.
ISBN 978-3-7315-1080-2

Band 11 Schnurr, Christoph Xaver
Ein Verfahren zur lexikographischen modellprädiktiven Regelung mit der Anwendung auf eine permanenterregte Synchronmaschine.
ISBN 978-3-7315-1095-6

Band 12 Schwab, Stefan
Guaranteed Verification of Dynamic Systems.
ISBN 978-3-7315-0965-3

Band 13 Pfeifer, Martin
Automated Model Generation and Observer Design for Interconnected Systems: A Port-Hamiltonian Approach.
ISBN 978-3-7315-1135-9

Band 14 König, Alexander
Absicherung hochautomatisierten Fahrens durch passiven virtuellen Dauerlauftest.
ISBN 978-3-7315-1141-0

Band 15 Stark, Oliver
Parameter- und Ordnungsidentifikation von fraktionalen Systemen mit einer Anwendung auf eine Lithium-Ionen-Batteriezelle.
ISBN 978-3-7315-1187-8

Band 16 Köpf, Florian
Adaptive Dynamic Programming: Solltrajektorienfolgeregelung und Konvergenzbedingungen.
ISBN 978-3-7315-1193-9

Die Bände sind unter www.ksp.kit.edu als PDF frei verfügbar oder als Druckausgabe bestellbar.

Band 17 Kölsch, Lukas
Dynamic Incentives for Optimal Control of Competitive Power Systems.
ISBN 978-3-7315-1209-7

Band 18 Schwartz, Manuel
Topologie-Optimierung eines radselektiv angesteuerten Fahrzeugs basierend auf einer optimalen Fahrzeugführungsregelung.
ISBN 978-3-7315-1222-6

Band 19 Rothfuß, Simon
Human-Machine Cooperative Decision Making.
ISBN 978-3-7315-1223-3

Band 20 Gellrich, Thomas Christoph
Dynamical Modeling and Control of Multiphase Heat Transport Systems Based on Loop Heat Pipes.
ISBN 978-3-7315-1231-8

Band 21 Maurer, Jona
Transactive Control of Coupled Electric Power and District Heating Networks.
ISBN 978-3-7315-1276-9

Band 22 Puccetti, Luca
Self-Learning Longitudinal Control for On-Road Vehicles.
ISBN 978-3-7315-1290-5

Band 23 Varga, Bálint
Limited Information Shared Control and its Applications to Large Vehicle Manipulators.
ISBN 978-3-7315-1325-4

Band 24 Walz, Fabian
Modellbasierte prädiktive Längsdynamikregelung für künftige Fahrerassistenz- und Automatisierungssysteme.
ISBN 978-3-7315-1331-5

Band 25 Strehle, Felix
A Framework for Decentralized Stabilization in Networked Energy System: A Passivity-Based Approach
ISBN 978-3-7315-1358-2

Band 26 Bischoff, Esther Sophie
Time-Extended Multi-Robot Task Allocation – A Reoptimization Framework with Provable Performance.
ISBN 978-3-7315-1449-7

Band 27 Malan, Albertus Johannes
Passivity-Based Stabilisation and Coordination in Networked Multi-Energy Systems.
ISBN 978-3-7315-1453-4